Engineered Bamboo Structures

Engineered Bamboo Structures

Yan Xiao

CRC Press
Taylor & Francis Group
Boca Raton London New York

CRC Press is an imprint of the
Taylor & Francis Group, an **informa** business

Cover image: Yan Xiao

First published 2022
by CRC Press/Balkema
Schipholweg 107C, 2316 XC Leiden, The Netherlands
e-mail: enquiries@taylorandfrancis.com
www.routledge.com – www.taylorandfrancis.com

CRC Press/Balkema is an imprint of the Taylor & Francis Group, an informa business

Library of Congress Cataloging-in-Publication Data
A catalog record has been requested for this book

ISBN: 978-1-032-06395-9 (hbk)
ISBN: 978-1-032-06917-3 (pbk)
ISBN: 978-1-003-20449-7 (ebk)

DOI: 10.1201/9781003204497

Typeset in Times New Roman
by Newgen Publishing UK

This book is dedicated to all who love bamboo.

Contents

Preface

Born and growing up in the Northern region of China, Suiyuan (Huhehaote), the first encounter I had with bamboo was probably the use of bamboo chopsticks, though my memory is blurry. What I remember clearly is that the first time I used bamboo to build something was to make a kite, with bamboo strips taken out from the bamboo curtain, sneaking them out while my mother was not looking. However, I never dreamed that one day I would study and construct buildings and bridges using bamboo and that often people would call me a bamboo engineer.

The whole thing can be traced back to 2001 when I took on the Chong Kong Chaired Professorship at Hunan University, in Hunan, one of the main bamboo growing provinces in China. One day, I was shocked and totally amazed by the beauty and immense waves of the so-called bamboo ocean, when I went to Yiyang, a town famous for its vast bamboo forest! I felt something warm in my heart then. Shortly afterwards, during a conversation with Prof. K.L. Liu of Hunan University, he showed me some furniture products made with bamboo. Immediately, I knew I had a vision for the future, and an idea, glubam, glued laminated bamboo, or bamboo-based glulam came to my mind.

My journey and that of my research group kicked off in 2005, with investigation into and development of engineered bamboo. We worked on the manufacture and testing of glubam as an alternative to timber structures, similar to those widely used in many industrialized countries. This book is mainly a summary of the research and demonstration design/construction projects over the past 16 years. Most information comes from the research findings of my own group; however, valuable references from published work by other researchers are also cited and discussed as far as possible. I hope to provide the academic and practicing community with an overall framework for working with engineered bamboo, particularly laminated bamboo.

Culturally, Chinese and oriental people like bamboo very much and consider bamboo as part of the spirit – elegant, strong, tough, flexible, and optimistic. In a famous poem, Su Shi (1037–1101) states that "residing without bamboo makes people vulgar."[1] I guess Su Shi's real meaning was that one should reside near to bamboo to connect with the natural beauty of a bamboo (an example is shown in Figure 0.1), not that one should live in a bamboo house – because, based on my limited knowledge of architectural history, there was actually no mainstream building technology using bamboo (though it is used as decoration, sub-system, etc.) developed or used by ancient Chinese society. Therefore, the work of this author and many of today's

Figure 0.1 "Residing without bamboo makes people vulgar," a quotation from the poems of Su Shi
(1037–1101).[1]

engineers is actually attempting to add new meanings to Su Shi's poem, "Living, in
Bamboo Buildings."

Reference

[1]   Jen, C. (Trans.). (2008). *The poetry of the Eastern Slope: A selected translation of poems, Ci
and prose works by Su Shi*. Fudan University Press, Shanghai.

Acknowledgements

One of the greatest joys I have had since studying bamboo is meeting so many enthusiastic people, throughout the world, who love bamboo and are passionate about building with bamboo. The author is fortunate to be associated with many excellent colleagues and students who worked with me on the investigation and design of bamboo structures. I would like to thank Professor B. Shan, of Hainan University, for his collaboration in many tasks involved in developing glubam. I appreciate the dedication and hard work of Drs. G. Chen, Z. Li, R.Z. Yang, Q. Zhou, R. Wang, C. Demartino, J.J. Xu, S.T. Deresa, and J. Wen, who took engineered bamboo as the subject of their doctoral or post-doctoral research at Hunan University, Nanjing Tech University, and Zhejiang University. Many students also worked with me and my group during their Master's or undergraduate thesis research, and their efforts are warmly appreciated.

The author has also been fortunate in receiving several research grants, without which the research could not be continued. These include the National Key Project of the National Natural Science Foundation of China; Distinguished Scholarships from Hunan University, Nanjing Tech University, and Zhejiang University; support from the Blue Moon Fund, and INBAR (International Bamboo and Rattan Organisation). The ongoing Key Project from the China Ministry of Science and Technology enables the author and his research group to validate the technology developed throughout the years in larger- scale real-life applications. The Engineering School and the Department of Civil and Environmental Engineering at the University of Southern California provided the author with many in-kind supports and helped significantly in publicizing the author's work during his tenure there. HunanTaohuajiang Bamboo Science and Technology, Co. Ltd., and Hangzhou Bamboo Technologies, Ltd. are warmly thanked for their kind supply of some raw and processed bamboo materials used in the author's research.

I must acknowledge two gentlemen who have sadly passed away: Professor Masahide Tomii and Professor Nigel M.J. Priestley. Though I was working with them on steel and concrete structures, they planted the seeds in my academic career for my shift in interest toward bio-based timber and bamboo. While I was pursuing my MS and Dr. of Engineering degrees at Kyushu University, Professor Tomii provided me with many opportunities to examine ancient and modern timber structures, so I was able to gain some understanding about ancient oriental architecture. When I was leaving the University of California, San Diego, to start my faculty career at the University of Southern California, I went to seek advice from Professor Priestley. To my surprise, as a world-renowned concrete structure expert, he suggested that I should keep an eye on timber structures. I was lucky enough to follow his suggestion and shortly after joining the University of Southern California, I was able to secure my first externally supported research grant as a junior professor, and the project was to study the seismic behaviors of clay and concrete tiles on timber truss roof systems.[1]

The author thanks Professor E.C. Zhu of Harbin University of Technology who reviewed part of the manuscript, particularly Chapter 4, and provided many useful suggestions. Ms. C.Q. Chen and Ms. C.Y. Mo of the Zhejiang University–University of Illinois Joint Institute (ZJUI) kindly spent considerable time in helping the author to prepare the manuscript. The encouragement and patience of editorial staff at Taylor & Francis are warmly appreciated.

Finally, and most importantly, I would like to thank my wife Connie Q. Liao, son Jeffrey, and daughter Nancy. Their patience and support throughout the years have made it possible for the author to devote himself to research.

The book is dedicated to all who love bamboo.

Permissions

Except where other sources are specified, the author including his research group has created all the tables and figures or has taken the photos for the figures.

About the Author

Yan Xiao is a Distinguished Chaired Professor of Civil Engineering and Director of the Energy, Environment and Sustainable Systems Sciences Department at the Zhejiang University–University of Illinois Joint Institute (ZJUI), and a professor in the Sonny Astani Department of Civil Engineering, University of Southern California. He directs the Zhejiang University (Ninghai) Joint Research Center for Bio-based Materials and Carbon Neutral Development. He received his Bachelor of Engineering degree from Tianjin University, China, and his Master and Doctor of Engineering degrees from Kyushu University, Japan. His academic experience includes a post-doctoral fellowship at the University of California, San Diego, where he was later also Lecturer and Assistant Research Scientist. He was Dean of the Civil Engineering Colleges at Hunan University and Nanjing Tech University. He is an associate editor for the American Society of Civil Engineers (ASCE) *Journal of Structural Engineering*, and *Journal of Bridge Engineering*, and editorial board member for the Elsevier *Journal of Constructional Steel Research*. He is an elected fellow of the ASCE and the American Concrete Institute (ACI). He holds a Professional Engineer license in California. Xiao's research and academic publications cover a broad spectrum of areas related to the design of structures against extreme loads, such as earthquakes and impacts. His recent research and industrial development efforts focus on modern bamboo structures with the goal of promoting environmentally and eco-friendly construction. Xiao has many patents to his name, and was responsible for the development of the award-winning technology GluBam (Glued Laminated Bamboo).

Introduction

As a Chinese set-phrase describes it, the author puts out this book to "throw away a brick in order to get a gem." This book contains nine chapters. Chapter 1 provides a brief discussion of bamboo resources, the progress of engineered bamboo research, and the roles of bio-based materials in sustainable development and carbon neutrality. Chapter 2 introduces the manufacturing process of engineered bamboo; specifically, glued laminated bamboo (Glubam). Chapter 3 provides details of the material behaviors of engineered bamboo. Chapter 4 attempts to establish the design values of engineered bamboo following existing design methodologies; specifically, the Chinese GB code, the U.S. National Design Specification (NDS), and the European Code EC-5. Chapters 5, 6, 7, and 8 discuss experimental performances of connections, components, trusses, and walls of engineered bamboo, respectively. Some design recommendations are also provided. Finally, Chapter 9 summarizes the design procedures for three major types of building: bamboo mobile buildings; lightweight frame buildings; and heavy frame multi-story buildings. Design examples abstracted from actual design projects are also provided.

Reference

[1] Xiao, Y., & Yun, H.W. (1998). Dynamic testing of full-scale concrete and clay tile roof models. *ASCE Journal of Structural Introduction Engineering*, 124(5), 482–489.

Chapter 1

Bamboo for Carbon Neutral Development

Bamboo is in the limelight as a bio-based green material for potential mass applications in modern construction. In this first chapter, bamboo resources are briefly discussed, followed by the definition of various bamboo structures, and discussions about the structural systems. In particular, the relationships between modern timber structures and bamboo structures are discussed. Based on the discussion, the framework is set for this book on developing engineered bamboo structures with close reference to modern timber engineering.

1.1 Bamboo Resources

As evergreen perennial flowering plants, bamboos are in the subfamily Bambusoideae of the grass family Poaceae, and are often referred as weedy grasses. There are more than a thousand species of bamboos widely distributed in the world.[1,2] This important non-wood forest resource has also seen a favorable increase in the world, particularly in China, as shown in Table 1.1. In a recent study, Canavan et al. examined the origin of more than 1662 bamboo species,[3] and Figure 1.1 shows the native and introduced species in different countries (only 20 countries extracted from Canavan et al. are shown). As exhibited in Figure 1.1, bamboos are native to countries in Asia, South America, and Africa; however, they have also been introduced in many parts of the world. It is particularly interesting to notice that the bamboo species in Australia, the United States, and some European countries are mainly introduced species. The author has also had a first-hand, surprising, and exciting encounter with vast bamboo forest bushes, shown in Figure 1.2, in a town named Belli Park, Queensland, Australia.

Some of the bamboos are called giant bamboo due to the size and the height they can reach. Among them are the Moso bamboo and the Guadua bamboo. Moso bamboo, Phyllostachys edulis, is a bamboo species native to China, also widely spread in East Asia. Moso is an evergreen arboreal bamboo of a genus of the Poaceae subfamily, with uniaxial dispersion, a large stalk which can reach more than 20 meters and a breast diameter as large as 20 centimeters. Guaduo, or Guadua angustifolia, is a species of clumping bamboo found from Central to South America. These two bamboo species form the main raw materials for the bamboo structures discussed in this book.

Bamboo resources are abundant in China. The unique geographical environment of the middle and lower reaches of the Yangtze River is extremely suitable for the growth of bamboo, particularly in the Hunan, Jiangxi, Zhejiang, and Fujian provinces. The quality of bamboo in these regions is very good, with relatively large culm diameter

DOI: 10.1201/9781003204497-1

Table 1.1 Bamboo area by country and region (1990–2010)

Country/region	Area of bamboo (1000 ha)			
	1990	2000	2005	2010
China	3856	4869	5426	5712
India	5116	5232	5418	5476
Total Asia (incl. China and India)	15,412	16,311	16,943	17,360
Total Europe	0	0	0	0
Total North and Central America	37	37	37	39
Total Africa	3688	3656	3640	3627
Total Oceania	23	38	45	45
Total South America	10,399	10,399	10,399	10,399
World	29,560	30,442	31,065	31,470

Source: FAO (2010).[2]

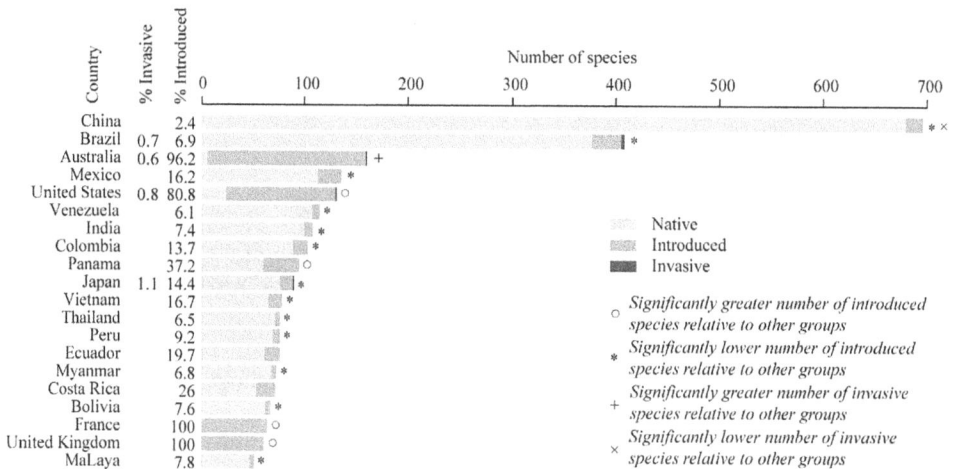

Figure 1.1 Native and imported bamboo species in different countries.
Source: Adopted from Canavan et al. (2017). [3]

and wall thickness. In addition, the mature cycle of the bamboo is relatively short, with three to five years of reproduction cycle. The main bamboo species used for buildings is Moso; this has been the case for a long time.[4,5]

Varieties of bamboo products are being developed and produced, including traditional and modern hand crafts, household tools, furniture, boards and panels, bamboo floorings, etc. In China, based on the 2018 statistics,[6] the total revenue of the bamboo industry exceeded 200 billion renminbi (RMB). However, despite the widely distributed bamboo forest resources, exploitation and logging management are not exercised sufficiently in many regions.

The levels of bamboo resource management and industrial development are not even in different regions. The differences in the development level of bamboo resources in the main bamboo producing areas have led to great differences in the management of

Figure 1.2 Bamboo bushes in Belli Park, Queensland, Australia.

bamboo resources. Some local governments regard the bamboo industry as one of their pillar industries, and have nurtured large-scale bamboo processing enterprises with a variety of products. Most enterprises have their own bamboo forest bases, and thus can achieve orderly logging management. The local government also vigorously promotes the cultivation of bamboo forests, increasing the area of bamboo forest. However, the author also examined some communities where the bamboo-related industry is relatively weak with only small workshops, mostly for small quantities of traditional household goods, and output of round bamboos, with low added values.

Innovations related to bamboo usage in other industrial products, such as clothing fabrics, medicine, paper, etc. are developing fast, but are beyond the scope of this book.

1.2 Bamboo as a Modern Building Material

As a natural material, bamboo has been used for thousands of years for numerous purposes.[7,8] In recent years, following the trend towards the sustainable and environmentally friendly development, there is a renewed interest for using bamboo in construction.[9-13] Reflecting a new interest in using bamboo in construction in the light of sustainable and carbon neutral development, in October 2007, the author organized the First International Conference on Modern Bamboo Structures (1-ICBS), in Changsha, China. This milestone event attracted scholars, government and NGO staff, and entrepreneurs from all over the world. The outcome was summarized in the proceedings of the ICBS-2007, including 33 papers selected from the final submissions.[9] The majority of the collected technical papers are on engineered bamboo and its applications.

Since the first international conference on modern bamboo structures, there has been a steady increase in scientific journal publications of bamboo-related research. Figure 1.3 depicts the annual numbers of SCI (Science Citation Index) papers in the

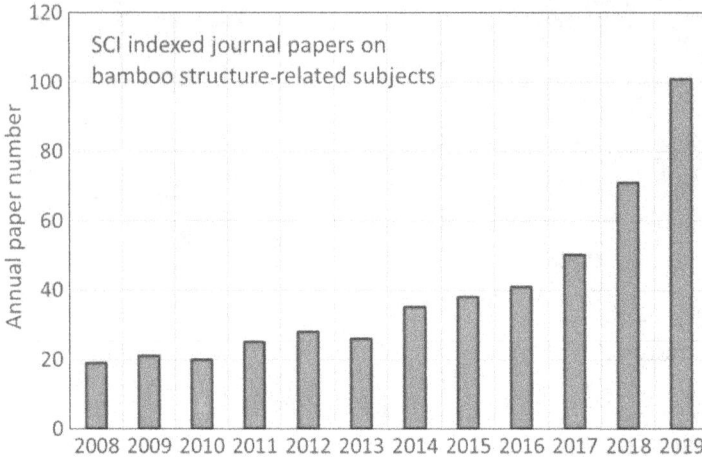

Figure 1.3 Annual journal papers related to bamboo structures in the period 2008–2019.

period 2008–2019. It should be clarified that most of the papers have been searched from the academic journals related to structural materials, structures, and construction; thus they may be limited in terms of the scope of the literature.

There is no doubt that research studies on bamboo and bamboo-based materials and structures have become a new academic field of research. The growing research will provide urgently needed background data, methodologies for testing, analysis, and design toward the development of specifications for engineered bamboo structures.

The merits of using bamboo as a basic structural material are quite apparent. First, bamboo as a natural resource is widely available in many parts of the world. Bamboo is essentially a kind of large grass, often nicknamed as giant grass, growing faster than most trees. Bamboo usually grows for about four years and can then be cut down and regenerated. Second, bamboo has good mechanical properties and is easy to process. Third, but certainly not last, the processing and manufacturing of bamboo has no substantial adverse impact on the environment. It basically can be processed without pollution and meets the requirements of sustainable development. Therefore, the development of modern structures with bamboo as the main structural material can effectively increase the added value of bamboo, increase the income of people in the bamboo growing areas, and promote sustainable development in the construction industry, which accounts for about 40% of carbon dioxide output.

In North America and many industrialized countries, another bio-based material, wood, has been widely used in the construction of bridges and houses.[14–16] Although forest coverage in North America and other places is wide and wood resources are abundant, it is still an urgent task for society to retain enough trees for future generations. The development and utilization of bamboo materials can provide new material options for bridge and housing construction on an international scale. As a direct return, the development of modern bamboo structures is not only beneficial to the developing countries with abundant bamboo, but also guides the cultivation and development of bamboo forests with fast growth and short maturity in other parts of the world and is conducive to environmental afforestation.

But the reality is that bamboo is not widely utilized in modern structures. The author believes that it is mostly due to the lack of supporting theory and research based on mechanics, material science, structural design, and experimental science, and the lack of establishment of modern design and construction procedures. In the next sections, a summary of the forms of modern bamboo structures is provided, followed by comparisons with modern timber structures.

1.3 Bamboo Structures

The usage of bamboo in buildings and bridges can certainly be traced back to ancient times; however, it should be re-examined in the context of modern engineering. In modern times, the forms for bamboo usage in structures can be categorized into the following:

1   round bamboo culm construction;
2   bamboo as reinforcement in combination with other materials, such as concrete and clay;
3   engineered bamboo composites.

It should be clarified that any bamboo structure designed and constructed using modern engineering concepts, process, methodology, etc. should be considered as engineered; however, the author adopts the above nomenclature to reflect the intensified engineering process in bamboo composites.

1.3.1 Construction Using Round Bamboo Culms

Bamboo buildings and bridges constructed directly from bamboo components are the most traditional forms of bamboo construction. In the 1950s, there was some research activity in China on round bamboo structures for the purpose of saving steel and timber.[17,18] However, systematic development and industrialized applications in construction do not seem to have followed. Most practices in construction using round bamboo in China are traditional in manner, represented by the so-called "Ganlan-style" bamboo buildings in the Southwest region of China. In recent years, buildings made of round bamboo are gradually returning to China, led primarily by architects, with some interesting buildings being constructed.[19,20]

Bamboo is often called the wood of the poor in Latin America and has a wide range of applications in buildings and bridges. Figure 1.4 shows photographs of some of the bamboo structures taken by the author during several conference trips to Colombia. Colombia may be the first country in the world to officially adopt the round Guadua bamboo structures in its design specifications.[21,22] In addition, the International Standards Organization (ISO) has also published design standards for round bamboo structures.[23]

Despite its natural beauty favored by many people, particularly architects and artists, round bamboo culm construction also delivers many engineering challenges, due

Figure 1.4 Round bamboo structures in Colombia: (a) the 45m-long Jenny Garzon Guadua bridge (Puente de Guadua) designed by S. Velez, and completed in 2003; (b) a complex of Guadua bamboo houses near the city of Armenia.

Figure 1.5 Examples of a batch of bamboo culms of different sizes.

primarily to its geometric nonuniformity and irregularity, as shown in Figure 1.5. The following are the main issues regarding the usage of round bamboo in structures:

1 Different from wood, the volume mass of one bamboo culm is normally not enough to serve as one structural element, although it can perform very well in its own natural mission. Thus, there is often a need to bind multiple culms together to form one structural element. Due to the geometric irregularity and nonuniformity in the

raw materials, such as differences of section shape, size and longitudinal variation, random degree of straightness, etc., such composition with multiple culms is difficult and can lead to waste of raw materials.

2   Due also to the geometric irregularity and nonuniformity, connections pose a particular difficulty in conforming to structural and constructional requirements. Specially designed complex connections are necessary to form connections between different structural elements, such as beams to columns.

3   The severe anisotropy of the bamboo itself includes extremely low lateral strength, since its fibers are all in the longitudinal direction. This may result in premature splitting failure when a lateral load is directly applied to the wall of the culm. Such laterally stressing situations are common particularly for the joints; thus special measures must be taken.

The author believes that one of the most basic characteristics of modern industry is its repeatability. Modern structures should be designed according to scientific principles and theories, constructed according to the design; the completed structure should meet the design requirements, and this "satisfaction" can be experimentally verified by modern scientific means. Nonetheless, although it is probably difficult for round bamboo to become a mainstream construction material due to its inherent irregular and nonuniform geometry, and extreme anisotropic mechanical properties, round bamboo culm still has an adequate share of the market and the pursuit of its inherent natural beauty will continue to enrich the built environment and structural inventory.

When it comes to round bamboo, one probably thinks about bamboo scaffoldings, such as the famous scene in the film *Rush Hour 2* (2001). Bamboo scaffoldings are still allowed and continue to be used successfully in the modern city Hong Kong (Figure 1.6), despite being banned in large cities in China. Experimental and analytical research work can be found in Chung & Yu's paper.[24]

1.3.2 Bamboo Reinforced Concrete Structures

Bamboo reinforced concrete used to be a major aspect of the application of bamboo in modern structural systems. The research and application of bamboo reinforced concrete have a long history.[25,26] In the 1950s, due to the lack of steel, bamboo reinforced concrete was used in concrete structures, and some buildings are still in service.[27] In the early 1960s, Brink and Rush[25] in the U.S. Navy Civil Engineering Laboratory systematically studied the mechanical performance of concrete beams with bamboo strip rebars and developed a calculation formula, design method, and table for the conversion of equivalent steel bars. The use of bamboo reinforced concrete in military engineering was targeted in order to avoid the need for electromagnetic shielding. However, this has been replaced by the fiber reinforced plastic (FRP) rebars that emerged in the 1960s and have been popularized since then. Compared with steel and FRP, the bamboo strip bar is too weak, and lacks deformability and bonding strength with concrete, making bamboo reinforced concrete not very suitable in modern structures with a higher demand for structural and functional performance. Low-cost buildings with lower strength concrete, masonry, or solidified soil with bamboo (including bamboo strips, fabrics, etc.) as reinforcement may still, however, be favored in many parts of the world.

Figure 1.6 Bamboo scaffolding in Hong Kong.

1.3.3 Engineered Bamboo Composites

This book is mainly concerned with engineered bamboo composites and their use in modern engineering structures. Based on the development in China and elsewhere, Liu et al. established comprehensive definitions of various bamboo products.[28] This book follows many of the definitions given in that reference work,[28] with some differences, however, to reflect many new developments in recent years. Table 1.2 lists some of the terminologies for typical timber and bamboo products.

The basic feature of essentially all engineered bamboo products is the integration of pieces of bamboo (mostly longitudinal strips) into a larger and timber-like mass that can have sizable and regular dimensions for applications based on engineered design. The industrialized process for such integration is mainly pressurized formation with the use of bonding agencies. Generally speaking, there are two types of pressurized formation to produce engineered bamboo composites: One is laminated bamboo; and the other is reconstituted bamboo or bamboo scrimber.

Laminated bamboo is a general term for bamboo products processed using adhesive lamination with layer-by-layer layups. Many of the products in this category were not originally targeted for construction, being used mainly for furniture, artwork, flooring, or concrete formwork, etc. In general, the laminated bamboo can be divided further based on the shapes of the preprocessed bamboo before lamination; typically, thin-strip (about 2 mm thick) lamination and thick-strip (about 5 to 8 mm thick) lamination. The planar laminated bamboo based on thin bamboo strips is also often referred to

Table 1.2 Nomenclatures of timber and bamboo products

Timber	Bamboo
Engineered timber	Engineered bamboo
Wood log	Round bamboo culms
Saw-cut lumber	Bamboo strips
Glulam	Glubam, Laminated bamboo lumber (LBL)
Plywood	Bamboo plywood or plybamboo
Scrimber	Bamboo scrimber
Parallel strand lumber (PSL)	Parallel strand bamboo (PSB)
Oriented strand board (OSB)	Oriented strand bamboo (bamboo OSB)
Laminated veneer lumber (LVL)	Laminated bamboo veneer lumber (bamboo LVL)
Nail laminated timber (NLT)	Nail laminated bamboo (NLB)
Cross laminated timber (CLT)	Cross laminated bamboo (CLB)
	Cross laminated bamboo and timber (CLBT or CLTB)[29]

as bamboo plywood or plybamboo; when based on thick bamboo strips, it is sometimes referred to as laminated bamboo lumber. Following the concept of wood-based laminated lumber (glulam), the author conceived a two-step lamination process, and the final product is termed as glued laminated bamboo or glubam, which is a bamboo-based glulam. The author originally registered "GluBam" as a trademark with the U.S. Patent and Trademark Office; however, glubam has now become a general technical terminology. Glubam has a clear technical definition and application purpose for structures and is the main subject of this book.

It should be pointed out that the lamination of bamboo can also be achieved by using mechanical connectors, such as steel nails, to form nail laminated bamboo (NLB), similar to nail laminated timber (NLT).

The reconstituted bamboo or bamboo scrimber is a product formed by squeezing bamboo strips into a rigid mold and activating the adhesive either in an elevated temperature or room temperature, similar to reconstituted wood.[30] Due to the "forced" squeezing process, the strips are hardened by adhesive under pressure from both vertical (the direction of the squeezing force) and lateral directions (by the passive pressure from the rigid mold). Due also to this "forced" squeezing process, the bamboo strips used can be irregular and can even be the smaller branches; thus there is an increased utilization of the raw materials. The higher strength and relatively high modulus of the final products are apparently favored for applications in structures; however, the relatively high density (typically higher than 1.0 compared with water) and increased usage of adhesives might be drawbacks. The research on bamboo scrimber has flourished in recent years,[31] and it seems to be particularly promising for its impact resistance.[32]

1.3.4 Engineered Bamboo versus Engineered Timber Structures

Engineered bamboo has been developed essentially following the industrial experiences and technology used in timber structures, with the targeted usage as an alternative to timber in modern construction industry. Therefore, it is useful to give a brief review of modern timber structures, which have mainly been developed in industrialized countries in the later half of the 20th century.

(a)   (b)

Figure 1.7 Wood log structures: (a) an ancient covered bridge (Xi-An Bridge) in Anhua, China; (b) the traditional way of building wood log frames in Hunan Province, China. Note that bamboo skin strips are used to tie the wood logs.

(a)   (b)

Figure 1.8 Different lightweight woodframe buildings under construction: (a) single family house in Arcadia; (b) a multi-story lightweight woodframe building under construction in Los Angeles.

Similar to round bamboo culms, round wood logs can also be directly used with minor modifications in buildings and bridges, as shown in Figure 1.7. Compared with round bamboo, wood can be cut into components with rectilinear shapes that more easily meet design specifications because of the large volume of wood cut from mature tree trunks. Such saw-cut wood lumbers are gradually standardized and are used in various wood structures. In North America, standard wood lumbers are the main structural elements in the construction of lightweight woodframe buildings (so-called 2 × 4 construction), as shown in some examples in Figure 1.8. For the skins of the woodframe buildings, plywood sheathing panels are typically used. Plywood is mass produced in a lamination process using rotary peeled wood veneers.[15]

Wood is also an anisotropic material with high strength and stiffness in the longitudinal direction; however, there are many natural defects caused by branches and rings,

Figure 1.9 Examples of glulam structures: (a) building frame with glulam girders under construction; (b) glulam girder room system in a shopping building in Los Angeles; (c) large-span curved glulam frame exhibition hall in Tacoma, Washington State.

so its usage efficiency is also limited as saw-cut lumbers, not mentioning the limitation in sizes. In the latter half of the 20th century, structural timbers made with smaller size saw-cut wood lumbers became a main structural form to enable timber to compete with steel and concrete in mainstream structures. Figure 1.9 shows some examples of glulam structures in large-scale buildings. For buildings, the long-time record holder was the Tacoma Dome with a diameter of 160 m, completed in 1983;[33] this was then overtaken by the 163 m span Superior Dome at Northern Michigan University, Marquette, Michigan, which was completed in 1991.[34] Most recently, timber buildings are receiving more and more attention in the applications of multi-story and tall buildings.[35–39] In particular, the advancement in materials and components manufacture, such as cross laminated timber (CLT), enables such development.

As regards bridges, the Flisa, Norway appears to be the record holder for its 70 m span as a roadway bridge made with wood.[40] The Xuhong footbridge (Figure 1.10) in Suzhou, China, appears to be the longest timber bridge with a main span length of 75.7 m and a total length of about 120 m.[41]

It is worthwhile mentioning that timber construction has had a large share in traditional buildings in China, as manifested by the 67 m tall 11th-century pagoda in

Figure 1.10 Record-setting footbridge built with glulam with a main span of 75.7 meters in Suzhou, China.

Source: Man et al. (2019). [41]

Yingxian. However, due to over-logging in the 20th century, development of modern timber structures in China was seriously delayed, until the beginning of the 21st century. On the other hand, timber structures continue to enjoy a fair share of the housing market in another industrialized Asian country, Japan. Traditional building methods and styles, originally influenced by China, are further matured and developed in combination with modern technologies. Figure 1.11 shows a timber house in Fukuoka, in which the author lived for several years as a student. Such traditional oriental style is well fused with modern ways of construction and living.

1.4 Code and Standard Issues

As a new material for modern construction, bamboo and structures using various forms of bamboo and bamboo composites are not fully specified for structural utilization, and there is only a limited number of technical documents available for testing, certification and as reference guidelines for design, manufacture and construction.[42] Since material similarity and the intended usage of engineered bamboo is in the same construction area as timber, standards and specifications for wood and timber design, manufacture and construction can be – and are currently being – followed with necessary modifications. Some standards and specifications have already been developed for round bamboo culms and laminated bamboo; however, not all of these target their

Figure 1.11 A typical Japanese-style family house in Fukuoka.

usage as structural members. In the next subsection , a summary on relevant standards and specifications is provided.

1.4.1 Relevant Wood Standards

It is essentially a natural choice to utilize the well-developed wood standards to set up the reference foundation for standardizing engineered bamboo. Several relevant wood and timber standards used in the studies of bamboo are listed in Table 1.3. First of all, studying and defining material properties is the most basic issue in structural engineering. The ASTM D143[43] "Standard test methods for small clear specimens of timber" provides the most important specifications to guide tests on the physical and mechanical properties of timber. ASTM D143 considers, especially, the standardized dimensions, controlled moisture content, temperature, hardness and bending, compression, toughness, shear, and tension strengths of small clear specimens. In China, the national standard (Guo Biao) GB/T 1928 "General requirements for physical and mechanical tests of wood"[44] was developed on the basis of referencing ISO 3129 which specifies the requirements for manufacturing samples, adjusting moisture, calibrations, calculations, and reporting.[45] In addition, several other GB standards regulate testing procedures of timber specimens, such as GB 1936 "Method for determination of the modulus of elasticity in static bending of wood";[46] GB 1937 "Method of testing in shearing strength parallel to grain of wood";[47] GB 1938 "Method of testing in tensile strength parallel to grain of wood";[48] GB 1939 "Method of testing in compression perpendicular to grain of wood";[49] GB 1940 "Method of testing in toughness of wood",[50] etc.

On the other hand, for wood structures, there are several specifications on engineered composite wood materials. Specifically, ASTM D3737 "Establishing allowable properties for structural glued laminated timber (Glulam)" provides procedures, and details of specimens to obtain the allowable design strengths of glulam.[51] ASTM D5457 covers

Table 1.3 Reference standards

Reference	Standards	Objectives	Notes
[43]	ASTM D143 "Standard test methods for small clear specimens of timber"	Engineering material, many aspects of tests	
[44]	GB/T 1928 "General requirements for physical and mechanical tests of wood"	Engineering material, basic requirements	[43] almost equals to [44 to 50]
[45]	ISO 3129 "Wood – Sampling methods and general requirements for physical and mechanical testing of small clear wood specimens"	Wood material testing	
[46]	GB 1936 "Method for determination of the modulus of elasticity in static bending of wood"	Wood properties in bending	
[47]	GB 1937 "Method of testing in shearing strength parallel to grain of wood"	Wood properties in shear	
[48]	GB 1938 "Method of testing in tensile strength parallel to grain of wood"	Tensile properties of wood	
[49]	GB 1939 "Method of testing in compression perpendicular to grain of wood"	Compressive properties of wood	
[50]	GB 1940 "Method of testing in toughness of wood"	Toughness of wood	
[51]	ASTM D3737 "Establishing allowable properties for structural glued laminated timber (Glulam)"	Glulam strength	
[52]	ASTM D5457 "Computing reference resistance of wood-based materials and structural connections for load and resistance factor design"	Glulam strength	
[53]	GB 19367 "Wood-based panels – Determination of dimensions of panels"	Wood-based panel properties	
[54,55]	GB 50005 Standard for design of timber structures; GB 50772 Code for construction of timber structures	Wood structures, and construction	
[56,57]	Design specifications for timber structures in the United States and Europe	Timber structures	

procedures for computing the reference resistance of wood-based materials and structural connections for use in the load and resistance factor design (LRFD).[52] Chinese national standards are also established for wood-based panels[53] and types of wood structures,[54] and relative construction procedures.[55] Specifications and standards in North America and Europe are also referenced in this book.[56,57]

1.4.2 Standards and Specifications for Bamboo

In the United States, the earliest reference to bamboo as a material can be found in an ICBO (International Conference of Building Officials, Inc.) technical evaluation report AC162, on the acceptance criteria for structural round bamboo released in 2000.[58] The document was developed by Bamboo Technologies, a Hawaii-based company. The document is not a code or standard, but rather a technical reference work that directly addresses the issue of code compliance, useful to both regulatory agencies and building-product manufacturers for implementing a non-code-regulated product (in this case, the round bamboo culms) into building practice. Agencies use evaluation reports to help determine code compliance and enforce building regulations, whereas manufacturers use the reports as evidence that their products (and this is especially important if the products are new and innovative) meet code requirements and warrant regulatory approval. The AC162 is now maintained by ICC ES (International Code Council, Evaluation Services), after the merger of the four major technical evaluation service agencies in the United States, including ICBO ES.

Internationally, the International Bamboo and Rattan Organisation (INBAR) sponsored the development of a model standard for the International Organization for Standardization (ISO).[23] In Latin America, Guadua bamboo has been widely used in dwellings for a long time. Colombia can probably be credited as the first country to implement the chapters for the use of bamboo in the national codes for seismic design of structures.[21] In China, the GB 2690 "Bamboo timber" defines and stipulates several visual aspects of round bamboo culms.[59] All these documents are related to the design of buildings made with round bamboo. The author was also commissioned by INBAR and drafted an engineered bamboo standard; however, unfortunately it was not submitted to ISO nor published. The issues related to developing design codes for engineered bamboo structures will be discussed in Chapter 4, and in this section, the existing standards, related to material testing, evaluation, and manufacture of engineered bamboo products are reviewed.

Properties of Bamboo Culms

Although bamboo-related materials are mainly used as a type of decorative or supplemental material to wood, the testing methods relating to the physical and mechanical properties of bamboos are developed comprehensively.[60] The Chinese standard GB 15780 "Testing methods for physical and mechanical properties of bamboos" provides thorough guidelines for testing bamboo, which incorporate initial bamboo culm sampling preparation, classic experimental procedures, and post processing as well as the report format. Explicitly, the measurements of the moisture content, shrinkage, and density and the testing programs of parallel-to-grain compressive, shear, tensile, and bending strength, are specified.[60]

Bamboo Boards, Plybamboos

In China, the development of various engineered bamboo products propels the establishment of standards. For example, the bamboo is processed and laminated for the product of bamboo flooring, which is regulated in GB 20240 "Bamboo flooring".[61] National standard GB 21179 "Decorative bamboo veneered panel" lays out several physical criteria including formaldehyde emission.[62] Laminated bamboo strip lumber is one of the most significant applications of bamboo-based materials in general. Chinese Forestry

Standard LY 1072 "Laminated bamboo strip lumber" provides the technical guidelines for grading, shapes, mechanical analyses, and testing methodologies.[63] The methods to obtain typical mechanical properties of MOR (modulus of resistance), MOE (modulus of elasticity), impact toughness, thermostable performance, and low temperature resistance are provided in LY 1073 "Experimental methodologies of physical mechanical properties of laminated bamboo strip lumber".[64] The utilization of strip plybamboo for floorboards of trucks and buses is standardized, inclusive of specifications of sampling shapes and necessary mechanical experimental designs.[65] On the other hand, bamboo-mat plywood has been considered on several criteria. GB 13123 "Bamboo-mat plywood" introduces the plybamboo material with its definitions, classifications, and inspection requirements.[66] The particular testing methods for bamboo-mat plywood are specified, covering devices and testing procedures in GB 13124 "Methods of testing bamboo-mat plywood".[67] Apart from the ornamental function, bamboo-based panels are employed to serve as subsidiary materials. Hence, the Chinese Machinery Industry Standard JB/T 6564 "Plybamboo for export machine packing boxes",[68] Construction Industry Standard JG 3059 "Plybamboo form with steel frame,"[69] and Forestry Standard LY 1574 "Plybamboo for concrete-form"[70] all provide certain procedures to regulate production, inspection of commodified products, and to set fundamental mechanical requirements.

Despite the fact that many of the above-mentioned standards are not developed for structural usage with bamboo, these relevant specifications, standards, and guidelines provide valuable references, and are useful for future development of standards and codes for bamboo structure design and construction.

1.5 Bamboo for a Sustainable Future

Following the increased severe changes in worldwide weather, global warming has become a common challenge to human society. Greenhouse gas (GHG) emission from all types of human activities in modern society, due to significant progress in industrialization, has come to the point of exceeding the limit that nature can bear. As shown in Figure 1.12, the GHG transferred to carbon dioxide (simply referred as carbon) has seen a significant upward trend, particularly starting from the end of World War II in 1945.[71]

In the fall of 2021, the Chinese government made a monumental announcement of its plan for reducing carbon emissions dramatically. As shown in Figure 1.13, the country will make all possible efforts to reach its peak of carbon emission by 2030, and to achieve carbon neutrality by 2060.

Among all the suggested measures to cut down carbon emissions, the utilization of bio-based materials in energy, construction, and perhaps all aspects of social activities may be crucial. For China and many of the developing countries in Asia, Africa, and Latin America, the fast-maturing bamboo is a rich and readily available resource.

Interest in and attention toward bamboo have already been growing in the world, as part of the trend of carbon neutral and sustainable development. In a joint international conference held in Costa Rica in 1998, participants from all over the world expressed hope for the use of bamboo for sustainable development.[72] Indeed, as a natural plant,

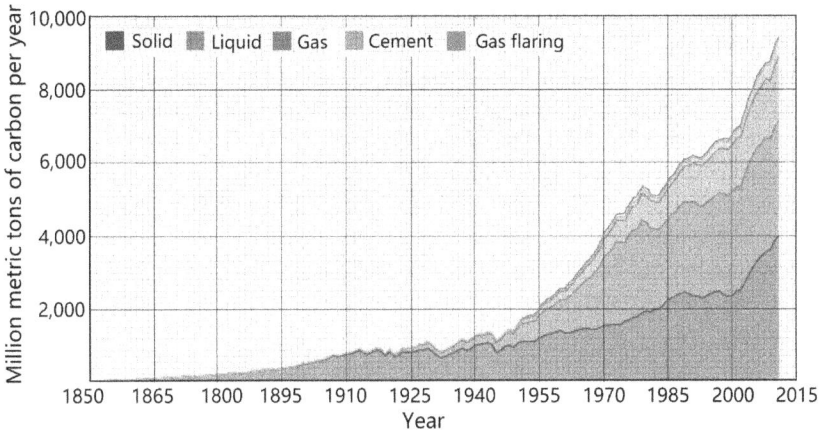

Figure 1.12 Carbon emission trend.

Source: Boden et al. (2015).[71]

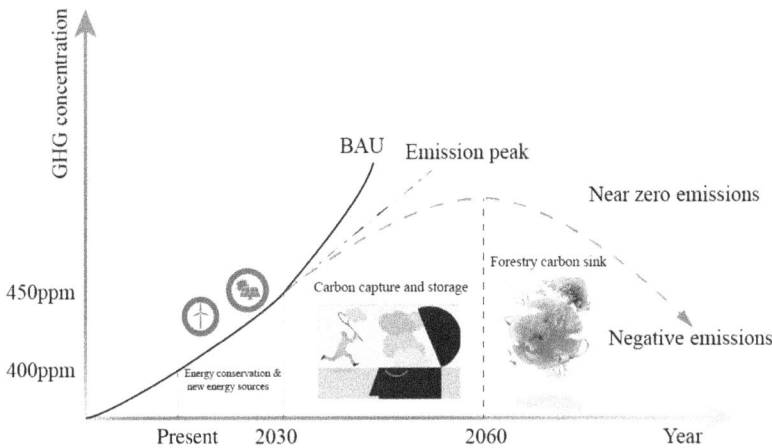

Figure 1.13 China's 2030/2060 carbon emission plan.

bamboo has many potential uses that can be advantageous to society. The following may be worthy of concern for future development:

1   Improvement of the efficiency of traditional bamboo usages: For example, bamboo shoots have been a delicacy in Asia. The yield and nutrition quality should be improved through improved management of the bamboo forest.[73] Similarly, for construction, using bamboo culms will continuously be an affordable building material for housing in many developing countries. Further improvement of yield, quality, even dimensions/shapes (for example, square section culms) may be expected from biological engineering research.

2   Utilization of bamboo will no doubt be a potential way to increase income for people in areas that have been left behind in terms of development, yet are rich in natural resources.

3   Adoption of advanced technology: Modern technology will help to improve the efficiency and reduce the end cost. For example, through the use of digital design and smart manufacture, round bamboo elements can be shaped to be more suitable for prefabrication in building structures.[74]

4   Bio-based material development: Engineered bamboo will continuously evolve as a bio-based composite material. In construction, and also in the furniture industry, engineered bamboo will be used as an alternative to engineered wood products. As a bio-based material, bamboo has the potential to replace plastics in daily life, etc.

Last, but not least, the formation of an engineered bamboo industry will create more jobs in bio-based material production. The author believes that we are on the verge of a new era of a bio-based sustainable society, in which bamboo will certainly play a significant role.

References

[1] Lobovikov, M., Paudel, S., Piazza, M., & Wu, H.R. (2007). *World bamboo resources: A thematic study prepared in the framework of the global forest resources assessment 2005*. Food and Agriculture Organization of the United Nations, Rome.

[2] FAO. (2010). *Global forest resources assessment 2010*. Food and Agriculture Organization of the United Nations, Rome.

[3] Canavan, S., Richardson, D.M., Visser, V., Le Roux, J.J., Vorontsova, M.S., & Wilson, J.R.U. (2017). The global distribution of bamboos: Assessing correlates of introduction and invasion. *AoB PLANTS*, 9(1), plw078.

[4] Liu, W.Y., Hui, C.M., Wang, F., Wang, M., & Liu, G.L. (2018). Review of the resources and utilization of bamboo in China. In H.P.S. Abdul Khalil (Ed.), *Bamboo – Current and future prospects*. IntechOpen, DOI: 10.5772/intechopen.76485

[5] Zhu, Z.H., & Jin, W. (2018). *Sustainable bamboo development*. CABI International, Boston, MA.

[6] Fei, B.H. (2020). Establishment of the national storage mechanism for bamboo material in China. *World Bamboo and Rattan*, 17(6), 1–4.

[7] Janssen, J.J. (2000). *Designing and building with bamboo*. Technical Report 20. International Network for Bamboo and Rattan (INBAR), pp. 130–133.

[8] Buckingham, K.C., Wu, L., & Lou, Y. (2014). Can't see the (bamboo) forest for the trees: Examining bamboo's fit within international forestry institutions. *AMBIO*, 43, 770–778, https://doi.org/10.1007/s13280-013-0466-7.

[9] Xiao, Y., Inoue, M., & Paudel, S.K. (Eds.). (2008). *Modern bamboo structures: Proceedings of First International Conference on Modern Bamboo Structures (ICBS-2007), Changsha, China, 28–30 October 2007*. CRC Press/Balkema, Leiden, The Netherlands.

[10] Xiao, Y., & Shan, B. (2013). *Modern bamboo structures, GluBam* (in Chinese). China Architecture and Building Press, Beijing.

[11] Xiao, Y., Yang, R.Z., & Shan, B. (2013). Production, environmental impact and mechanical properties of glubam. *Journal of Construction and Building Materials*, 44, 765–773.

[12] Sharma, B., Gatóo, A., Bock, M., & Ramage, M. (2015). Engineered bamboo for structural applications. *Construction and Building Materials*, 81, 66–73.

[13] Sun, X.F., He, M.J., & Li, Z. (2020). Novel engineered wood and bamboo composites for structural applications: State-of-art of manufacturing technology and mechanical performance evaluation. *Construction and Building Materials*, 249, 118751.

[14] Ritter, M.A. (1990). *Timber bridges: Design, construction, inspection, and maintenance*. U.S. Department of Agriculture, Forest Service, Engineering Staff, Washington, DC.

[15] Forest Products Laboratory (U.S.). (1987). *Wood handbook: Wood as an engineering material*. General Technical Report FPL-GTR-113. U.S. Department of Agriculture, Forest Service, Forest Products Laboratory, Madison, WI.

[16] Thelandersson, S., & Larsen, H.J. (Eds.). (2003). *Timber engineering*. John Wiley & Sons, Chichester, UK..

[17] Chen, Z.Y. (1958). Several issues about bamboo roof trusses (in Chinese). *Tsinghua Science and Technology*, 4(2), 269–286.

[18] Huang, X. (1959). *Roof bamboo structures* (in Chinese). Architectural Engineering Press.

[19] Tan, G.Y., & Yang, L. (2014). *Tectonics of bamboo* (in Chinese). Southeast University Press, Nanjing.

[20] Liu, K.W., Xu, Q.F., Wang, G., Chen, F.M., & Leng, Y.B. (Eds.). (2019). *Modern bamboo buildings in China* (in Chinese). China Architecture & Building Press, Beijing.

[21] Asociación Colombiana de ingeniería Sísmica (AIS). (2010). *Reglamento Colombiano de Construcción Sismo. Resistente NSR-10 Titulo G-Estructuras de Madera y Estructuras de Guadua*. Bogotá, Colombia, www.unisdr.org/campaign/resilientcities/uploads/city/attachments/3871-10684.pdf

[22] Correal J. (2016), Bamboo design and construction. In K.A. Harries & B. Sharma (Eds.), *Nonconventional and vernacular construction material*, Elsevier, London, Chapter 14.

[23] International Organization for Standardization (ISO). (2021). ISO 22156-2021: Bamboo – structural design. ISO, Geneva, Switzerland, www.iso.org

[24] Chung, K.F., & Yu, W.K. (2002). Mechanical properties of structural bamboo for bamboo scaffoldings. *Engineering Structures*, 24(4), 429–442.

[25] Brink, F.E., & Rush, P.J. (1966). *Bamboo reinforced concrete construction*. U.S. Naval Civil Engineering Laboratory, Port Hueneme, CA.

[26] Ghavami, K. (1995). Ultimate load behaviour of bamboo-reinforced lightweight concrete beams. *Cement and Concrete Composites*, 17(4), 28288.

[27] Rong, B.S. (2008). Opening speech. In Y. Xiao, M. Inoue, & S.K. Paudel (Eds.), *Modern bamboo structures: Proceedings of First International Conference on Modern Bamboo Structures (ICBS-2007), Changsha, China, 28–30 October 2007.*. CRC Press/Balkema, Leiden, The Netherlands.

[28] Liu, X., Smith, G.D., Jiang, Z., Bock, M.C.D., Boeck, F., Frith, O., Gatóo, A., Liu, K., Mulligan, H., Semple, K.E., Sharma, B., & Ramage, M.H. (2015). Nomenclature for engineered bamboo. *BioResources*, 11(1), 1141–1161.

[29] Xiao, Y. (2018).A type of CLBT for wall and slab panels, Chinese patent, ZL201720561674.8, 2018-03-30.

[30] Hutchings, B.F., & Leicester, R.H. (1988). Scimber. In R. Itani (Ed.), *Proceedings of the 1988 International Conference on Timber Engineering, September 19–22, 1988, Seattle, WA*, Vol. 2, pp. 525–533.

[31] Huang, Y., Ji, Y., & Yu, W. (2019). Development of bamboo scrimber: A literature review. *Journal of Wood Science*, 65, Article 25, https://doi.org/10.1186/s10086-019-1806-4

[32] Chen, F., Deng, J., Cheng, H. et al. (2014). Impact properties of bamboo bundle laminated veneer lumber by preprocessing densification technology. *Journal of Wood Science*, 60, 421–427, https://doi.org/10.1007/s10086-014-1424-0

[33] Western Wood Structures. (n.d.). Clear-span timber domes. Retrieved from: www.westernwoodstructures.com/projects

[34] TMP Architecture. (n.d.). Northern Michigan University Superior Dome. Retrieved from: www.tmp-architecture.com/project/superior-dome

[35] Buchanan, A., Pampanin, S., Newcombe, M., & Palermo, A. (2009). *Multi-storey timber buildings using post-tensioning*. NOCMAT, Bath, UK.

[36] Pei, S.L., van de Lindt, J.W., Popovski, M., & Berman, J.W. (2014). Cross-laminated timber for seismic regions: Progress and challenges for research and implementation. *ASCE Journal of Structural Engineering*, 142(4), E2514001.

[37] Cornwall, W. (2016). Would you live in a wooden skyscraper?, Science, September.

[38] Tollefson, J. (2017). The wooden skyscrapers that could help to cool the planet. *Nature*, May.

[39] Green, M., & Taggart, J. (2020). *Tall wood buildings: Design, construction and performance*, Second and Expanded Edition. Birkhäuser, Berlin and Boston, MA.

[40] Ekeberg, P.K., & Søyland, K. (2005). Flisa Bridge, Norway: A record-breaking timber bridge. *Bridge Engineering*, 158(BE1).

[41] Man, Z., Zhuang, H.Y., & Lin, A. (2019). Xuhong Bridge: A long span wooden arch bridge with glued laminated timber. *Structural Engineering International*, 31(1), DOI: 10.1080/10168664.2019.1679061

[42] Xiao, Y., Shan, B., & Li, Z. (2019). Recent progress in engineered bamboo development. In *Modern engineered bamboo structures: Proceedings of the Third International Conference on Modern Bamboo Structures (ICBS 2018), June 25–27, 2018, Beijing, China*. CRC Press, Leiden, The Netherlands.

[43] ASTM International. (2014). ASTM D143-14: Standard test methods for small clear specimens of timber. ASTM International, West Conshohocken, PA, www.astm.org

[44] National Technical Committee 41 on Timber of Standardization Administrator of China. (2009). GB/T 1928-2009: General requirements for physical and mechanical tests of wood (in Chinese). China Building Industry Press, Beijing.

[45] International Organization for Standardization (ISO). (1975). ISO 3129: Wood – Sampling methods and general requirements for physical and mechanical testing of small clear wood specimens. ISO, Geneva, Switzerland, www.iso.org

[46] National Technical Committee 41 on Timber of Standardization Administrator of China. (2009). GB 1936-2009: Method for determination of the modulus of elasticity in static bending of wood (in Chinese). China Building Industry Press, Beijing.

[47] National Technical Committee 41 on Timber of Standardization Administrator of China. (2009). GB 1937-2009: Method of testing in shearing strength parallel to grain of wood (in Chinese). China Building Industry Press, Beijing.

[48] National Technical Committee 41 on Timber of Standardization Administrator of China. (2009). GB 1938-2009: Method of testing in tensile strength parallel to grain of wood (in Chinese). China Building Industry Press, Beijing.

[49] National Technical Committee 41 on Timber of Standardization Administrator of China. (2009). GB 1939-2009: Method of testing in compression perpendicular to grain of wood (in Chinese). China Building Industry Press, Beijing.

[50] National Technical Committee 41 on Timber of Standardization Administrator of China. (2009). GB 1940-2009: Method of testing in toughness of wood (in Chinese). China Building Industry Press, Beijing.

[51] ASTM International. (2018). ASTM D3737-18: Establishing allowable properties for structural glued laminated timber (Glulam). ASTM International, West Conshohocken, PA, www.astm.org

[52] ASTM International. (2019). ASTM D5457-19: Computing reference resistance of wood-based materials and structural connections for load and resistance factor design. ASTM International, West Conshohocken, PA, www.astm.org

[53] National Technical Committee 198 on Wood-based Panels of Standardization Administration of China. (2009). GB 19367-2009: Wood-based panels: Determination of dimensions of panels (in Chinese). China Building Industry Press, Beijing.

[54] Ministry of Housing and Urban–Rural Development of the People's Republic of China. (2017). GB 50005-2017: Code for design of timber structures (in Chinese). China Architecture & Building Press, Beijing.

[55] Ministry of Housing and Urban–Rural Development of the People's Republic of China. (2012). GB 50772-2012: Code for construction of timber structures (in Chinese). China Building Industry Press, Beijing.

[56] American Wood Council. (2018). *National design specification for wood construction.* American Wood Council, Leesburg, VA.

[57] European Committee for Standardization (CEN). (2016). *Timber structures – calculation and verification of characteristic values: EN 14358-2016 [S].* European Committee for Standardization, Brussels.

[58] Bamboo Technologies. (2000). *Acceptance for structural bamboo.* Technical Evaluation Report AC162. ICBO Evaluation Service, Inc., Whittier, CA.

[59] National Technical Committee 263 on Bamboo and Rattan of Standardization Administration of China. (2000). GB 2690-2000: Bamboo timber (in Chinese). China Building Industry Press, Beijing.

[60] National Technical Committee 263 on Bamboo and Rattan of Standardization Administration of China. (1995). GB 15780-1995: Testing methods for physical and mechanical properties of bamboos (in Chinese). China Building Industry Press, Beijing.

[61] National Technical Committee 263 on Bamboo and Rattan of Standardization Administration of China. (2006). GB 20240-2006: Bamboo flooring (in Chinese). China Building Industry Press, Beijing.

[62] National Technical Committee 198 on Wood-based Panels of Standardization Administration of China. (2007). GB 21179-2007: Decorative bamboo veneered panel (in Chinese). China Building Industry Press, Beijing.

[63] National Forestry Administration. (2002). LY 1072-2002: Laminated bamboo strip lumber (in Chinese). National Forestry Administration, Beijing.

[64] National Forestry Administration. (1992). LY 1073-92: Experimental methodologies of physical mechanical properties of laminated bamboo strip lumber (in Chinese). National Forestry Administration, Beijing.

[65] National Forestry Administration. (2000). LY 1575-2000: Strip plybamboo for bottom boards of trucks and buses (in Chinese). National Forestry Administration, Beijing.

[66] General Administration of Quality Supervision, Inspection and Quarantine of the People's Republic of China. (2003). GB 13123-2003: Bamboo-mat plywood (in Chinese). China Building Industry Press, Beijing.

[67] General Administration of Quality Supervision, Inspection and Quarantine of the People's Republic of China. (1991). GB 13124-91: Methods of testing bamboo-mat plywood (in Chinese). China Building Industry Press, Beijing.

[68] National Industry Administration. (1993). JB/T 6564-93: Plybamboo for export machine packing boxes (in Chinese). National Industry Administration, Beijing.

[69] National Urban–Rural Development Administration. (1999). JG 3059-1999: Plybamboo form with steel frame (in Chinese). National Urban–Rural Development Administration, Beijing.

[70] National Forestry Administration. (2000). LY 1574-2000: Plybamboo for concrete-form (in Chinese). National Forestry Administration, Beijing.

[71] Boden, T.A., Marland, G., & Andres, R.J. (2015). *Regional and national fossile-fuel CO2 emissions.* Carbon Dioxide Information Analysis Center, Oak Ridge National Laboratory, U.S. Department of Energy, Oak Ridge, TN.

[72] Kumar, A., Sastry, C., & Ramanuja Rao, I.V. (1998). *Bamboo for sustainable development: Proceedings of the Vth International Bamboo Congress and the VIth International Bamboo Workshop, San Jose, Costa Rica, 2–6 November 1998.*

[73] Zhen, R., Zheng, W.P., Fang, W., Huang, Y.H., & Zhang, S.S. (2010). Shapes and nutrients of dendrocalamopsis oldhami bamboo shoots in 12 production areas. *Journal of Zhejiang Forestry College,* 27(6), 845–850.

[74] Lorenzo, R., Godina, M., Mimendi, L., & Li, H. (2020). Determination of the physical and mechanical properties of moso, guadua and oldhamii bamboo assisted by robotic fabrication. *Journal of Wood Science,* 66(20).

Chapter 2

Production of Engineered Bamboo

As discussed in the previous chapter, engineered bamboo can have a broad definition: bamboo products that are designed, manufactured, and constructed using modern engineering concepts, procedures, including inspection, etc. Even many round bamboo culm structures are designed and built with modern engineering methodology and therefore can be considered as engineered (round culm) bamboo structures. This book, however, mainly focuses on the introduction and discussion of laminated bamboo, and specifically, glued laminated bamboo or glubam. Therefore, this chapter first discusses the manufacture of engineered bamboo.

2.1 Definition of Glubam

Since the production of engineered bamboo for construction is essentially a new industry, it is important to define the industrial manufacture process. Glubam is an engineered bamboo manufactured with the intended application of serving as an alternative to wood-based laminated lumber, or glulam.[1–3] Glubam is manufactured using two-step lamination processes.[1,4] Since bamboo is geometrically nonuniform and irregular in its original form, the first process is to convert it into a more regular shape, through a hot-pressure process. The end product at this stage is typically in the form of boards with a specific dimension, and can be called glubam board, laminated bamboo board, or plybamboo. Of course, many of the board products can have their own usages, such as floorings, shipping container floorings, concrete formwork, etc. The typical bamboo boards used for making glubam have a thickness of 25 mm to 40 mm. The second lamination process is essentially similar to manufacturing glulam using a cold-pressing lamination process.

It might be interesting to point out the distinct difference between glubam and glulam, in terms of the first manufacturing process. For glulam, the first process is to saw-cut the wood logs into standardized lumbers, a kind of dis-integration process. However, for glubam, the first process is to integrate the bamboo pieces (strips) into larger-size and standardized boards. Also, the first process (for glubam) is typically completed in factories near bamboo growing areas (partially for transportation efficiency to avoid long-distance transfer of bamboo culms which contain a significant amount of air), while the second process can be executed at fabrication facilities close to the construction site. This is actually quite logical as it can be analogically compared with the process of producing steel structures, as shown in Table 2.1.

DOI: 10.1201/9781003204497-2

Table 2.1 Analogical comparison of steel and glubam structures

Steel	*Engineered wood – glulam*	*Engineered bamboo – glubam*
Mining iron ores	*Foresting, logging*	*Planting, nursing and logging*
Further refining iron using a heating/combustion process to separate/mix materials	Various treatment procedures	Drying, carbonization, impregnation of chemicals for durability, etc.
Hot rolling process to produce sheets, shapes, etc.	Manufacture of standard lumbers, plywood, etc.	Hot-pressing process to produce bamboo boards, plybamboo
Transportation in mass quantities		
Cutting, welding, drilling to produce structural elements, at the steel structure fabrication facilities	Cold-pressing process to produce glulam elements*	Cold-pressing process to produce glubam elements
Site installation and connection of steel components	Site installation and connection of glubam components	Site installation and connection of glubam components

Note: *Photo courtesy of Mr. J.J. Zhou, Suzhou CROWNHOMES, China.

In the following two sections, the process of making bamboo boards is first introduced, followed by descriptions of the final process for manufacturing glubam.

2.2 Manufacturing Process of Laminated Bamboo Boards

As defined in the previous section, glubam is made with two processes. The first process is the manufacture of laminated bamboo boards. Typically, there are mainly two types of laminated bamboo boards made, and the author defines them as the thick-strip bamboo board and the thin-strip bamboo board. The final glubams can then be termed as thick-strip glubam and thin-strip glubam, respectively.

2.2.1 Laminated Thick-strip Bamboo Board

The thick-layer laminated bamboo boards or sheets are made by pressure gluing a few layers (typically three to five layers) of relatively thicker (about 5–8 mm) bamboo strips with a width of about 20–25 mm, often referred as laminated bamboo lumber (LBL)[5] or glued bamboo laminates (GBL).[6–10] The top-of-the-line products are made for flooring plates, as an alternative to wood flooring.

As shown in Figure 2.1, the three- to five-year-old Moso bamboo (Phyllostachys edulis) culm is first split into arc segments, then the bark and the inner layers are removed, and rectangular bamboo strips or slats are shaped. Since the thick- strips are required to have accurate dimension, the round culms need first to be selected to reduce the irregularity. Furthermore, due to the fact that the wall thickness of a culm may vary along the height, the length of the strip may be limited. Once the strips are made (also called rough planing), they need to be cleaned, boiled, steamed, and dried. For further improving durability, carbonization can be applied. The carbonization on the one hand can destroy worms, their eggs, and fungi in the bamboo; on the other hand, it can eliminate the nutritional source for the borers, avoiding mildew and worm damage

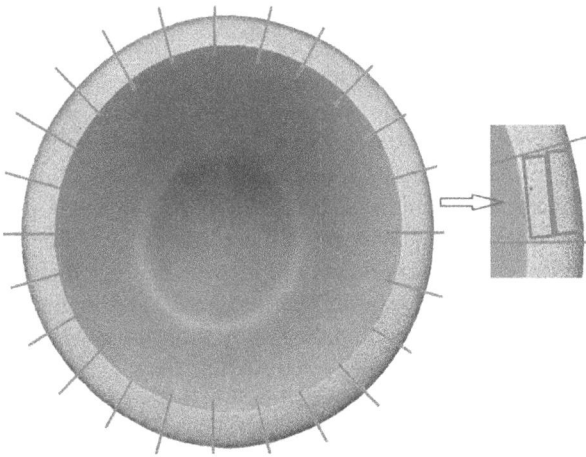

Figure 2.1 Splitting bamboo culm for relatively thick strips.

Figure 2.2 Examples of thick bamboo strips without carbonization (a); with medium carbonization (b); and deep carbonization (c).

in the bamboo during its later use. The bamboo strips are placed in a carbonization chamber and steamed at a temperature of 150–300°C for hours and then cooled down. Afterwards, the strips need to be planed again to form a precise dimension. Figure 2.2 shows the bamboo strips with different levels of carbonization. The medium carbonization process is 125°C for 80 min., whereas the deep carbonization is 140°C for 110 min.

In the first lamination process, the precisely planed bamboo strips are glued with urea-formaldehyde (UF) or phenol-formaldehyde (PF, can be used for outdoor structures), aligned to form a plane board, and then pressed under elevated temperature at 80–90°C, pressure 2–3 MPa (mega pascal), for about 60 min. The board can be configured with the strips aligned either flatwise or sidewise, as shown in Figure 2.3 (a) and (b), respectively; or in a combination of the two, as shown in Figure 2.3 (c). In order to avoid regular alignment of glue lines, the strips should be staggered for flatwise lamination, as shown in Figure 2.3 (d). If necessary, the strips can also be cross laminated to form CLB, cross laminated bamboo. For the hot-pressed bamboo boards, the overall thickness is typically limited to 50 mm, due to the fact that the heat transfer efficiency in bamboo can generate a temperature gradient and affect the hardening of the adhesive.

Note that since the wall thickness of a bamboo culm is thinning from the bottom to the top, cultivating only long pieces of strips may result in more waste of bamboo layers. Therefore, the strips need to be lengthened for production of longer boards. For thick-strips, the joining ends of the two thick-strips can first be shaped into hooks and then clamped together when aligning the strips before pressure gluing. Normally, the hooked joints should be alternated to avoid concentration of weak sections, when aligning the strips.

It should be pointed out that the thick-strip laminated bamboo boards are typically more expensive, due primarily to: i) removal of more materials when cutting strips from the bamboo culm; ii) the stringent requirements for the dimension of the strips; iii) and the need for accurate alignment in lamination.

Figure 2.3 Configurations for thick-strip laminated bamboo boards: (a) flatwise pressed; (b) multiple layers, staggered flatwise pressed; (c) sidewise pressed; (d) "I-shape" combination pressed.

Source: Photos courtesy of Ms. T. Li, Hunan Taohuajiang Bamboo Science & Technology Co., LTD.

2.2.2 Laminated Thin-strip Bamboo Board – Plybamboo

The thin-strip laminated bamboo boards were originally developed for concrete form-work, floor plating of ship container boxes, etc. Two types of cutting methods are used in producing thin bamboo strips; i.e., the circumferential cutting and the radial cutting, as shown in Figure 2.4 (a) and (b) respectively. For circumferential cutting, the Moso bamboo culms are first split into arc segments and then the segments are sliced into thin-strips approximately 20 mm wide and 1.8–2.4 mm thick, either by machine or manually. The bamboo bark and the inner layer are typically removed. For radial cutting, the strips are cut about 2.0 mm thick along the radial direction without removing the bamboo bark and the inner layer. For the same bamboo culms, the usage of materials for thin-strips is higher than that for thick-strips; particularly in the case of radial cutting thin-strips, there is almost no waste left.

The thin-strips are then netted into unidirectional, single-layer curtains – 2.6 m long for the longitudinal direction (long curtain) and 1.3 m long for the transverse direction (short curtain). The netted strips are dried first in a drying chamber to reduce their water content below 10±2%. The dried nets are saturated in adhesives for about 2 to 3 min. and then dried again to control the water content ratio in a range of 16±2%. Phenol-formaldehyde adhesive is used for the hot-pressing process and the typical composition is approximately 1.0:1.17 for phenol and formaldehyde. If necessary, other ingredients such as insect repellent agents can be added.

The thin-layer bamboo boards for glubam typically have a nominal thickness of 30 mm which contains 21 layers of thin-strip curtains with a longitudinal to transverse

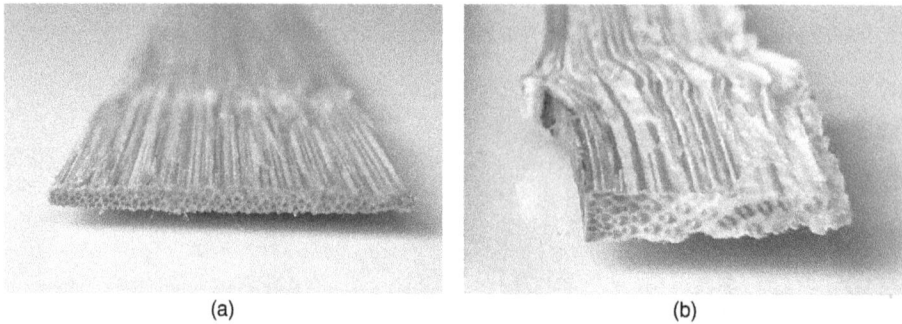

(a)   (b)

Figure 2.4 Different cutting methods for thin-strips: (a) circumferential cutting (sample width about 25 mm, average thickness about 2.0 mm); (b) radial cutting (sample width about 11.0 mm, average thickness about 1.8 mm).

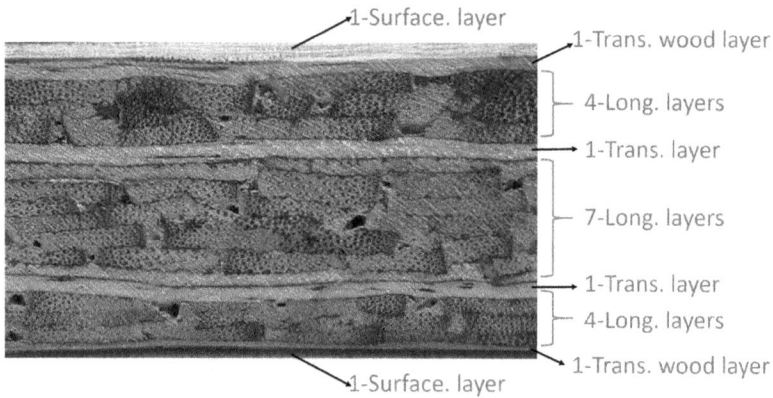

Figure 2.5 Example of configuration of thin-strip curtains.

strip ratio of about 4:1. As shown in Figure 2.5, the curtain configuration is typically (from one side to the other): surface layer, 1 layer of wood transverse curtain, 4 layers of longitudinal curtains, 1 layer of transverse curtain, 7 layers of longitudinal curtains, 1 layer of transverse curtain, 4 layers of longitudinal curtains, 1 layer of wood transverse curtain, and another surface layer. The wood-strip curtains serve as a kind of cushion during the hot-pressing process to control the uniformity of the board thickness.

The multi-layer curtains form a mat, about 2.6 m long and 1.3 m wide and this is first pressed at room temperature with a pressure of about 2–5 MPa to reduce its thickness; then it is placed in the hot-presser machine that can produce multiple boards at one time. The hot presser can apply 18.5–22 MPa pressure to the mats under an elevated temperature of 145°C for about 60 min. (including 10 min. to raise the temperature and 10 min. cooling stage). The time required for pressing is roughly 1.5 min. for every 1 mm thickness. After trimming, the bamboo boards typically have a dimension of 2.44 m long and 1.22 m wide, which is the typical dimension of plywood in the United States; i.e., 8 ft by 4 ft.[11] Bamboo boards up to 6.0 m long can also be produced; however, the

Figure 2.6 Production process of glubam boards.

Note: ①: Raw bamboo, ②:Weaving, ③: Drying, ④: Dipping, ⑤: Paving, ⑥: Hot-press forming, ⑦: Sawing, ⑧: Products.

pressing machine required for pressing longer pieces becomes unnecessarily costly, inefficient, and less flexible for production variations, and it is more difficult to pack and transport the end products. Figure 2.6 shows the overall process of thin-strip bamboo board production.

2.3 Cold-pressing Process of Glubam

In the first industrial process, the standardized laminated bamboo boards are mass produced, as commodities. Some of them have their own market of applications, but for engineered bamboo structures, they are semi-finished products. The bamboo boards are then used to produce the glubam structural components, based on structural design.

The standardized laminated bamboo boards only have a specific size, and the thickness is typically limited to 20 to 40 mm, therefore they need to be sized by cutting, spliced for longer length, and laminated again by a cold-pressing process to form structural components based on a specific design, similar to that for glulam.[12]

Since typical structural components are longer than the standardized bamboo boards, one of the key technical details is how to extend the limited length to produce longer components to satisfy design need. Actually, similar extension is also always needed to manufacture glulam for timber structures. A technique called finger jointing is typically

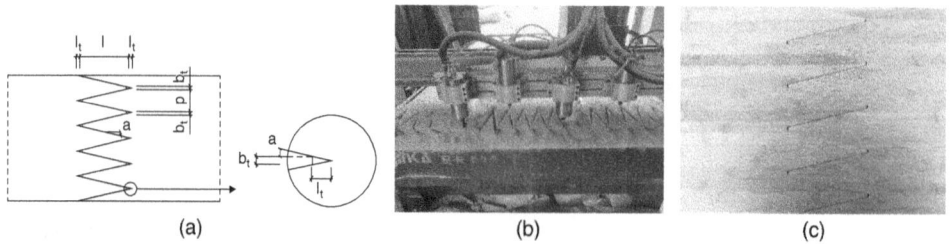

Figure 2.7 Finger joint: (a) dimension details; (b) processing using CNC cutting machine; (c) abutting figure joint.

Figure 2.8 Cold lamination pressing machine.

adopted, where the joining ends of timber elements are shaped into a zigzag shape and glued together with pressure applied along the axis of the joining elements. A similar technique is used for glubam during the cold glue process to make up the length required for structural components,[13] as exhibited in Figure 2.7. The finger joint teeth can be cut using a computer numeric control (CNC) cutting machine or waterjet cutting machine. It should be pointed out that for components primarily used for bearing compression, a butted joint is acceptable

In the cold-pressing lamination process, the elements sized from laminated bamboo boards are applied with adhesives and aligned based on design. The adhesive is typically epoxy,[2,4] or resorcinol resin adhesive.[11] Figure 2.8 shows the combined multi-function cold press machine developed by the author and his colleague Shan.[14] The length of the cold presser can be adjusted according to the requirements of the glubam components, and the pressure can be applied in two directions by the synchronous hydraulic system.

Figure 2.9 Glubam structural components: (a) thick-strip glubam girder; (b) thin-strip glubam columns.

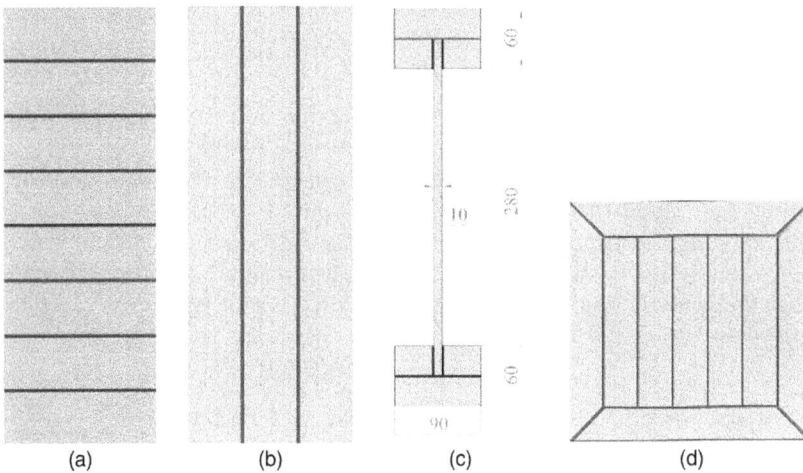

Figure 2.10 Configurations of glubam elements in sections: (a) flatwise layup; (b) sidewise layup; (c) I-shape joist section; (d) example of possible column section.

Figure 2.9 shows examples of glubam components made with thick-strip and thin-strip bamboo, respectively. It is important to note that for a laminated girder member, it has two different configurations; i.e., flatwise layup and sidewise layup, as shown in Figure 2.10 (a) and (b), respectively. The flatwise layup and cold-pressing formation are similar to the configuration for wood-based glulam; however, this is not as efficient as the sidewise layup. As can be seen in later chapter discussions, the stiffness and capacity of the glubam beams and girders with sidewise layup are better than those with flatwise layup.[15] For the same size of glubam beam to be manufactured using bamboo board with the same thickness, the flatwise layup configuration not only requires more cutting work, but also results in a larger total lamination contact area compared with the sidewise layup lamination; thus it may cost more in terms of adhesive. However, for

Figure 2.11 Segmented connection of glubam girders.

the sidewise layup, larger pressing equipment is needed since each of the lamination surfaces is larger, compared with the flatwise layup.

In addition, the I-shape joist[16] can also be made as shown in Figure 2.10 (c). Columns can be made with solid sections and covered by a layer of bamboo plates as illustrated in Figure 2.10 (d). The glubam structural members can also be made as hollow sections to reduce the weight and to enlarge the section modulus.

For very long components, such as long-span glubam girders which may be restrained by transportation limits, it is recommended that segments are manufactured and then connected at the construction site; an example of this type of connection is shown in Figure 2.11.

2.4 Environmental Assessment of Glubam Production

As a structural material, the environmental impact of the production of glubam is also an important aspect to be considered. It is well known that wood is a low-carbon, environmentally-friendly green building material compared with concrete, steel, and aluminum. Bamboo is called a "green" material because of its carbon sequestration capacity and lower production energy consumption. Similarly, we need to study glubam's production energy consumption and carbon emissions to assess its overall environmental impact.

The author's team conducted a survey on the glubam production base of the manufacturers in Yanling County, Hunan Province of China, and inspected the whole process of producing glubam.[3]

2.4.1 Analysis of Total Energy Consumption of Production

According to the actual production conditions of the manufacturer, the relevant data for calculating the energy consumption of glubam production per cubic meter were obtained. Production energy consumption mainly comes from the following

Table 2.2 Energy consumption value during main production process of glubam

Process	Transportation	Splitting	Weaving Drying Cutting	Manufacturing adhesive	Hot pressing	
Raw material or energy consumption	Diesel oil	Bamboo	Electric energy	PF Phenolic glue	Hydraulic oil	Bamboo scrap
Material or energy use	——	672,000 plants (The average diameter is 35 cm)	145,400 kWh	1,050,000 kg	4000 kg	4200 t
Raw material consumption per cubic meter of glubam	6.5136 kg	56 plants	121 kWh	87.5 kg	0.33 kg	350 kg×70%
Coefficient of material energy consumption	1.4571 kg Standard coal/kg	——	0.400 kg Standard coal/ kW h	22.5 MJ/kg	1.2 kg Standard coal /kg	21.16 MJ/kg
Energy consumption GJ/m³	0.2783	——	1.4185	1.9687	0.0116	-5.184GJ

Note: MJ = megajoule; GJ = gigajoule.

aspects: transportation of materials, power consumption, preparation of adhesives, consumption of hydraulic oil for hot-pressing equipment, water consumption for heating and cooling, consumption of boiler fuel, energy consumption of plant and equipment, etc. In order to save energy, the manufacturer uses the residuals (rods, leaves, etc.) of the bamboo as boiler fuel, which can be used instead of coal. In calculating the energy released by the burning of bamboo waste, it is assumed that 70% of the energy is used for glubam board production, and the other 30% is used for other heating facilities and energy loss in the factory. Since the energy combustion is derived from raw bamboo, this value is added as a negative value when calculating the coal consumption. In the calculation of glubam energy consumption, the energy loss caused by water circulation during the heating and cooling of the hot-press procedure is not included in the total value. The main production processes and corresponding energy consumption calculations are shown in Table 2.2.

After research and calculation, the energy consumption per cubic meter of glubam is 2.67 GJ/m³ which includes the summation of data from Table 2.2 and the energy consumption nested in the production equipment and the factory facilities.[3] It can be seen from Table 2.2 that the electrical energy consumed by the production and the preparation of the adhesive are the main contributors of energy consumption. Therefore, improving the hot-pressing efficiency and reducing the amount of adhesive used are important means to reduce the energy consumption of glubam.

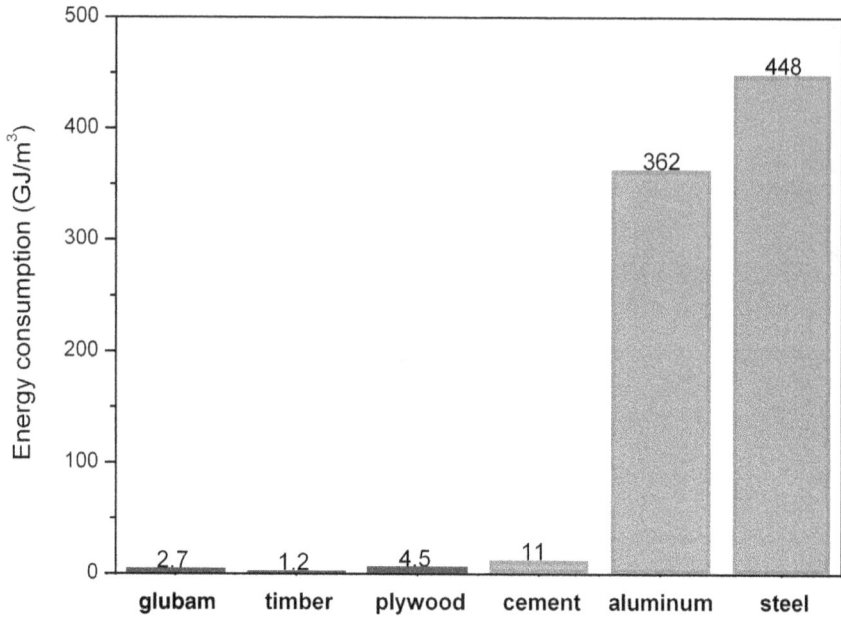

Figure 2.12 Comparison of production energy consumption between glubam and other building materials

Source: Yiao et al. (2013). [3]

Figure 2.12 compares the energy consumption of glubam materials with several commonly used building materials such as timber, plywood, cement, aluminum, and steel. The energy consumption data for cement come from Hammond and Jones,[17] and the data for other materials come from Buchanan and Honey.[18] As can be seen from the figure, the energy consumption of aluminum and steel is much higher than that of other materials. Glubam consumes far less energy than cement and steel, but slightly more than timber.

2.4.2 Analysis of Carbon Emissions

The low carbon performance of bamboo products has been widely recognized; therefore glubam is also expected to be potentially widely used in future low carbon buildings. The main carbon emission processes and corresponding values produced in the manufacturing of glubam are analyzed using information in literature[19–21] and shown in Table 2.3.[3] A large number of studies have shown that bamboo has a higher carbon sequestration ability than woody tree species. The amount of solidified carbon dioxide per cubic meter of bamboo can be calculated by this equation:[22]

$$C = QY\frac{N}{M}$$

(2.1)

Table 2.3 Carbon emissions from main production processes of glubam

Process	Bamboo solidified carbon	Carbon emissions during processing								
		Cutting and transportation	Weaving	Drying	Cutting	Performance testing	Manufacturing adhesive	Hot pressing		
Degree of automation	Natural	Semiautomatic		Automatic and manual			Manual	Automatic		
Material requirement	Natural environment	Diesel oil		Electric energy			PF Phenolic glue	Hydraulic oil	Water for cooling	Bamboo scrap (in boiler)
Annual material consumption		—		145,400 Kwh			1,050,000 kg	4000 kg	3000 t	4200 t
Energy consumption per cubic meter of glubam (energy consumption)		6.5136 kg (0.2783 GJ)		121 Kwh (1.4185 GJ)			87.5 kg (1.97 GJ)	0.33 kg (0.012 GJ)	250kg	350 kg (7.406GJ)[18]
CO_2 emission coefficient		74.1 kg/GJ[19]		0.975 kg/Kwh = 271 kg/ GJ[20]			0.356 kg/GJ	2.1 t/t Standard coal = 73.3 kg/GJ[19]	0	3.3 t/t Standard coal = 112,000 kg/T]
CO_2 emission kg/m^3	-2166 Eq.2.1	482.66		384.41			31.15	0.85	0	829.47

Note: t/t = ton of CO_2 per ton of standard coal; T] = terajoule.

where C is the amount of CO_2 fixed per cubic meter of bamboo (tCO_2/m^3); Q is the amount of CO_2 that can be fixed per hectare per year of bamboo forest, which is about 36.44 $tCO_2/(hm^2)$ based on Zhou and Jiang's research;[23] N is the number of bamboos required per cubic meter of glubam material, and assumed to be about 52; M is the number of bamboos that can be accommodated per hectare of bamboo forest, taken as 3500; Y is the mature age, taken as four years. According to the above situation and the actual investigation, the amount of carbon dioxide that can be fixed during the bamboo growth process required for one cubic meter of glubam is calculated as 2.166 t.

Table 2.3 shows the main emission data during the process.[3] Based on the research and calculation of carbon dioxide emissions from the overall process of glubam production, the CO_2 emissions from glubam were -261 kg/m^3 including data from Table 2.3 and carbon emissions counted for the equipment, factory inventories, etc.[3] Obviously, the glubam material is a carbon negative structural material, similar to timber.

Figure 2.13 shows the carbon emissions of glubam and other commonly used construction materials such as timber, plywood, cement, aluminum, and steel. Compared to other materials, cement, aluminum, and steel release more carbon dioxide during production; glubam, timber, plywood, etc. can fix carbon dioxide. The comparison shows that glubam has a negative carbon emission and its carbon sequestration is greater than that of timber and glulam.

The study by the author's research group on the carbon emissions of glubam might be preliminary as the first pioneer attempt. However, it basically reflects the environmental impact of laminated bamboo products in the context of current manufacturing processes and technology.

2.5 Emerging Technologies

Corresponding to the development of sustainable construction around the world, newer technologies or revival of existing technologies to produce various bamboo products are developing rapidly. This section attempts to introduce some of the technologies for engineered bamboo production.

2.5.1 Flattened Bamboo

Recently, the so-called flattened bamboo is gaining some attentions in application.[24] A Chinese article published in 1988 by renowned scholar Prof. Q.S. Zhang of the Nanjing Forest University describes the basic concept, processing method, and machinery development.[25] The main advantage of flattened bamboo is the possibility of using almost the full thickness of the bamboo culms, through various processes. However, due to the very basic nature of the bamboo – its uneven thickness and diameter along the length – the length of the flattened bamboo strips is rather limited. Besides, its cost is still high, similar to the thick-bamboo strip, due to the need for added processing and high accuracy for dimensioning. In this sense, the laminar using flattened bamboo may be categorized as a special thick-strip glubam.

Flattening of the originally round bamboo requires special conditions of preparation of the culms, warming and heating, as well as applying stresses.[26] Flattened bamboo is currently used in the manufacture of floorings and cutting boards. Figure 2.14 (a) and (b) show the specially selected Moso bamboo culms and the short flattened bamboo

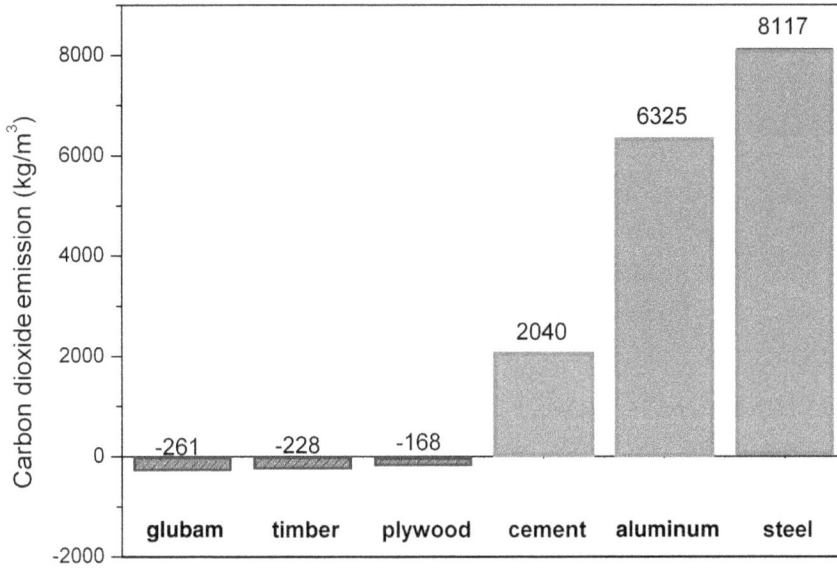

Figure 2.13 Carbon emissions from glubam and other building materials.
Source: Yiao et al. (2013). [3]

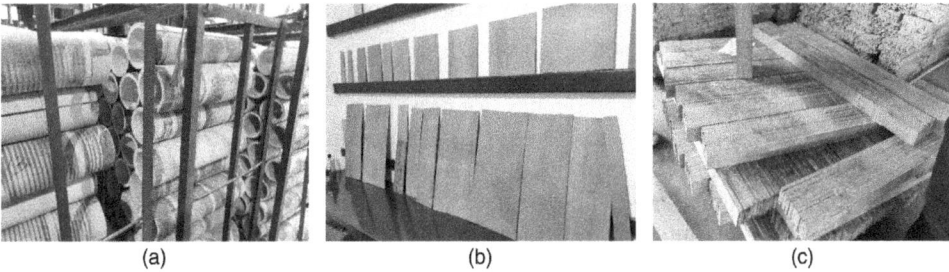

(a)   (b)   (c)

Figure 2.14 Flattened bamboo in processing: (a) selected and preprocessed raw material; (b) short flattened bamboo; (c) laminated flattened bamboo about 1.2 m long.

pieces for furniture in a factory in Longyou, Zhejiang, China. Figure 2.14 (c) exhibits some trial lamination studs with 1.2 m long flattened bamboo strips, at the Guizhu Corp., Ganzhou, China.

2.5.2 High Frequency Hot Press Lamination

The limitations in terms of achievable thickness during the hot-pressing process is one of the main hurdles for producing large-size structural components of engineered bamboo. Drawing on the experiences and development in timber-based glulam and plywood, high frequency hot-pressing technologies are being utilized in producing laminated bamboo, particularly LBL or thick-strip glubam.[27,28]

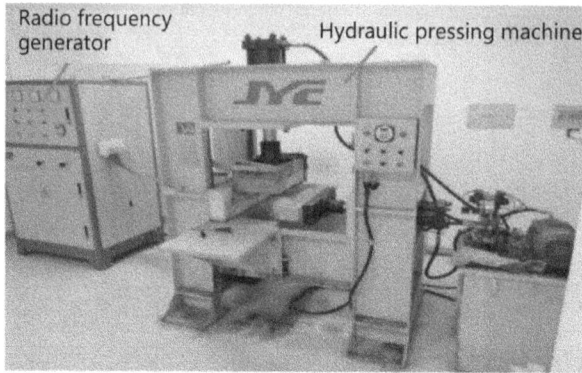

Figure 2.15 High frequency hot presser.
Source: Liu et al. (2020).[31]

The high frequency heating is also termed as high frequency dielectric medium heating. The concept is well known and similar to the working principle of micro-wave heaters. When a high frequency electric field is applied to a material, it repeatedly polarizes the polar molecules inside the material, causing the rapid movement of molecules, thus generating heat due to internal friction. This technology is widely used in processing wood for various purposes, such as drying, lamination, and pest elimination, etc., due primarily to the extensive understanding of the dielectric properties of wood.[29]

In Nguyen et al.'s research study,[28] the bamboo strips were bonded by phenol-formaldehyde (PF) resin using a high frequency hot press to prepare glued laminated bamboo (GLB). The GLB specimens were 500 mm long, 200 mm wide, and 20 mm thick with 4 layers of bamboo strips (5 mm thick) . Nguyen et al. investigated the influences of bamboo moisture content and adhesive spread on the physical and mechanical properties of GLB. During lamination, the normal pressure was 3.0 MPa and the side pressure was 5.0 MPa. The results showed that higher moisture content and spread of adhesive improved the performances of GLB. The optimal pressing was achieved by a combination of adhesive spread of 300 g/m², moisture content of 12% and pressing time of 6 min., under high frequency heating. Nguyen and Zhang also studied the temperature immediately after the pressing and showed that the temperature can reach 100°C during the short process.[30] Similar research has also been carried out recently by Liu et al.[31] using a high frequency hot pressing lamination machine, as shown in Figure 2.15.

In the literature to date, the thickness dimensions of the high frequency hot press laminated bamboo or thick-strip glubam was relatively small (20 mm composed of 4 layers of flat strip). Jiang et al. also reported making bamboo scrimber with a dimension of 3000 mm × 800 mm × 200 mm using high frequency hot press lamination.[27]

2.5.3 Bamboo Filament Winding

Another proprietary technology, named as "Zhu Chanrao" in Chinese or bamboo winding has been announced recently.[32] The technology appears to be essentially a

Figure 2.16 Bamboo winding products.

filament winding process in which the bamboo fiber strips are sorted through to be wound into layers of walls of tubular shapes, along with continuous fabrics. The developer claims that this bamboo-based composite can be used for manufacturing many industrialized products, including utility pipes for urban cities, train carriage structure, houses, etc. Figure 2.16 depicts the full-scale section model in the exhibition hall of the developer's company.

2.5.4 Cross Laminated Bamboo and Timber (CLBT)

Inspired by the great success and growing development of cross laminated timber (CLT) in Europe and elsewhere,[33–36] the author conceived the idea of combining bamboo and timber for developing cross laminated bamboo and timber (abbreviated as CLBT or CLTB and referred to as CLBT in this book).[37,38] Another motivation is to take the advantage of the rich natural resources of bamboo and the growing availability of the

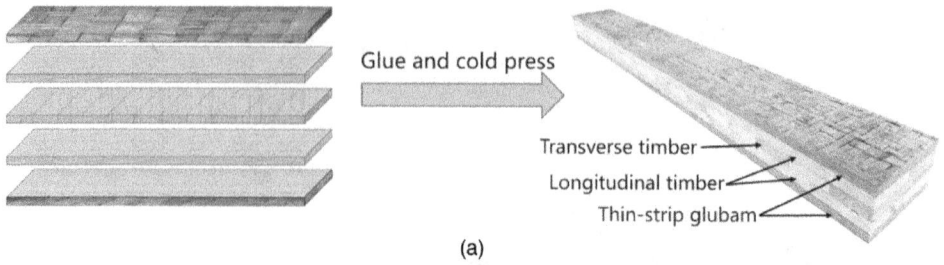

Glue and cold press

Transverse timber
Longitudinal timber
Thin-strip glubam

(a)

(b)

Figure 2.17 CLBT configuration (a); and 6 m long CLBT after lamination (b).

vast output from planted forests, such as the poplar, eucalyptus, and China fir forests. The overall goal is to aid in the push toward a carbon neutral society.

A pilot study was carried out on testing CLBT slab beams.[38] The testing parameters included the combination of two types of timber, spruce-pine-fir (SPF) and poplar, as inner layers, as well as two types of glubam, thin-strip and thick-strip , as surface layers. Figure 2.17 (a) shows the configuration of a typical CLBT flat beam. Recently, the 6 m long CLBT slabs have been produced using the CLT production line at the Ningbo Sino-Canada Low-Carbon Technology Institute Co. Ltd. (Figure 2.17 (b)). During the lamination process, a pressure of 0.8–1.2 MPa was applied for 1–1.5 h in the ambient environment at a room temperature of about 25°C for curing the one-component poly-urethane (PUR) adhesive. Following the room-temperature pressing process, the CLBT specimens were trimmed and excessive adhesives on the surfaces removed, for testing their bending and shear properties.[39]

2.5.5 Automation

Throughout history, human beings have invented countless tools, machines, and methods which have significantly enhanced the productivity of all types of goods to support the activities of human life and society. Certainly, this process is endless.

Automation has been very successful in many sectors of manufacturing industry, including the furniture industry, and is now spreading in the construction industry. Today, the overall manufacturing industry is moving fast toward the smart manufacturing phase, which will revolutionize the interaction among human and machines and different sectors of society. Behind the interaction are the data and information science and technology.[40] How to collect, organize, and utilize data is the key to smart manufacturing. Artificial Intelligence, Virtual and Augmented Reality, and advanced robotics all have a role to play in smart manufacture.

In the construction industry, Building Information Modeling (BIM) can serve as a platform to integrate or disintegrate digital models (digital twins) of all parts or systems of a building, a city, or an even larger system. Lorenzo et al. recently attempted to establish a platform for the design and detailing of building systems constructed with round bamboo culms.[41] They propose to use digital scanning to obtain the geometrical and material information of the natural bamboo culms, and then digitize and model them to enable robotic manufacturing, such as cutting and tailoring the joints, etc. This concept was also demonstrated by producing a large number of small clear specimens from bamboo culms.[42]

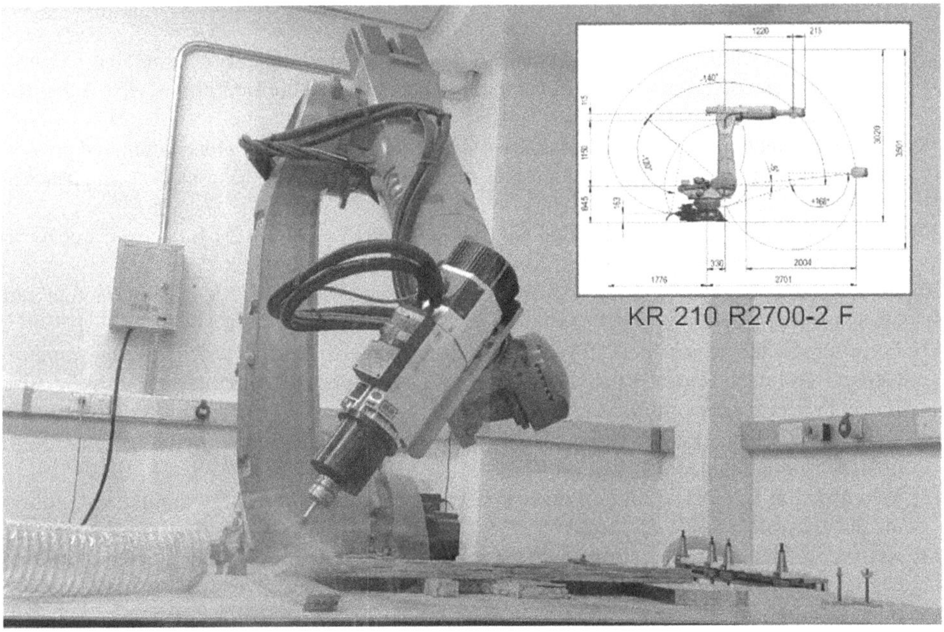

Figure 2.18 Robotic manufacture of glubam elements.

Ma et al. from the author's research laboratory have been working on developing digital design and manufacture of spherical domes with glubam elements.[43] A Grasshopper-based (software) program was developed for the parametric assessment and digital design of the dome, based on a given number of divisions of the dome using triangle elements. The information related to the shape and dimension of the elements and end joints can then be input to the robotic machine for manufacture. Figure 2.18 shows the cutting and drilling of the glubam element by the robotic machine.

References

[1] Xiao, Y., Shan, B., Chen, G., Zhou, Q., & She, L.Y. (2008). Development of a new type of Glulam – GluBam. In Y. Xiao, M. Inoue & S.K. Paudel (Eds.), *Modern bamboo structures: Proceedings of First International Conference on Modern Bamboo Structures (ICBS-2007), Changsha, China, 28–30 October 2007*. CRC Press/Balkema, Leiden, The Netherlands.

[2] Xiao, Y., Zhou, Q., & Shan, B. (2010). Design and construction of modern bamboo bridges. *ASCE Journal of Bridge Engineering*, 15(5), 533–541.

[3] Xiao, Y., Yang, R.Z., & Shan, B. (2013). Production, environmental impact and mechanical properties of glubam. *Journal of Construction and Building Materials*, 44, 765–773.

[4] Xiao, Y., & Shan, B. (2013). *Modern bamboo structures, GluBam* (in Chinese). China Architecture and Building Press, Beijing.

[5] Lee, A.W.C., Bai, X., & Bangi, A.P. (1998). Selected properties of laboratory-made laminated-bamboo lumber. *Holzforschung*, 52, 207–210.

[6] Wahab, R., Mohamed, A.H., Sulaiman, O., & Samsi, H. (2006). Performance of polyvinyl acetate and phenol resorcinol formaldehyde as binding materials for laminated bamboo and composite-ply from tropical bamboo species. *International Journal of Agricultural Research*, 1(2), 108–112.

[7] Mahdavi, M., Clouston, P.L., & Arwade, S.R. (2011). Development of laminated bamboo lumber: Review of processing, performance, and economical considerations. *ASCE Journal of Materials in Civil Engineering*, 23(7), 1036–1042.

[8] Luna, P., Takeuchi, C., & Cordón, E. (2014). Mechanical behavior of glued laminated pressed bamboo guadua using different adhesives and environmental conditions. *Key Engineering Materials*, 600, 57–68.

[9] Li, Y.J., Shen, Y.C., Wang, S.Q., Du, C.G., Wu, Y., & Hu, G. (2012). A dry-wet process to manufacture sliced bamboo veneer. *Forest Product Journal*, 62(5), 395–399.

[10] Ni, L., Zhang, X., Liu, H., Sun, Z., Song, G., Yang, L., & Jiang, Z. (2016). Manufacture and mechanical properties of glued bamboo laminates. *Bioresources*, 11(2), 4459–4471.

[11] Forest Products Laboratory (U.S.). (1987). *Wood handbook: Wood as an engineering material*. General Technical Report FPL-GTR-113. U.S. Department of Agriculture, Forest Service, Forest Products Laboratory, Madison, WI.

[12] Tang, H.H., Chen, H.B., & Wang, Z. (2004). Brief introduction about manufacture process of structural laminated bamboo (in Chinese). *A China Wood-based Panels*, 11, 16–19.

[13] Yeh, M.C., & Lin, Y.L. (2012). Finger joint performance of structural laminated bamboo member. *Journal of Wood Science*, 58, 120–127, DOI: 10.1007/s10086-011-1233-7.

[14] Xiao, Y., & Shan, B. (2015). Cold-pressing machine for glued laminated bamboo and timber, Chinese Innovation Patent, CN201210347267.9, Notification No. CN102862196B, applied on September 19, 2012, approved on October 28, 2015.

[15] Li, Z., Yang, G.S., Zhou, Q., Shan, B., & Xiao, Y. (2018). Bending performance of glubam beams made with different processes. *International Journal of Advances in Structural Engineering*, 22(2), 535–546.

[16] Tang, Z., Shan, B., Li, W.G., Peng, Q., & Xiao, Y. (2019). Structural behavior of glubam I-joists. *Construction and Building Materials*, 224, 292–305.

[17] Hammond, G., & Jones, C. (2008). *Inventory of carbon & energy (ICE), Version 1.6a*. University of Bath, UK, Bath.

[18] Buchanan, A.H., & Honey, B.G. (1994). Energy and carbon dioxide implications of building construction. *Energy and Buildings*, 20(3), 205–217.

[19] Zhou, F.C. (1991). Combustion values of 70 bamboo species. *Bamboo Research*, 10(1),18–21.

[20] Intergovernmental Panel on Climate Change (IPCC). (2006). *IPCC guidelines for national greenhouse gas inventories: Volume II: Energy*. IPCC, Geneva, Switzerland.

[21] Xia, D.J., Ren, Y.L., & Shi, L.F. (2010). Measurement of life-cycle carbon equivalent emissions of coal-energy chain. *Statistical Research*, 27(8), 82–89.

[22] Yang, R.Z. (2013). Research on material properties of Glubam and its application (in Chinese). Dr. of Engineering thesis supervised by Y. Xiao, Hunan University.

[23] Zhou G.M., & Jiang P.K. (2004). Density, storage and spatial distribution of carbon in phyllostachy pubescens forest (in Chinese). *Scientia Silvae Sinicae*, 40(6), 20–24.

[24] Lou, Z.C., Wang, Q.Y., Sun, W., Zhao, Y.H., Wang, X.Z., Liu, X.R., & Li, Y.J. (2021). Bamboo fattening technique: A literature and patent review. *European Journal of Wood and Wood Products*, https://doi.org/10.1007/s00107-021-01722-1

[25] Zhang, Q.S. (1988). Research on bamboo plywood – I: Softening and fattening of bamboo. *Journal of Nanjing Forest University* (Natural Science Edition), 4, 13–20.

[26] Liu, J., Zhang, H., Chrusciel, L., Na, B., & Lu, X.N. (2013). Study on a bamboo stressed flattening process. *European Journal of Wood Production*, 71, 291–296, DOI: 10.1007/s00107-013-0670-y.

[27] Jiang, S.X., Zhang, Q.S., Fu, W.S., & Mu, G.J. (2011). *Development and application of high frequency heating molding press for parallel bamboo sliver lumber* (in Chinese). China Forestry Science and Technology.

[28] Nguyen, T., Zhang, Q.S., & Jiang, S.X. (2014). *Technology of glued laminated bamboo laminated in high-frequency hot press and its properties* (in Chinese). China Forestry Science and Technology.

[29] James, W.L. (1975). *Dielectric properties of wood and hardboard: Variation with temperature, frequency, moisture content, and grain orientation*. USDA Forest Service Research Paper FPL 245, Forest Service, Forest Products Laboratory, Madison, WI.

[30] Nguyen T., & Zhang, Q.S. (2015). Temperature inside mats of high frequency hot pressed glued and laminated bamboo (in Chinese). *Journal of Zhejiang A&F University*, 32(2), 167–172.

[31] Liu, Y., Zhou, J., Fu, W., Zhang, B., Chang, F., & He, W. (2020). Preparation and mechanical property evaluation of glued laminated bamboo based on high frequency heating (in Chinese). *Scientia Silvae Sinicae*, 56(8), 131–140.

[32] Smith, L., Chen, F., Zhou, H., Wang, G., Ye, L., Li, M., & Wei, X. (2019). Development of bamboo winding composite pipe (bwcp) and its compression properties. *Bioresources*, 14, 5875–5882.

[33] Karacabeyli, E., & Douglas, B. (2013). *CLT handbook: Cross-laminated timber*. FP Innovations, Rexdale, ON.

[34] Brandner, R., Flatscher, G., Ringhofer, A., &Thiel, A. (2016). Cross laminated timber (CLT): Overview and development. *European Journal of Wood and Wood Products*, 74(3), 331–351, https://doi.org/10.1007/s00107-015-0999-5

[35] Pei, S.L., van de Lindt, J.W., Popovski, M., & Berman, J.W. (2014). Cross-laminated timber for seismic regions: Progress and challenges for research and implementation. *Journal of Structural Engineering*, 142(4), E2514001.

[36] Wei, P., Wang, B.J., Wang, L., Wang, Y., Yang, G., & Liu, J. (2019). An exploratory study of composite cross-laminated timber (CCLT) made from bamboo and hemlock-fir mix. *Bioresources*, 14(1), 2160–2170.

[37] Xiao, Y. (2018). *A type of CLBT for wall and slab panels*. State Patent Office of China, Beijing.

[38] Xiao, Y., Cai, H., & Dong, S.Y. (2021). A pilot study on cross-laminated bamboo and timber beams. *ASCE Journal of Structural Engineering*, 147(4), 06021002.

[39] Wei, P., Wang, B.J., Wang, L., Wang, Y., Yang, G., & Liu, J. (2019). Composite cross-laminated timber. *Bioresources*, 14(1), 2160–2170.

[40] Oracle. (2021). *The digital future of manufacturing*. Retrieved from: www.oracle.com/webfolder/assets/digibook/digital-manufacturing/index.html

[41] Lorenzo, R., Lee, C., Oliva-Salinas, J.G., & Ontiveros-Hernandez, M.J. (2017). Bim bamboo: a digital design framework for bamboo culms. *Proceedings of the Institution of Civil Engineers: Structures & Buildings*, 170(4), 295–302.

[42] Lorenzo, R., Godina, M., Mimendi, L., & Li, H. (2020). Determination of the physical and mechanical properties of moso, guadua and oldhamii bamboo assisted by robotic fabrication. *Journal of Wood Science*, 66(20).

[43] Ma, K., Li, Z.K., & Xiao, Y. (2020). General parametric glubam spherical grid shell design approach. Paper presented at The 2020 Structures Congress, GECE, Seoul, South Korea, August 26–28.

Chapter 3

Material Properties of Glubam

As introduced in Chapter 2, glued laminated bamboo (glubam) is a type of engineered bamboo, which resembles several features of timber-based glued laminated lumber (glulam). The glubam components are made by pressure laminating engineered bamboo board elements (20–40 mm thick); thus the performance is highly dependent on the quality and properties of the engineered bamboo boards. This chapter summarizes research findings of the author's research team and others, focusing on the basic behavior of engineered bamboo materials.

3.1 Basic Physical Properties of Glubam

The glubam and the engineered bamboo boards can be categorized into the thick-strip lamina and the thin-strip lamina.[1,2] In the first step of manufacture, the thick-strip laminated glubam is glued using a few layers of relatively thicker bamboo strips (5–8 mm thick and 20–25 mm wide), as illustrated in Figure 3.1 (a). By contrast, the thin layers of the laminated glubam are typically generated by laminating netted mats with bamboo strips approximately 2 mm thick and 20 mm wide, as shown in Figure 3.1 (b). Other researchers have also focused their studies on the thick-strip bamboo board.[3,4] The thick-strip glubam boards are generally made unidirectional with all the strips aligned in the longitudinal direction, but bidirectional laminated bamboo or CLB (cross laminated bamboo) can also be produced. As for the thin-strip glubam boards, application of a bidirectional structure with typical ratios of longitudinal and transverse bamboo strips is more prevalent. Specifically, a ratio of 4:1–7:1 is utilized for columns or girders, while 1:1 is used for making sheathing panels. The arrangement of bamboo strips is essentially the configuration of bamboo fibers in different directions within the engineered bamboo board.

To be consistent, in this book, the coordinate system illustrated in Figure 3.1 (c) is established. In the coordinate system, the x-axis represents the longitudinal direction which is aligned with the main orientation of the bamboo fibers, using 100% fibers for the thick-strip panel and 80% or more fibers for the thin-strip panel; the y-axis serves as the transverse direction with less fibers – namely, 0% for the thick-strip panel, and 20% or less for the thin-strip panel; and the z-axis represents the thickness or glue surface direction.

Engineered bamboo boards and sheets for glubam have been extensively tested. In a recent testing program, the oven-dry density of thick-strip bamboo is 631 kg/m³, with a standard deviation of 41.2 kg/m³; the oven-dry density for thin-strip bamboo

DOI: 10.1201/9781003204497-3

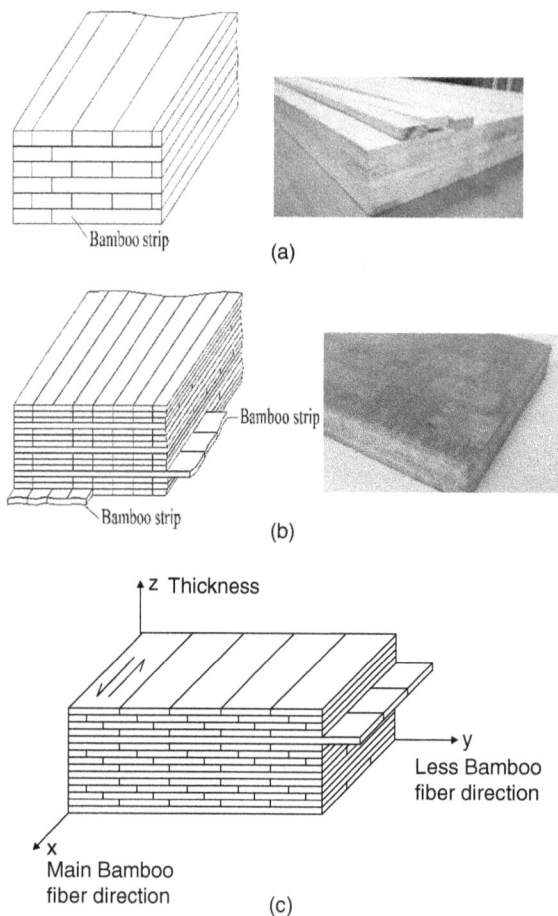

Figure 3.1 Schematic diagram of glubam board: (a) thick-strip glubam; (b) thin-strip glubam; (c) definition of coordinate system.

is 862 kg/m³, with 49.3 kg/m³ as its standard deviation.[5] Observations of old samples of engineered bamboo boards reveal that, under indoor conditions, the engineered bamboo has an advantageous quality and stability, with the average moisture content ranging from 7% to 12%. After storage for many years, there is no notable warpage deformation or mildew, and the fluctuations are negligible amongst all basic mechanical properties.

3.2 Research Background of Mechanical Properties of Engineered Bamboo

As discussed in Chapter 1, in recent years, a growing number of academic studies on engineered bamboo has been produced by researchers and practitioners around the world, particularly by scholars in China, South America, and elsewhere. Sharma et al. considered the basic mechanical properties (e.g., tension, compression, shear, and bending strengths) of bamboo scrimber and laminated bamboo sheets following the timber standards.[4] The comprehensive properties, including the physical and mechanical aspects, of glued laminated Guadua (GLG) were studied by Correal et al. They studied the basic physical characteristics such as moisture content, density, shrinkage, and hardness as well as the mechanical properties including the compressive, tensional, shear, and bending strengths, and Poisson's ratio.[6]

Both analytical and experimental analyses were undertaken for laminated (Moso) bamboo lumber (LBL) with thick strips by Li et al., based on a timber testing standard.[3] The experimental side is significant for carrying out both material performance tests and component tests. Consequently, the mean ultimate tensile and compressive strengths were obtained to suggest a tri-linear stress-strain relationship in the parallel direction of the bamboo strip. Li et al. paid attention to the bending directions and the presence of internal joints as well, revealing that the performance of the beam is influenced by these factors to a certain degree. Analytically, regardless of the bending orientations, three bending failure modes were presented with their calculation formulas and the ultimate bending deflections, which also illustrated correlation with the experimental results. However, research by the author's team[2,7] focused more on the bending performance of full-scale beams made with different thicknesses of bamboo strips; i.e., the thick-strip glubam and thin-strip glubam. The results showed that the two types of glubam manifested an elasto-brittle behavior, similar to wood-based glulam. Li et al. also discovered that the ultimate loading capacities are different depending on the bending direction, with the capacity of edgewise bending (strong axis bending for the strips) being greater than that of flatwise bending (weak axis bending of the strips).[2]

Considering the shear strength, Xiao et al. have tested glued laminated bamboo (glubam), manufactured using three types of bamboo strip configuration.[8] It is notable that the probability models of in-plane and out-plane shear strength values project to normal and Weibull distributions respectively, which provides a strong foundation for the follow-up experimental and numerical analyses. Takeuchi et al.'s research compared numerical analysis on finite elements for the shear strength of laminated Guadua bamboo (LGB) with experimental outputs.[9] It is incredible that the analysis of of failure cracks from three various orthogonal directions of fibers are exceptionally close to the testing results.

Additionally, by considering glubam as bidirectional bamboo fiber lamina, Yang et al. reported experimental insights regarding the off-axis tensional strength of glubam material.[10] The complete testing results were correlated with theoretical criteria based on Hankinson's equation and the Tsai–Wu failure theory. Also, the interaction coefficient became the most significant parameter for the difference from the angles of off-axis tests.

Due to the intended applications of engineered bamboo and its similarity to timber, most studies on bamboo materials are conducted following the standards and specifications developed for timber. One important difference between bamboo and wood which has to be noticed is that the distribution of wood knots mainly determines the geometric sizes of standard wood specimens. Bamboo has no knots, but it has nodes which usually form the weakest part of the bamboo strips. The definition of the most reasonable specimen size for structural bamboo panels should be based on the probabilistic distribution of bamboo nodes and subsequently, fibers. However, in most cases, only the average performance is considered. The size of specimens is decided in accordance with the panel thickness recommended in current research. Many Chinese scholars have examined the applicability of existing timber standards for bamboo products. Liu et al.[11] conducted a comparative study on GB/T 15780 "Testing methods for physical and mechanical properties of bamboos"[12] and J G/T 199 "Testing methods for physical and mechanical properties of bamboo used in building".[13]

Gao et al. studied the properties and applications of bamboo composites for building structures.[14] The form and properties of bamboo slabs are analogous to those of structural veneer lumber. The long and thick structural profile seems related to that of plywood. Consequently, the Japanese Ministry of Agriculture, Forestry and Fisheries Notice (JAS) No. 237 (February 27, 2003) "Structural veneer lumber"[15] and European standard EN1194-1999 "Wood structure. Glulam – Determination of performance grading and eigenvalues"[16] were employed to build the foundation of the standards for bamboo structures. Hence, the performance of bamboo slabs for building was judged on the basis of those standards. However, the results obtained by Gao et al.[14] indicated that, due to the high level of toughness and low degree of rigidity of bamboo, the elastic modulus and static bending strength grade produces mismatches between bamboo and cemented wooden materials. In conclusion, the experimental specifications for glued laminated wood should be the appropriate reference work.

Zhang and He[17] explored the physical and mechanical properties of bamboo laminated timber and bamboo finger jointed laminated timber, which were compared with the mechanical properties of general building timbers, perforated masonry brick, and concrete. Specifically, the mechanical properties of bamboo laminated timber comprised tensile, compressive, and shear strength and elastic moduli. Except for the drawback that the fiber ratios in different directions of the bamboo sheets are not considered, the outcomes have a reference value for the testing of engineered bamboo.

In addition, Ye et al.,[18] Wang et al.,[19] Zhang et al.,[20] Jiang et al.,[21] and other researchers have devoted themselves to analyzing the mechanical properties of bamboo and bamboo composites. It can be deduced from these studies that research on engineered bamboo in China increasingly emphasizes construction applications.

Many researchers have explored experimental methodologies for the mechanical properties of engineered bamboo from the perspective of their independently selected materials. However, due to the lack of uniform industry standards or specifications for engineered bamboo, the approaches and outputs of research which has focused on mechanical properties are not comparable. The lack of unified conclusions generates an uncertainty regarding the performance of engineered bamboo, hindering its application in building structures.

The author's research team recently completed a comprehensive program of material testing with a considerable number of samples. The tests were conducted following the existing standards for wood and timbers; however, the researchers made necessary modifications to the dimensions of the specimens to suit the typical dimension of engineered bamboo. In the following section, the results are described and research by other scholars is also summarized.

3.3 Basic Mechanical Properties of Engineered Bamboo

3.3.1 Relevancy of Existing Specifications

Due to the similarity with timber and the intended usage of bamboo, research on bamboo typically follows the existing well-developed timber standards; however, some adjustment and modifications are needed – for example, with regard to the number of sampling specimens. In ASTM D143, there is no specific requirement for the number of specimens.[22] According to the Chinese national standard for testing wood properties,[23] the number of samples is determined according to a confidence level of 0.95. Therefore, the test is calculated according to the minimum number of samples required to achieve an accuracy index of $P = 5\%$:

$$n_{min} = \frac{v^2 \eta^2}{P^2} \tag{3.1}$$

In Eq.3.1, v is the coefficient of variation of the property to be determined; η is the reliability index of the result, according to a confidence level of 0.95, the value is 1.96. Chinese standard GB/T 1928 "General requirements for physical and mechanical tests of wood"[24] also gives the average coefficient of variation of the main properties of commonly used wood. For wood compressive strength, this value is 13%. Because of the lack of relevant data for glubam, the authors refer to the relevant values of wood to determine the number of samples. According to the calculation result of Eq.3.1, the minimum number of samples required for a glubam sheet in the grain compression test is 26. Generally speaking, as regards the shear behavior of glubam and woodlike materials, the scattering of shear strength values is relatively large. In a shear testing program, Xiao et al. studied the sensitivities of specimen numbers.[8] Based on an initial coefficient of variation, $v = 20\%$, Eq.3.1 gives a minimum specimen number of 62. In the testing program, 62 specimens of the 4:1 thin-strip glubam were first tested for each loading condition, and then the sensitivity of the specimen numbers was evaluated. The authors randomly selected 30 specimens from the 62 specimens and the statistical characteristics of different testing numbers were investigated, as shown in Figure 3.2. From the comparative results shown in Figure 3.2 (a), it is found that the mean of 30 specimens is in line with that of 62 specimens, and their standard deviations are also very close. The P-Values for both cases are greater than the significance level $\alpha = 0.05$, indicating that the test results follow the normal probability distributions. The comparisons shown in Figure 3.2 indicate that tests of specimens with a total number 62 or 30 both yield approximately similar statistical results. This suggests that for man-made materials, such as glubam, the scattering of material properties is much smaller than for natural wood.

In the Chinese code GB 50005 (2017), Appendix F,[25] the characteristic values of man-made woody materials are calculated based on tests with different specimen numbers, and the smallest number of sampling is specified as 10.

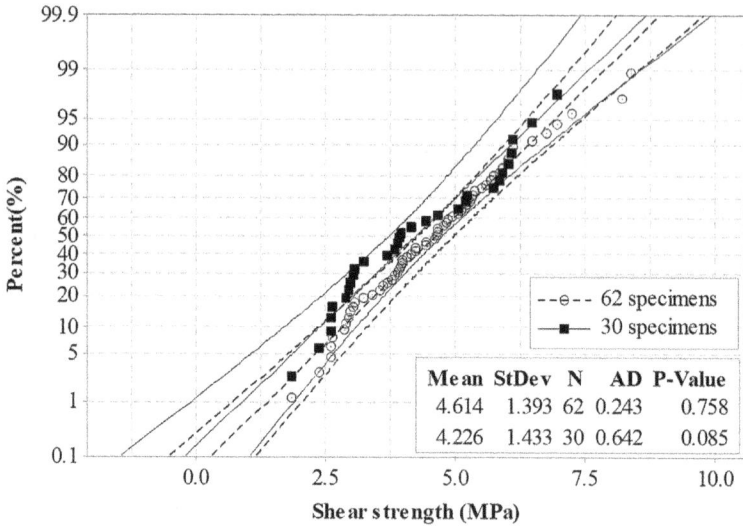

(a)

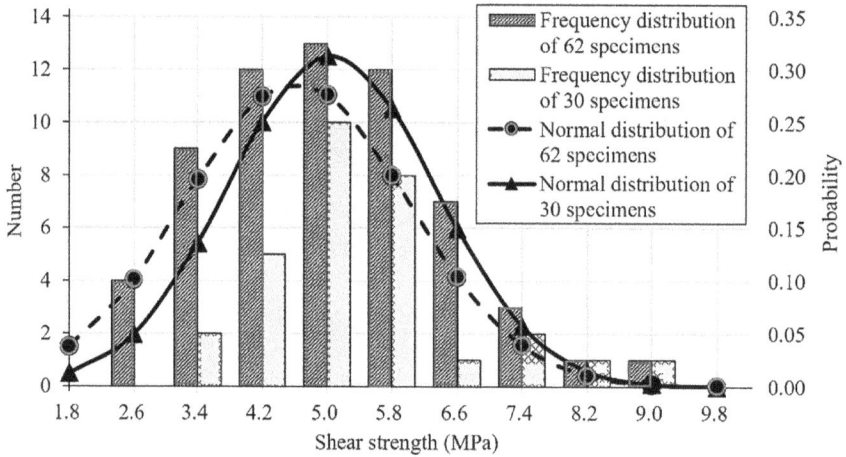

(b)

Figure 3.2 Probability graph of GB4-1ZL test series: (a) Normality test of 62 specimens and 30 specimens; (b) Frequency distribution and normal distribution.

3.3.2 Tensile Properties

3.3.2.1 Tensile Test Specimens

One of the main differences between engineered bamboo and wood is the hardness, making the specimens difficult to fabricate. The author's research group suggested adjusting the test specimens recommended in ASTM D143 (2014)[22] to better conform to the characteristics of glubam. Since the typical thickness of engineered bamboo

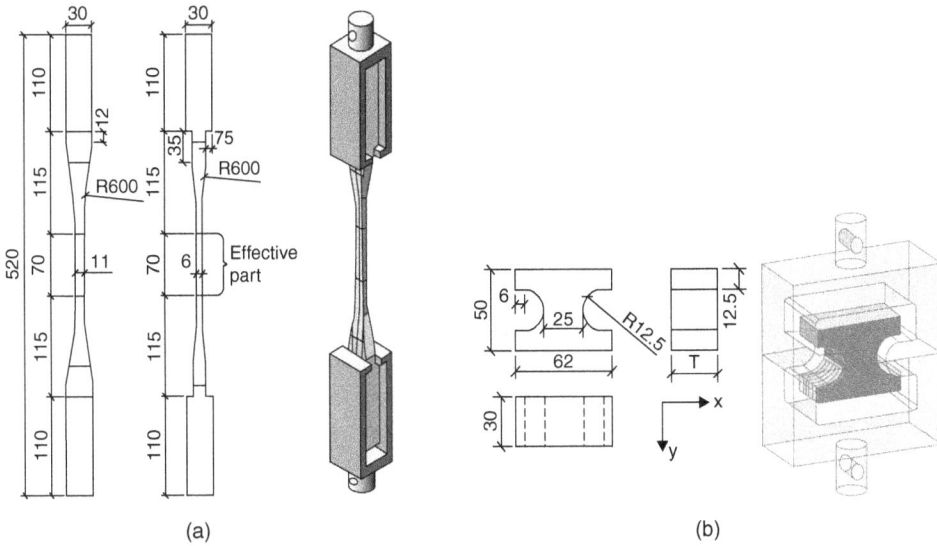

Figure 3.3 Glubam tensile tests (unit: mm): (a) tensile test along main fiber direction; (b) tensile test along transverse direction.

boards for glubam fabrication is 30 mm, the dimensions of the standard test specimen and the details of the holding jigs were modified further, as shown in Figure 3.3 (a) and (b) for the longitudinal and transverse directions, respectively.[5] The specimen can be cut using a computer numeric control (CNC) machine or waterjet cutting machine, with high precision.

In a tension test, the stress of laminated bamboo along its bamboo strip direction is calculated as:

$$\sigma_t = \frac{F}{bt} \tag{3.2}$$

where, F is the measured force during the test, the tension strength f_t is obtained when this force reaches its maximum value F_{max}; b and t are the measured effective width and thickness of the specimen in the middle testing portion, and are designed as 11 mm and 6 mm, respectively. The loading rate was kept as 1 mm/min., during testing.

A strain gauge is attached in the effective middle portion of the specimen and the strain is measured and recorded at the same time as the applied force. The tensile modulus is calculated by:

$$E_t = \frac{\Delta F}{bt\Delta\varepsilon} \tag{3.3}$$

where, ΔF is the load equal to the difference between the lower and upper limits of loading, $\Delta\varepsilon$ is the strain increment corresponding to the lower and upper limits of loading, using the arithmetic means of the results. According to the linear portion of load-deflection curves of tension tests, 1 kN (kilonewton) and 2.5 kN are typically selected as the lower and upper limits of loading for determination of the tensile modulus of elasticity E_t. The loading and unloading are typically repeated five times in the range of lower and upper limits to calculate the tensile modulus.

3.3.2.2 Tensile Behavior of Engineered Bamboo along Fiber Directions

Figures 3.4 and 3.5 show the failure patterns of typical specimens recently conducted by Li et al.[5] For specimens under main-fiber direction tension (denoted as x-direction in Figure 3.1), there are no apparent differences in failure modes between thick-strip and thin-strip bamboo specimens. The tension rupture of bamboo specimens in the effective region was the typical failure mode noticed during the test, as shown in Figure 3.3 (a). Only failures located in the effective part would be accepted for the strength calculations. For thick-strip bamboo under y-direction tension, the splitting rupture of bamboo strips was sudden and instant, without any resistance remaining, due to the fact that there is no fiber in the direction. For thin-strip bamboo specimens under less-fiber direction tension, which is denoted as y-direction herein, the outer-layer surface bamboo strips were broken first, while the remaining bamboo strips inside could still contribute some resistance until the end of the test.

The tensile stress and deformation relationships obtained from the tensile tests are shown in Figure 3.6 and Figure 3.7, for the main fiber direction (x-direction) and for the secondary direction (y-direction), respectively. Deformations for the x-direction tests shown in Figure 3.6 are measured using extensometers attached in the middle effective testing zone of the specimens. However, the deformations for y-direction tests are based on the relative displacement of the loading platens; thus these only serve as reference, but do not reflect accurate deformation. Relatively large variations are noticed for tensile curves as shown in Figure 3.6 (a) and (b). The coefficient of variance (COV) for tension strength along the main bamboo fiber direction (x-direction) is 18.4% for uni-directional thick-strip glubam and 21.5% for bidirectional thin-strip glubam. Although the tension strength of thin-strip bamboo boards along main fiber x-direction (80.5 MPa) is less than that of thick-strip ones (119.3 MPa), the tension strength of thin-strip

Figure 3.4 Failure patterns of glubam in tension along main fiber direction.

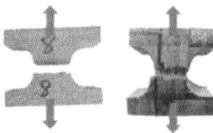

Figure 3.5 Failure of specimens subjected to tension in the transverse direction.

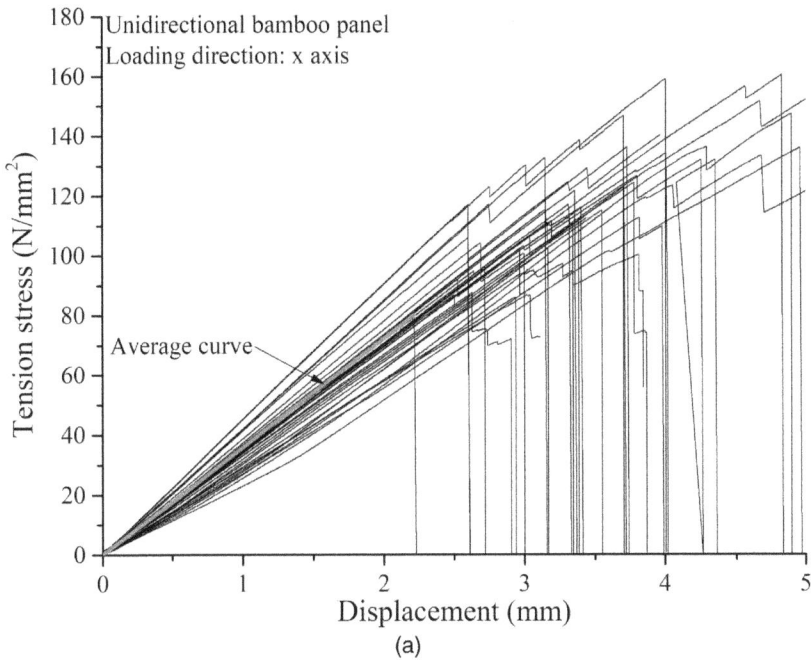

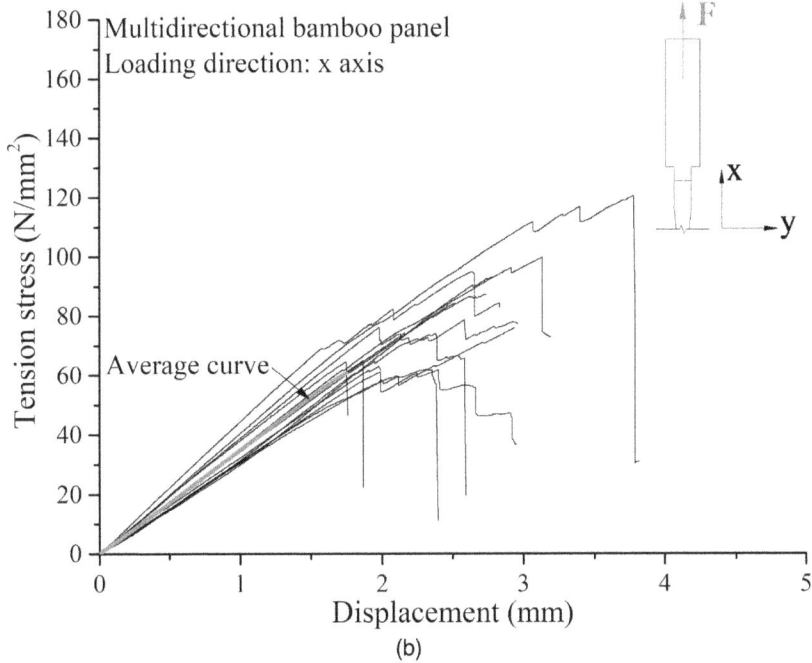

Figure 3.6 Stress-displacement relationships: (a) thick-strip glubam subjected to tension in main fiber direction x; (b) thin-strip glubam subjected to tension in main fiber direction x.

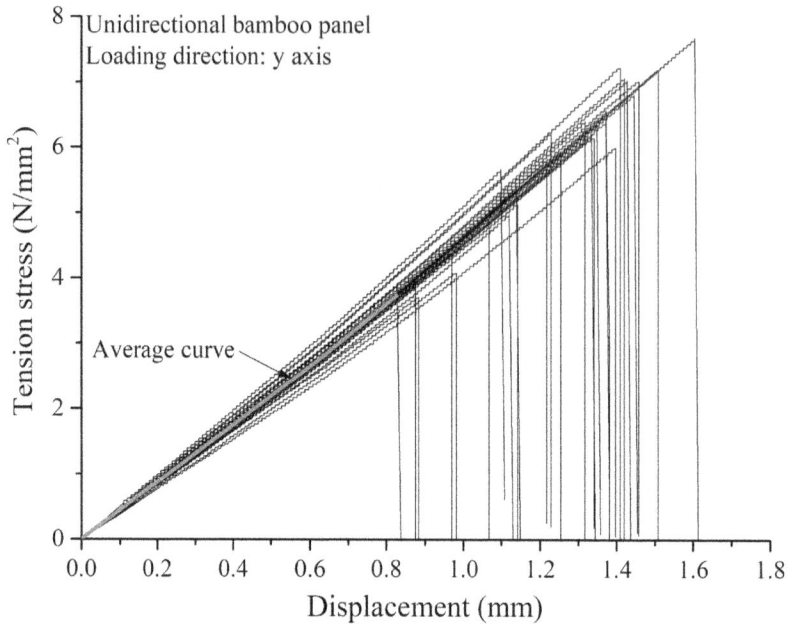

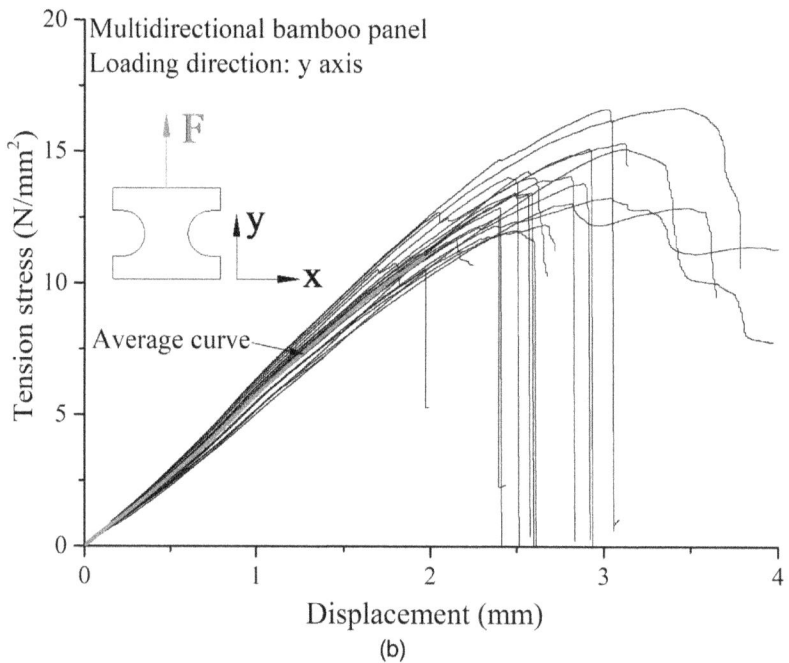

Figure 3.7 Stress-displacement relationships: (a) thick-strip glubam in y-direction tension; (b) thin-strip glubam in y-direction tension

bamboo boards along less fiber y-direction (13.3 MPa) is more significant than that of thick-strip bamboo panels (5.9 MPa).

The strengths of the thin-strip bamboo panels shown in Figure 3.6 and 3.7 have been recently obtained,[5] and are slightly lower than corresponding values obtained in Xiao et al.[8] with different specimen details, and a different batch of production of engineered bamboo boards. The recent tests may be considered more conservative than the previous studies. On the other hand, it also implies that for each batch of engineered bamboo boards to be used for fabricating glubam components, sampling and tests should be carried out to certify the strengths.

3.3.2.3 Tensile Behavior Relevant to Fiber Orientations

For isotropic materials such as steel, the elastic modulus and tensile strength are the same in all directions, but for anisotropic or orthotropic materials such as wood, bamboo, or various fiber composites, the elastic properties are quite different in all directions. Unlike wood and raw bamboo, glubam has bamboo fibers distributed in both the longitudinal and transverse directions, and the fiber content in both directions can be configured differently, so the strength and elastic modulus in different directions are inevitably different. In order to fully reveal the tensile properties of glubam in different directions, seven types of test specimens were prepared at different angles from the longitudinal direction of the sheet.[10] The sampling of the test piece is exhibited in Figure 3.8 (a), and the CNC waterjet cutting tool is shown in Figure 3.8 (b). Except for the 0° test piece, the number of test pieces at each angle was 10. In order to more accurately determine the material properties in the direction of the grain, the number of 0° test pieces was selected to be 16.

Figure 3.9 shows the results of the tensile test of the specimens at various angles. It can be seen from the figure that the bearing capacity of the 0° specimens is significantly higher than that of the specimens with other fiber angles. Figures 3.10 (a) and (b) show the relationship between angle and tensile strength and tensile modulus, respectively. The two relationship curves are similar in shape and show the same trend. The tensile strength and elastic modulus of the glubam plate in the direction of the 0° angle are 83 MPa and 10.3 GPa (giga pascal), respectively; and the tensile strength and elastic

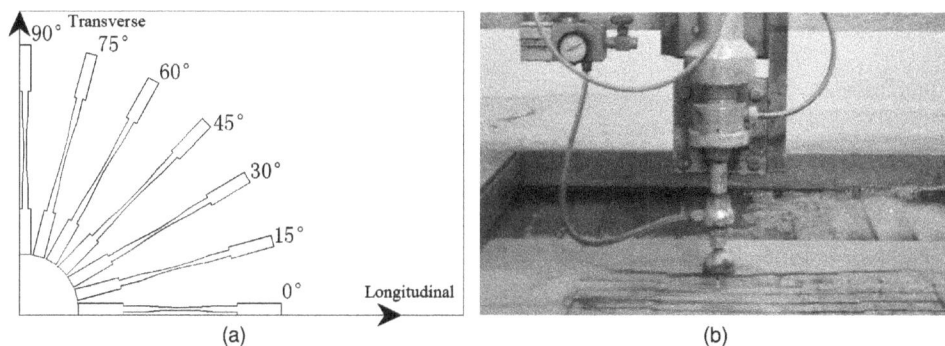

Figure 3.8 Tensile test specimens: (a) cutting angle; (b) waterjet cutting.

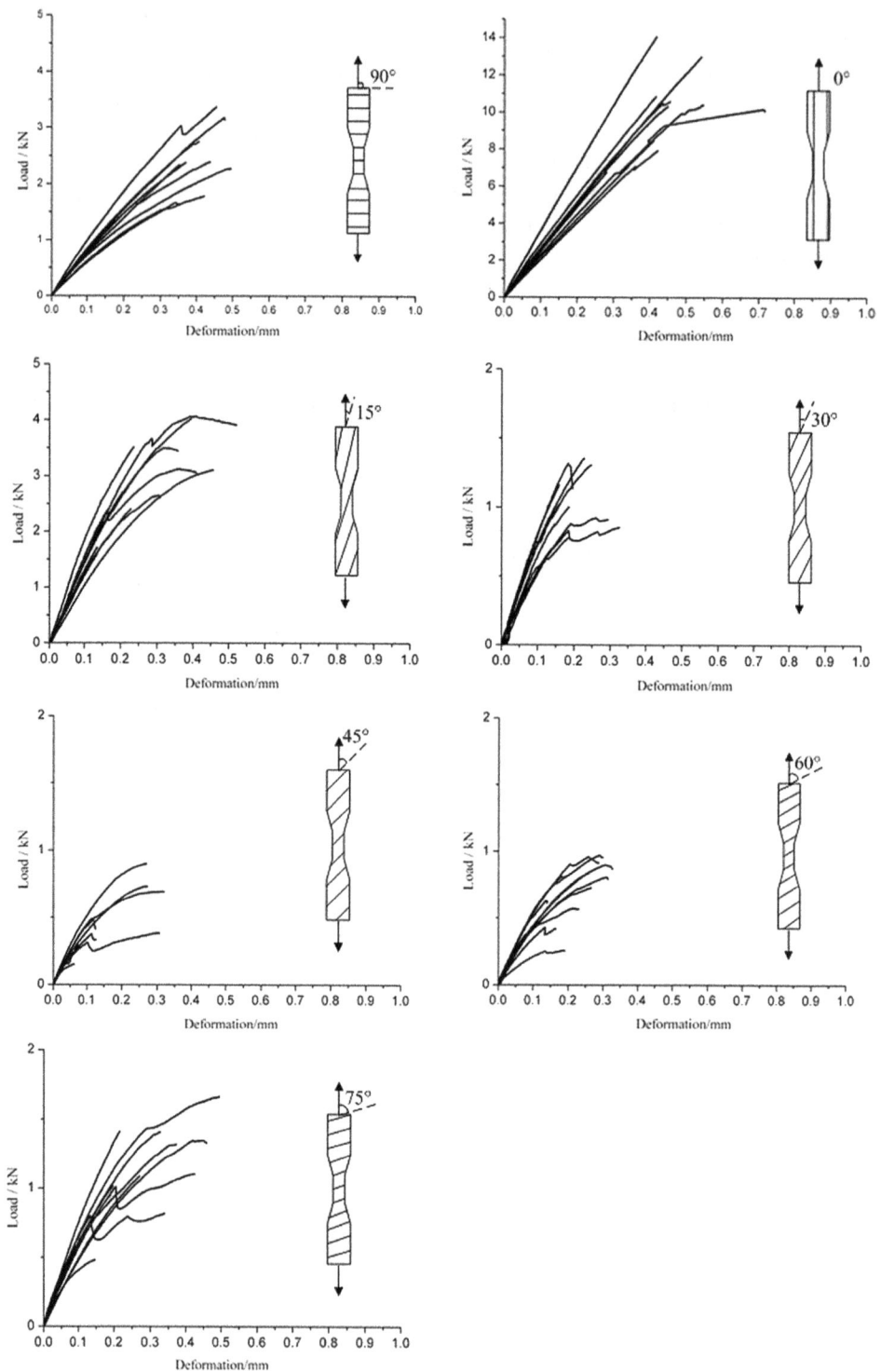

Figure 3.9 Displacement load curve of specimens with different fiber angles.

$f_t / f_{t,0} = \exp(3.19\theta^2 - 6.01\theta)$

(a)

$E/E_0 = \exp(1.74\theta^2 - 3.64\theta)$

(b)

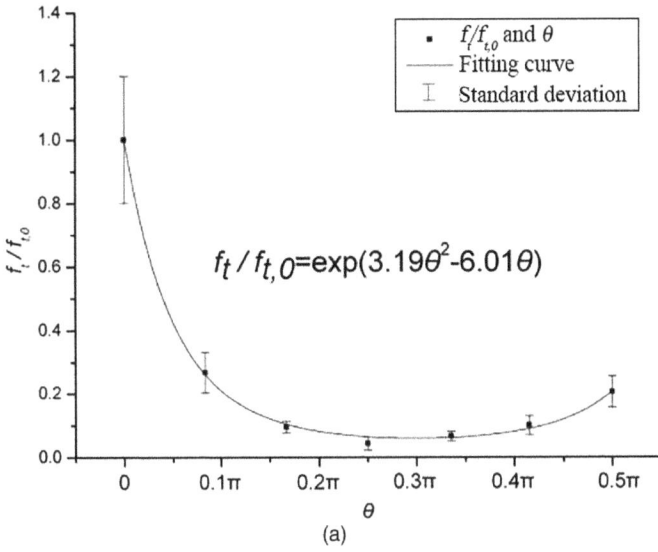

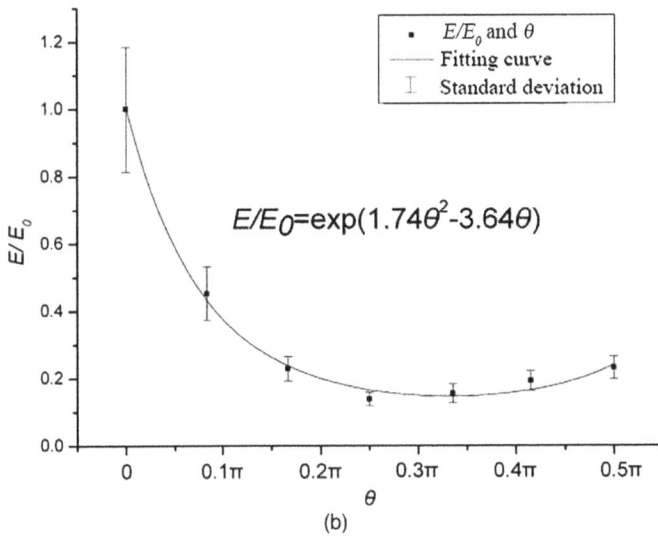

Figure 3.10 Mechanical properties corresponding to fiber angles: (a) relationship between $\frac{f_t}{f_{t,0}}$ and angle θ; (b) relationship between $\frac{E}{E_0}$ and angle θ.

modulus in the transverse direction – that is, the 90° direction are 17 MPa and 2.4 GPa. It can be seen that the ratio of the performance parameters in both directions is around 4:1, which is basically consistent with the fiber ratio of 4:1 of the grain and the grain of the plate itself.

Kwan et al. studied the mechanical properties of bamboo as a natural material in 1987; in particular, they investigated the relationship between bamboo fiber quantity and elastic modulus.[26] The results show that the relationship between Young's modulus E and experimental fiber content is linear. This is very similar to the mechanical behavior of a typical fiber-reinforced composite. This property of bamboo itself provides the basis for the design, production, experimentation, and application of glubam. However, it is more noteworthy that, unlike natural bamboo, when the fiber and load angle is 45° in a glubam sheet, the strength value and elastic modulus of the glubam material reach a minimum, as shown in Figure 3.10. The phenomenon is closely related to the arrangement and interaction of fibers in glubam.

There is no formula for the influence of the direction of the applied force and the angle of the fiber direction on the tensile properties of glubam with a certain ratio of longitudinal and transverse fibers. Based on the tensile test values at various angles, Yang et al. obtained the relationship between the strength of each direction and the angle of the elastic modulus.[10] According to the principle of obtaining as high a goodness of fit as possible, the exponential curve of the strength and elastic modulus of a glubam sheet with fiber angle θ is obtained:[10]

$$\frac{f_t}{f_{t,0}} = exp(3.19\theta^2 - 6.01\theta) \; 0 \le \theta \le \pi/2 \tag{3.4a}$$

$$\frac{E}{E_0} = exp(1.74\theta^2 - 3.64\theta) \; 0 \le \theta \le \pi/2 \tag{3.4b}$$

In Eq.3.4: f_t represents the tensile strength, $f_{t,0}$ representing the tensile strength in the 0° direction; E represents the elastic modulus, E_0 representing the elastic modulus in the 0° direction. The determination coefficients of the above two fitting curves are 0.998 and 0.995, respectively. Further, in the case of the angle in the Eq.3.4 $\theta = 0$, the specific strength and the specific modulus value are 1.0; when $\theta = \pi/2$, it is about 0.2.

3.3.3 Compressive Behavior

3.3.3.1 Compression Test Specimens and Setup

Compressive behavior and strength are the most important mechanical properties for engineering materials such as rock, concrete, bricks, and timber. For bamboo materials, many researchers have studied the compressive strengths of round bamboo culms. For engineered bamboo, a significant number of compressive tests have been carried out by the author's research team.[1,5] Figure 3.11 (a) shows the specimen size with a length to widths ratio of 4.0 based on ASTM D143, but adjusted to the thickness of engineered bamboo board. Figure 3.11 (b) and (c) describe the test setup of compression loading along the main fiber x-direction (grain direction for timber) and transverse directions, respectively. The loading rate was 1 mm/min. Tests were also carried out by Yang et al.[10] based on Chinese guidelines.[23] Based on the Chinese code, the test specimen has a length to width ratio of 3:2, different from the ASTM standards.[22] The compression stress is calculated using Eq.3.2, by replacing σ_t with compressive stress σ_c.

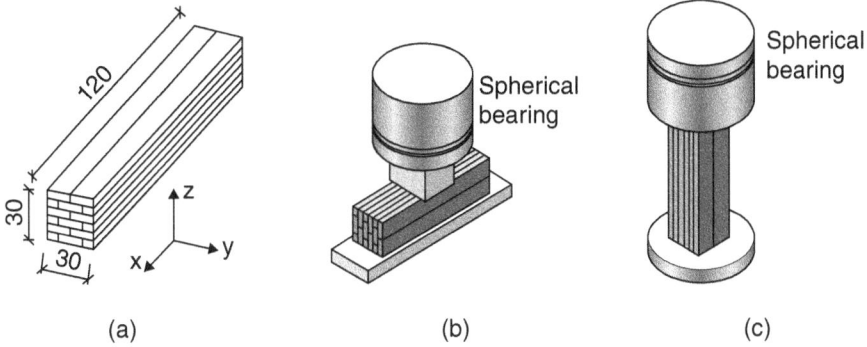

Figure 3.11 Specimen details and setups for compression tests: (a) specimen; (b) main fiber direction; (c) secondary or transverse direction with less or no bamboo fibers.

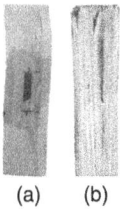

Figure 3.12 Typical failure patterns in x-direction compression: (a) thick-strip glubam compressed in x-direction; (b) thin-strip glubam compressed in x-direction.

The compression strength f_c is obtained when the compressive force reaches its maximum value F_{max}. For compressive specimens, the width b and thickness t are both designed to be 30.0 mm. For y- and z-loading directions, F_{max} is the maximum value of the linear-elastic portion (proportional limit) of the load-displacement curve, which is obtained through test curves. Strain gauges were affixed in the middle of two sides of the compression surface for strain measurement for tests along the x-direction. The lower and upper limits of loading in the x-direction are 6 kN and 20 kN, respectively. The elastic modulus E_c is calculated using Eq.3.3, by replacing E_t with E_c.

3.3.3.2 Behavior of Engineered Bamboo under Compression

Engineered bamboo specimens with thick and thin strips after failure under compression in the x-direction are shown in Figure 3.12 (a) and (b), respectively. As shown in Figure 3.12 (a), for thick-strip glubam specimens, the out-of-plane buckling failure under x-direction loading was noticed during the tests, but no delamination was noticed for the lamination layers among the bamboo strips. For thin-strip bamboo specimens, a different failure mode was observed with delamination of bamboo layers under compression in the x-direction (longitudinal direction), as shown in Figure 3.12 (b). Local crushes occurred when the loading directions were perpendicular to the bamboo strips, in y- and z-directions, and both thick- and thin-strip engineered bamboo specimens

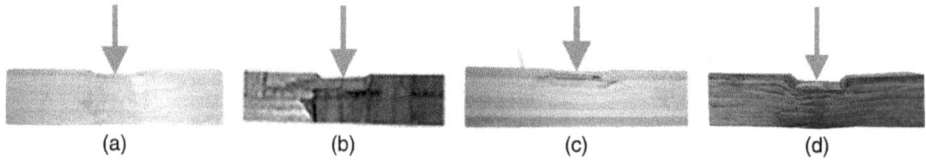

Figure 3.13 Typical failure patterns: (a) thick-strip glubam compressed in y-direction; (b) thin-strip glubam compressed in y-direction; (c) thick-strip glubam compressed in z-direction; (d) thin-strip glubam compressed in z-direction.

exhibited similar results under compression in these two directions, as shown in Figure 3.13.

Stress-deformation curves obtained through compression tests in the x-direction, as shown in Figure 3.14, indicate that one failure of engineered bamboo is that it is quite ductile under compression. As shown in Figure 3.14, a plastic plateau can be noticed for the engineered bamboo specimens under x-direction compression. The behaviors of the engineered bamboo specimens are even more ductile and stable when subjected to compression in the transverse directions (y and z directions), as shown in Figure 3.15. This might be attributed to the testing method, in which the loading is actually the bearing type with a localized load causing a local confinement effect, particularly for the loading in the direction of thickness (z direction). Experimental tests are needed to establish the material behavior of engineered bamboo in transverse compression.

Experimental research was also carried out by Yang[10,27] on the compressive behavior of thin-strip glubam specimens with and without a cold glue interface. Testing results indicate that the compressive strength is reduced when there is a cold glue interface. Figure 3.16 depicts one example of the final failure of a specimen with a glue interface.

3.3.4 Bending Resistance of Glubam

3.3.4.1 Bending Test Specimen

Static bending is one of the most critical tests for timber and woodlike materials, as well as many brittle materials, as the measurement of modulus of elasticity (MOE) and modulus of rupture (MOR) is the basic criterion for quantifying the properties of timber and other brittle materials. For static bending tests, four-point and three-point loading methods are typically used; however, the three-point loading method seems to be more typical for timbers.

As shown in Figure 3.17, following the three-point loading test method given by ASTM D143 (2014)[22] and ISO 13061 (2014),[28] the size of the specimen for static bending is 20 × 20 × 300 mm for thick-strip bamboo boards and 30 × 30 × 450 mm for thin-strip bamboo boards, whereas the distance between the lower supports is 240 mm and 360 mm, respectively. This keeps the ratio of the span and the section height of the test piece at 14, in order to minimize the influence of shear force introduced by the

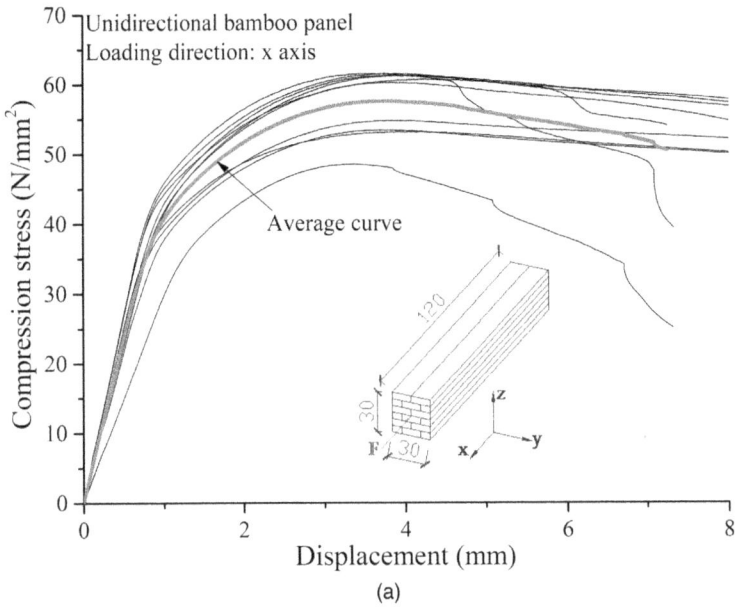

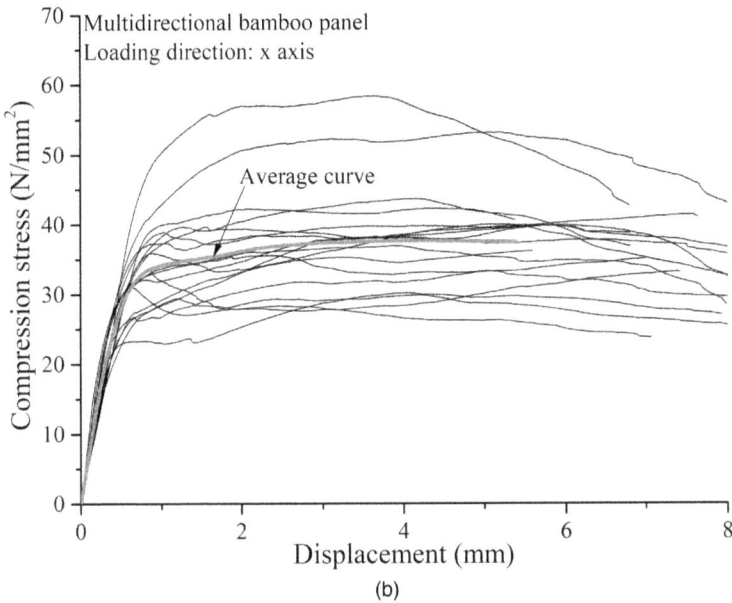

Figure 3.14 Stress-displacement relationships: (a) Thick-strip glubam compressed in main fiber x-direction; (b) Thin-strip glubam compressed in main fiber x-direction.

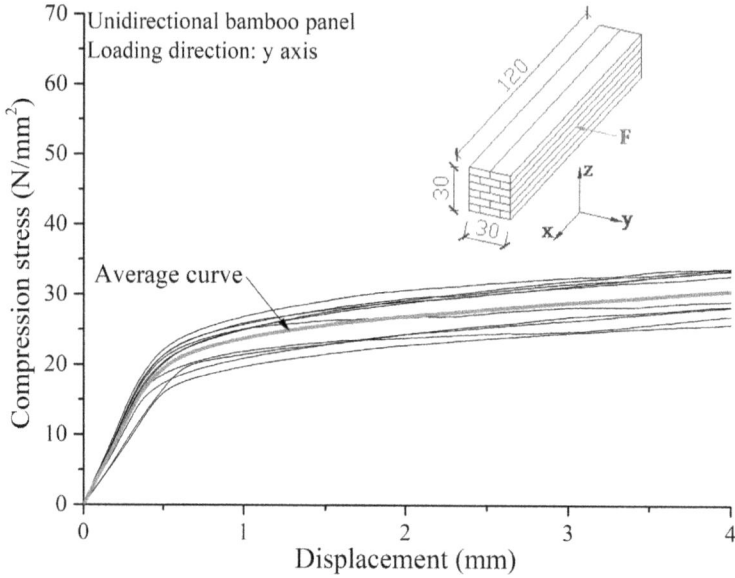

(a)

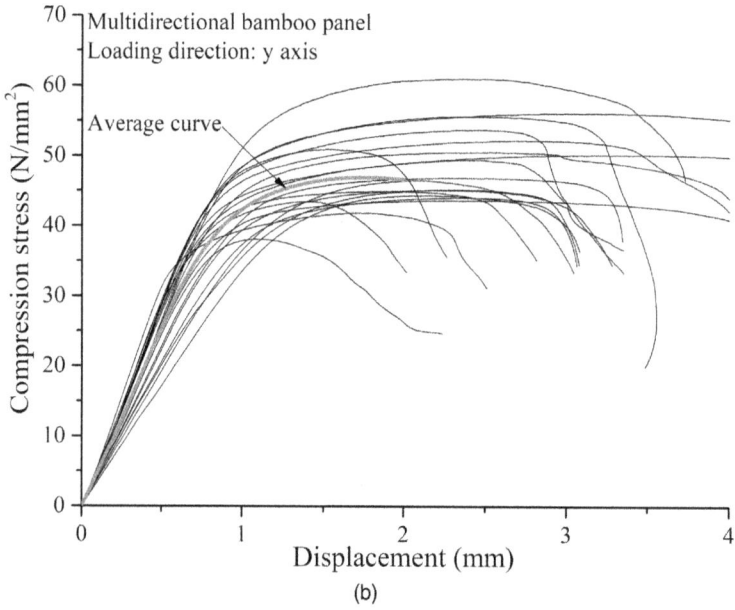

(b)

Figure 3.15 Stress-displacement relationships: (a) Thick-strip glubam in *y*-direction compression; (b) Thin-strip glubam in *y*-direction compression; (c) Thick-strip glubam in *z*-direction compression; (d) Thin-strip glubam in *z*-direction compression.

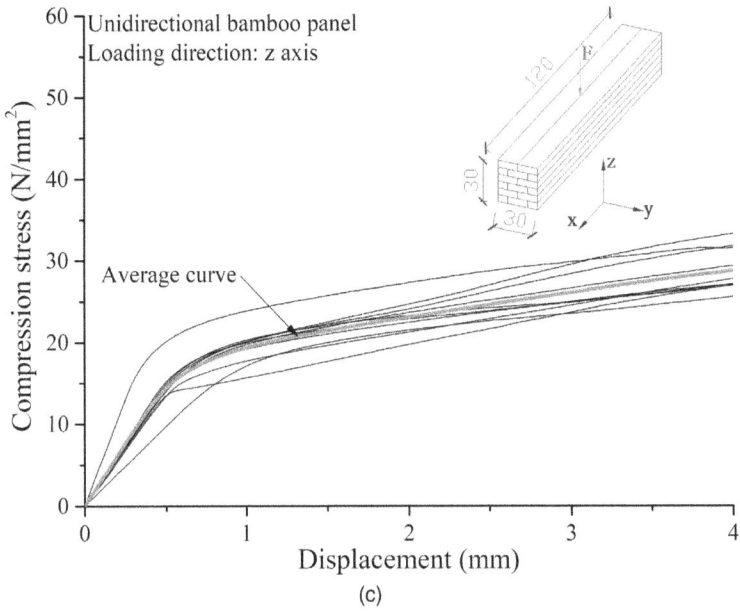

(c)

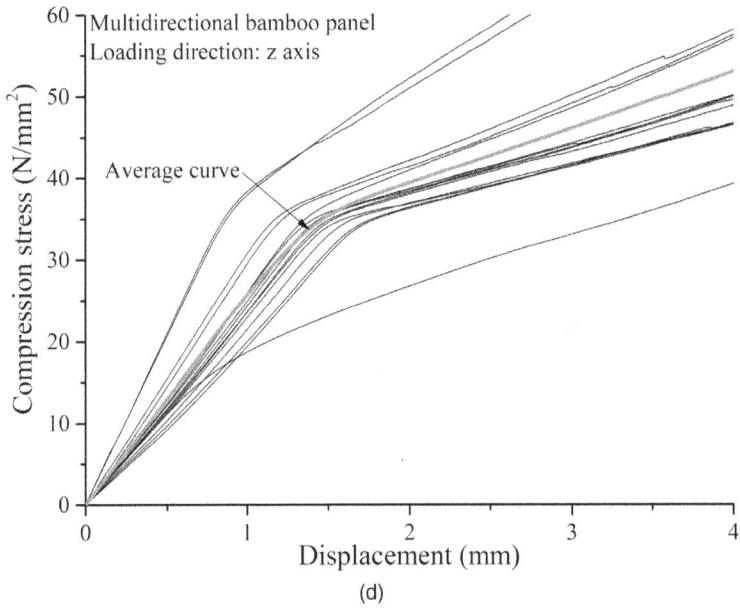

(d)

Figure 3.15 Continued

Figure 3.16 Failure of a thin-strip glubam specimen with interface glueline.

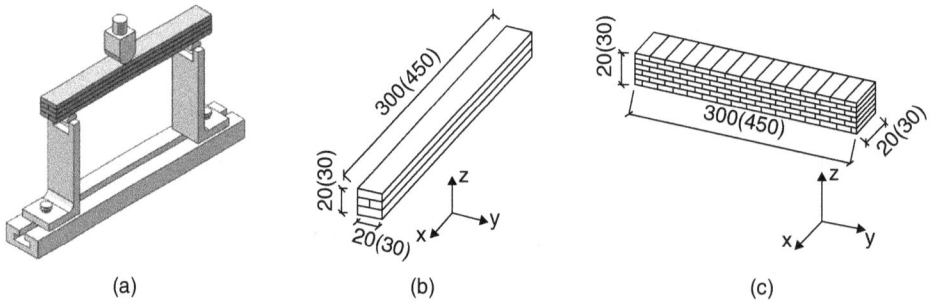

(a)   (b)   (c)

Figure 3.17 Bending testing: (a) Specimen for bending in longitudinal direction aligned with main bamboo fibers; (b) Specimen for bending in transverse direction aligned with less or no bamboo fibers; (c) Test setup.

Note: For bending specimens made with thin-strip multidirectional laminated bamboo panels, the cross-section of the bending specimen is 30 × 30 mm, and the length is 450 mm.

loading head. The radius of all the supports and the loading heads[29] was 30 mm. The loading rate was 0.5 mm/min. for the measurement of modulus of elasticity (MOE) and 2 mm/min. for the measurement of modulus of rupture (MOR). A digital dial gauge, with an accuracy of 0.001 mm was used to measure the center line deflection at the mid-span. All specimens were first loaded from 200 N (newton) to 800 N three times, to measure the modulus of elasticity (MOE); they were then loaded until failure for the calculation of ultimate strength, or modulus of rupture (MOR). The modulus of elasticity (MOE), E_w, of each test piece is calculated as:

$$\text{For three – point bending:} \quad E_w = \frac{Pl^3}{4bh^3 f} \tag{3.5a}$$

For four − point bending: $E_w = \dfrac{23Pl^3}{108bh^3 f}$ (3.5b)

where, P is the load equal to the difference between the upper and lower limits of loading, which is 200 N and 800 N in this research; l is the distance between the centers of the supports, which is 240 mm or 360 mm in this research; b is the breadth of the test piece, which is 20 mm or 30 mm; h is the height of the test piece, which is also 20 mm or 30 mm herein; f is the mean deflection, equal to the difference between the results obtained in measuring the deflection at the upper and lower limits of loading mentioned above, in mm.

The stress on the top/bottom line of the section in static bending is calculated as:

For three − point bending: $\sigma_{b,w} = \dfrac{3Pl}{2bh^2}$ (3.6a)

For three − point bending: $\sigma_{b,w} = \dfrac{2Pl}{bh^2}$ (3.6b)

The modulus of rupture (MOR),[28] or the ultimate strength f_m is obtained when the load reaches the maximum value P_{max}. Altogether, for the cases with bending stress occurring in the longitudinal (x-) direction and transverse (y-) direction, four loading directions are considered, which are $f_{m,xz}$ and $f_{m,xy}$, $f_{m,yz}$ and $f_{m,yx}$. In the expression for bending stress $f_{m,ij}$, the subscript i indicates that the induced stress is acting in the direction of the i-axis; j is the loading direction.

3.3.4.2 Bending Behaviors

According to the summary in ASTM D143,[22] there are generally six failure modes of wood specimens under static bending: simple tension; cross-grain tension; splintering tension; brash tension; compression; and horizontal shear. The actual bending behavior of glubam specimens is complicated and may depend on the tension and compression behavior along the bamboo fibers. The conventional linear elastic model is only acceptable for stresses smaller than the proportional limit. The stiffness of the compression zone of the bamboo beam is reduced when the stress is over the proportional limit; therefore, non-linear deformations would develop, and lead to a shift of the neutral axis. During the bending tests, the glubam specimens typically do not show a visible crack in the compression zone under a stress level corresponding to $f_{m,xz}$ and $f_{m,xy}$, as shown in Figure 3.18 (a) to (d). The bending failure is governed by the tension strength of the bamboo fiber as well as the appearance of non-linear deformations in the compression zone of the beam-type specimen. The failure pattern is essentially the same as the simple tension failure categorized in ASTM D143,[22] except for the case of the thick-strip specimen under $f_{m,xy}$, shown in Figure 3.18 (c). Damage to flatwise bending specimens (xz bending) is slightly more substantial than that with edgewise specimens (bending in xy), as can be seen by comparing Figure 3.18 (a) and (c), as well as Figure 3.18 (b) and (d). This is consistent with conclusions obtained from results of

Figure 3.18 Typical failure patterns in glubam bending tests: (a) thick-strip glubam in *xz* bending; (b) thin-strip glubam in *xz* bending; (c) thick-strip glubam in *xy* bending; and (d) thin-strip glubam in *xy* bending; (e) thick-strip glubam in *yz* bending; (f) thin-strip glubam in *yz* bending; (g) thick-strip glubam in *yx* bending; (h) thin-strip glubam in *yx* bending.

experiments using a full-scale glubam beam (9 m long).[2] For specimens under $f_{m,yz}$ and $f_{m,yx}$, brittle failures were noticed, and the resistance for thick-strip (Figure 3.18 (e) and (g)) glubam can basically be neglected, since there is no bamboo fiber in the *y*-direction. Such a failure pattern can be categorized as brash tension failure as described in ASTM D143.[22] The failure modes of thin-strip bending specimens under $f_{m,xz}$ and $f_{m,xy}$ (Figure 3.18 (f) and (h)) are similar to the thick-strip ones; however, improved bearing capacities can be noticed under $f_{m,yz}$ and $f_{m,yx}$, due to the existence of bamboo strips in the *y*-direction.

For thick-strip glubam, the behaviors and measured MOE and MOR are almost the same for the specimens with the loading direction parallel to the laminated surface and the specimens with the loading direction perpendicular to the laminated surface, as shown in Figure 3.19 (a) and (c). For thin-strip bamboo panels, the specimens with loading directions parallel to the laminated surface have higher stress and MOE and MOR values than specimens with loading directions perpendicular to the laminated surface, as can be seen by comparing Figure 3.19 (b) and (d). Stress-displacement curves with four different loading directions inducing stresses in the *y*-direction (with less bamboo fiber for thin-strip glubam or no bamboo fiber for thick-strip glubam) are given in Figure 3.19 (e) to (h). Clearly, thin-strip glubam specimens out-perform the thick-strip counterpart specimens due to the existence of bamboo fibers in the *y*-direction.

3.3.5 Shear Performance

3.3.5.1 Specimen and Testing Method

Based on ASTM D143,[22] the test specimen for wood materials is shown in Figure 3.20. The standard thickness of the test piece is 60 mm, which is twice the nominal thickness of glubam board. The author's research team has carried out a significant number of tests strictly based on the size required as per the ASTM standard and on specimens with a practical dimension – a thickness of about 30 mm.[5,8]

During a shear test, the specimen is placed in the fixture first, and the baffles in both directions are tightened to ensure that the specimen is snugly held in position. Then the fixture on which the test piece is mounted is placed on the lower loading platen of a

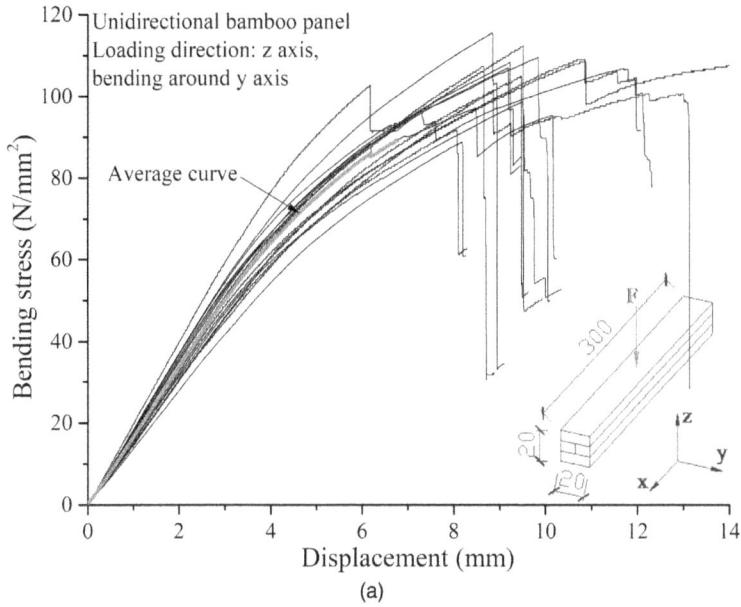

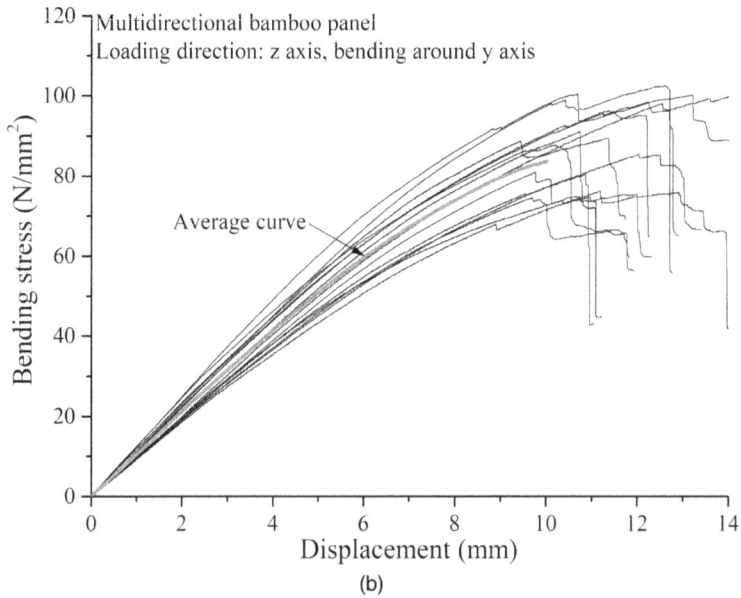

Figure 3.19 Stress-displacement relationships: (a) Thick-strip glubam in *xz* bending; (b) Thin-strip glubam in *xz* bending; (c) Thick-strip glubam in *xy* bending; (d) Thin-strip glubam in *xy* bending; (e) Thick-strip glubam in *yz* bending; (f) Thin-strip glubam in *yz* bending; (g) Thick-strip glubam in *yx* bending; (h) Thin-strip glubam in *yx* bending.

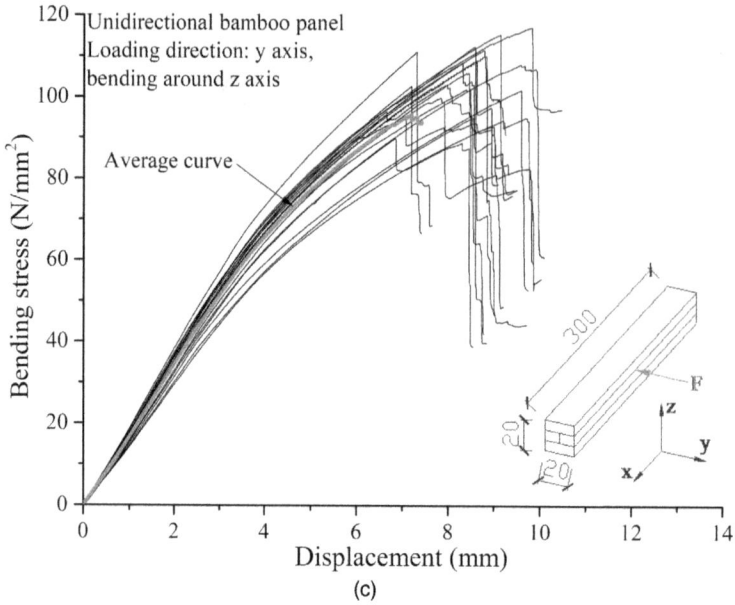

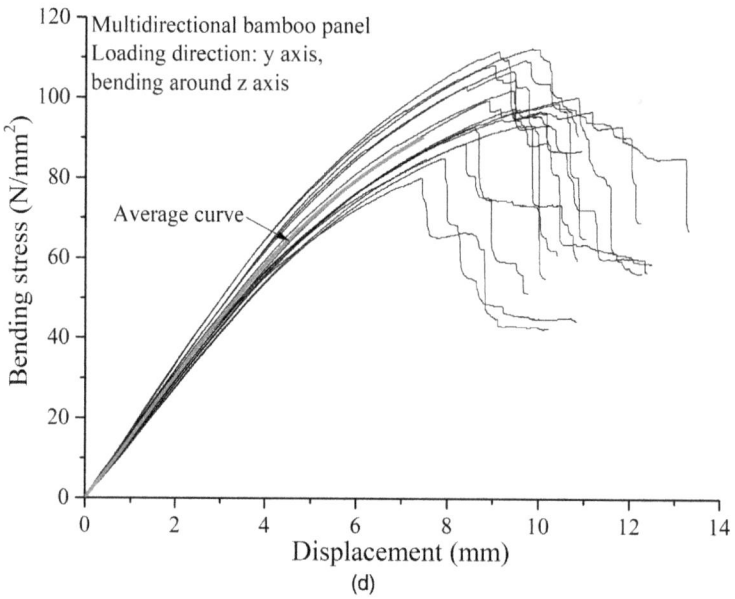

Figure 3.19 Continued

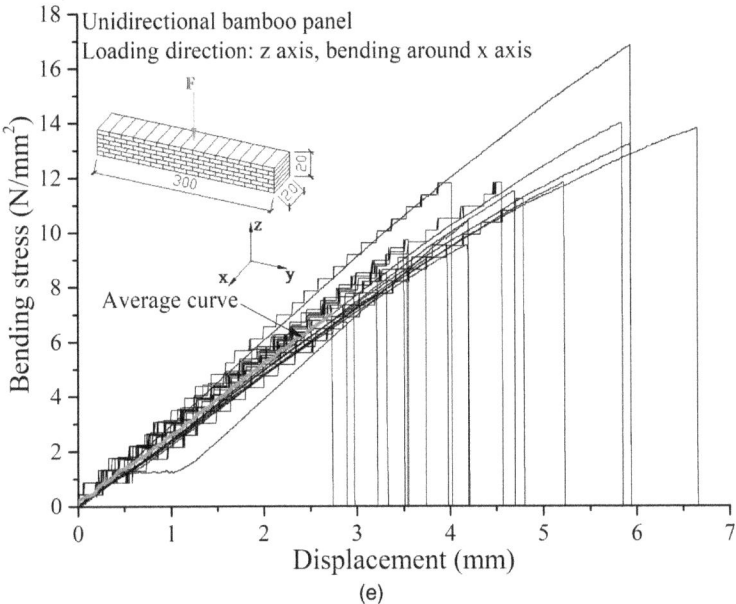

(e)

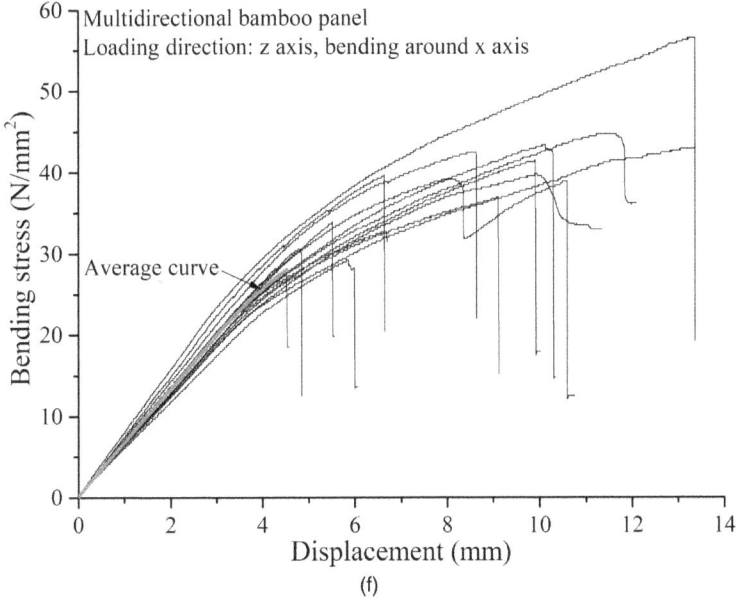

(f)

Figure 3.19 Continued

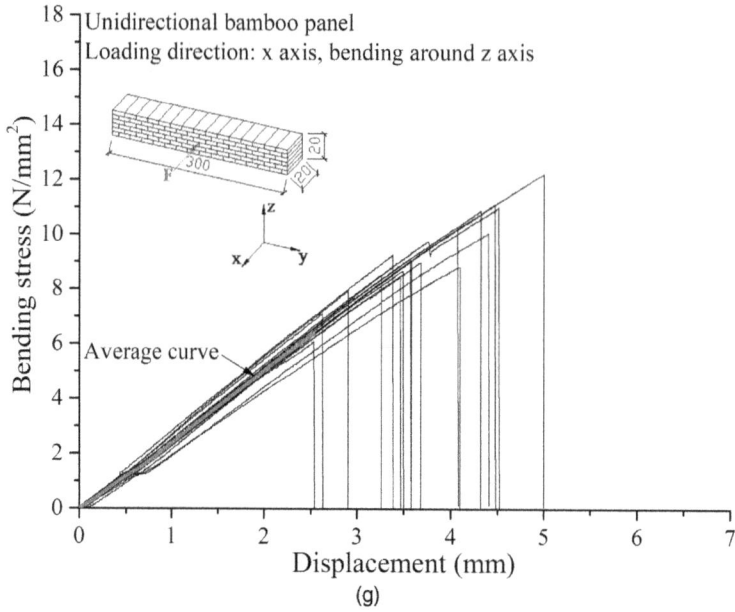

(g)

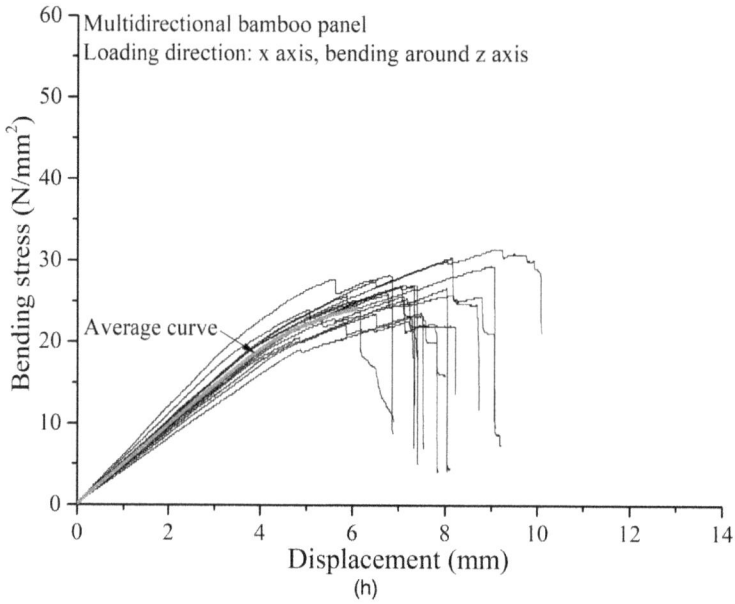

(h)

Figure 3.19 Continued

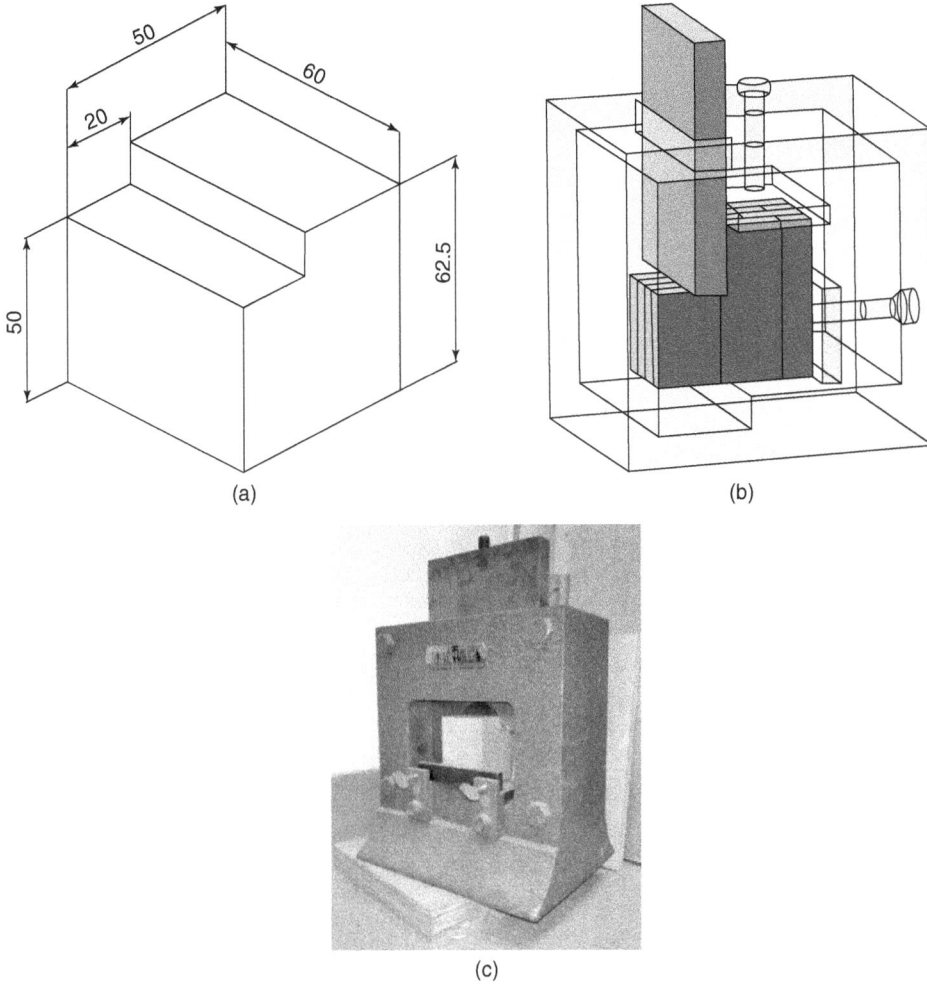

Figure 3.20 Shear test: (a) specimen; (b) loading method; (c) loading jig.

universal testing machine. The loading speed should be kept as 2 mm/min., and the test is normally finished within 2 min.

Due to its complex configuration of bamboo fiber in strip layers, glubam has distinct characteristics in different directions; thus, shear tests of the glubam need to be conducted on six possible shearing directions. These shear directions are denoted as τ_{yx}, τ_{xy}, τ_{xz}, τ_{zx}, τ_{yz}, and τ_{zy}. For τ_{ij} ($i = x, y, z; j = x, y, z$), the first subscript i indicates the direction normal to the shear plane, j is the loading direction, as shown in Figure 3.21.

The shear stress is calculated as:

$$\tau = \frac{F}{bh} \tag{3.7}$$

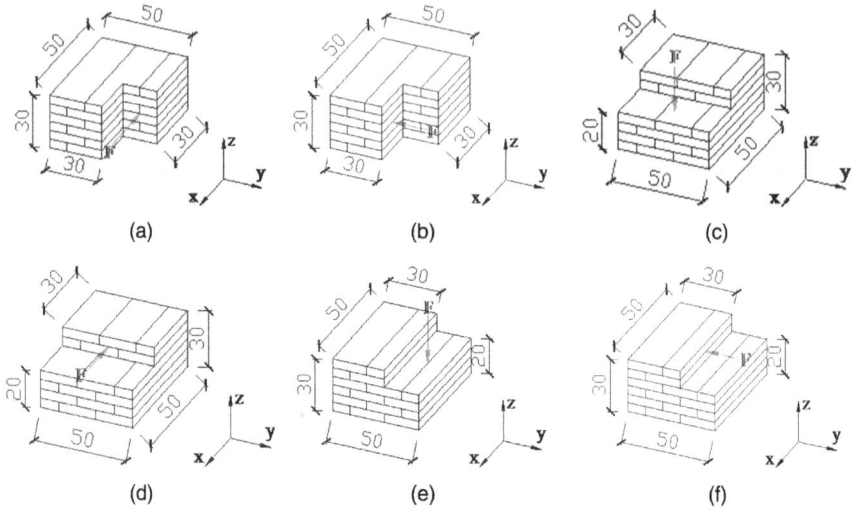

Figure 3.21 Shear tests and specimens: (a) shear plane xoz, along x-direction; (b) shear plane yoz, along y-direction; (c) shear plane yoz, along z-direction; (d) shear plane xoy, along x-direction; (e) shear plane xoz, along z-direction; (f) shear plane xoy, along y-direction.

where F is the measured force during the shear test, the shear strength f_v is obtained when this force reaches its maximum value F_{max}; however, for a test without a clear peak force, F_{max} is defined as the load recorded when the displacement is 2 mm;[5] $b \times h$ is the measured area to resistant shear force during the test. The loading rate was 0.6 mm/min.

3.3.5.2 Shear Behaviors

The failure patterns of thick-strip and thin-strip glubam specimens subjected to shear in various directions are shown in Figure 3.22 and Figure 3.23, respectively. When the transverse section of the glubam specimen (y-plane) is subjected to the shearing force along the longitudinal direction (x-direction), the failure appears to be the shearing off of the glubam portion along the shear force application plane, as shown in Figure 3.22 (a) and (d) for thick-strip glubam, and Figure 3.23 (a) and (d) for thin-strip glubam. The failure of the two types of glubam is quite different when the section crossing the longitudinal direction (x-direction) is subjected to shear in the transverse direction (y-direction). As shown in Figure 3.22 (b), the thick-strip glubam failed with a shearing off along the direction perpendicular to the shear force, whereas the thin-strip glubam ruptured along the shear application plane, Figure 3.23 (b). This is evidence of the effects of bamboo fiber configured in the transverse direction for thin-layer glubam. When the cross section perpendicular to the longitudinal direction is subjected to the shear stress along the thickness direction (z-direction), as shown in Figure 3.22 (c) and Figure 3.23 (c), both types of glubam had similar failure patterns with shear and tearing off along the plane perpendicular to the shear stress direction, in which no bamboo

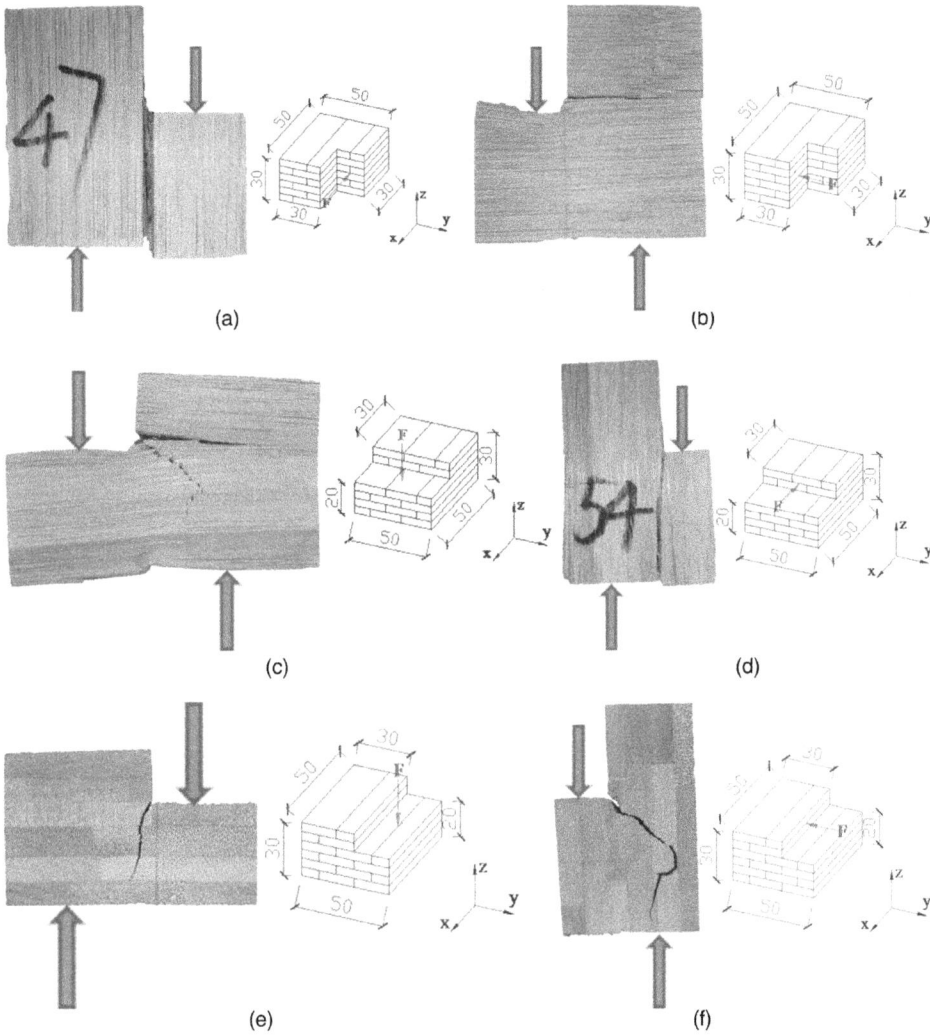

Figure 3.22 Failure patterns of thick-strip specimens subjected to: (a) shear plane xoz, along x-direction, τ_{yx}; (b) shear plane yoz, along y-direction, τ_{xy}; (c) shear plane yoz, along z-direction, τ_{xz}; (d) shear plane xoy, along x-direction, τ_{zx}; (e) shear plane xoz, along z-direction, τ_{yz}; (f) shear plane xoy, along y-direction, τ_{zy}.

fiber is provided for either glubam. For the shearing loading case where the shear force is applied in a plane parallel to the bamboo fibers (along the section's cross y-direction), failure was initiated by a main crack at the inside corner of the specimen in a direction of about 45°, and then the crack turned to run approximately parallel to the applied shear force and then penetrated through the specimens, as shown in Figure 3.22 (e) and Figure 3.23 (e), for thick-strip and thin-strip glubams, respectively. Similar failure can also be observed as shown in Figure 3.22 (d) for the thick-strip glubam subjected

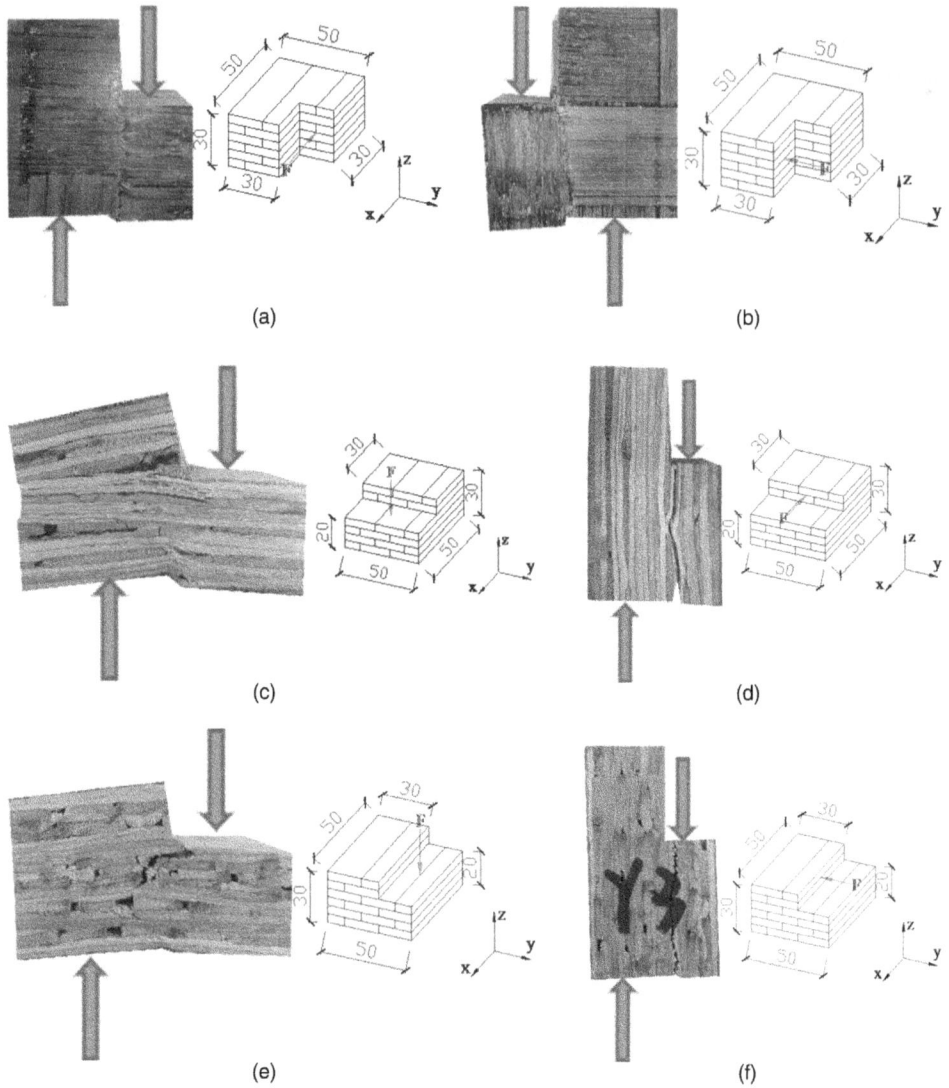

Figure 3.23 Failure patterns of thin-strip specimens subjected to: (a) shear plane xoz, along x-direction, τ_{yx}; (b) shear plane yoz, along y-direction, τ_{xy}; (c) shear plane yoz, along z-direction, τ_{xz}; (d) shear plane xoy, along x-direction, τ_{zx}; (e) shear plane xoz, along z-direction, τ_{yz}; (f) shear plane xoy, along y-direction, τ_{zy}.

to shear parallel to the bamboo fibers, but along the section across the thickness or z-direction. However, the direct shearing fracture failure is observed for the specimen of thin-strip glubam as shown in 3.23 (d).

For the shear stress-displacement curves shown in Figure 3.24, the displacement only provides a qualitative description of the shear behavior for the specimens, as it

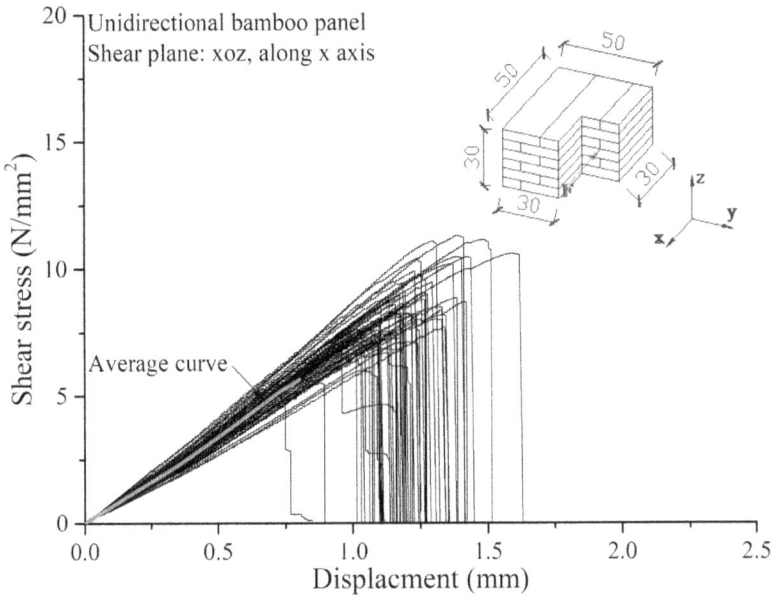

(a)

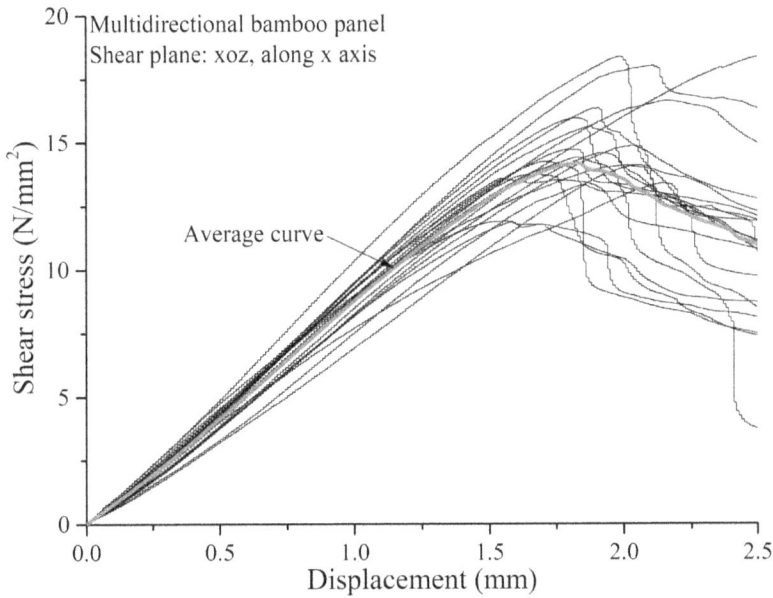

(b)

Figure 3.24 Stress-displacement relationships: (a) Thick-strip glubam τ_{yx}; (b) Thin-strip glubam τ_{yx}; (c) Thick-strip glubam τ_{xy}; (d) Thin-strip glubam τ_{xy}; (e) Thick-strip glubam τ_{xz}; (f) Thin-strip glubam τ_{xz}; (g) Thick-strip glubam τ_{zx}; (h) Thin-strip glubam τ_{zx}; (i) Thick-strip glubam τ_{yz}; (j) Thin-strip glubam τ_{yz}; (k) Thick-strip glubam τ_{zy}; and (l) Thin-strip glubam τ_{zy}.

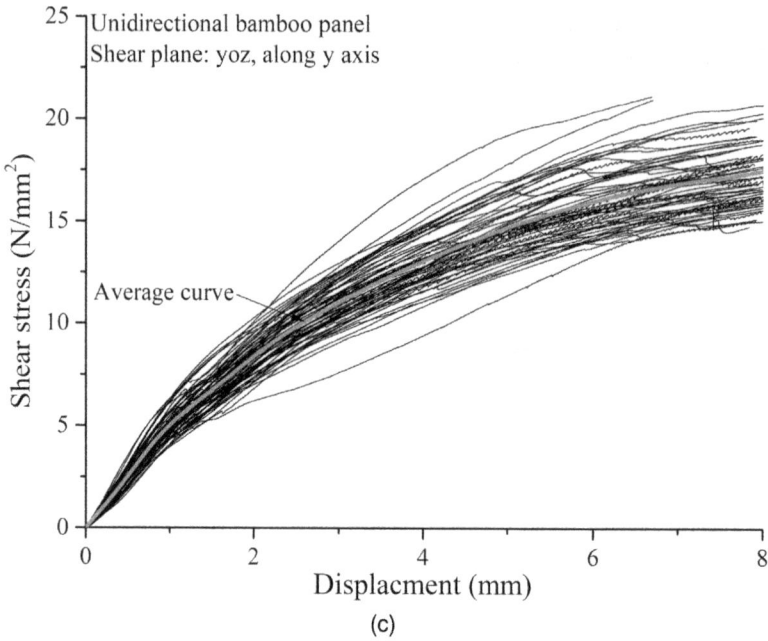

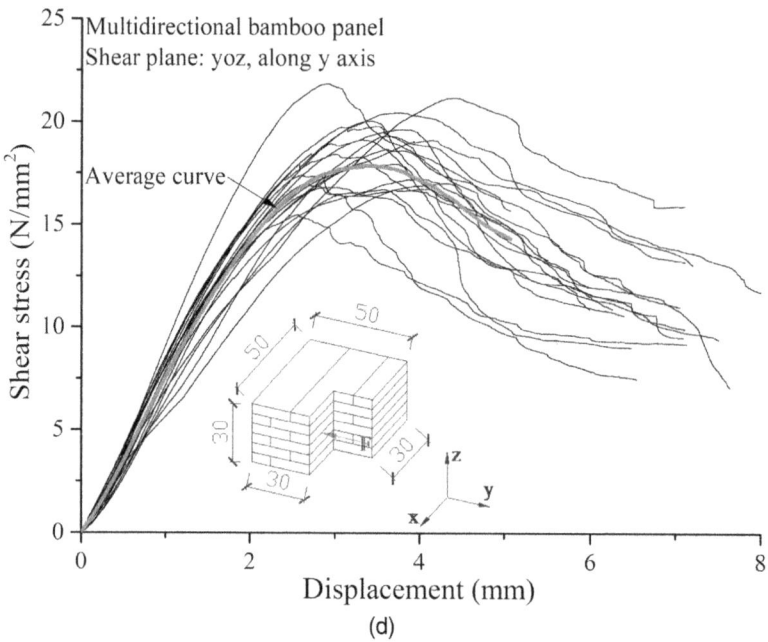

Figure 3.24 Continued

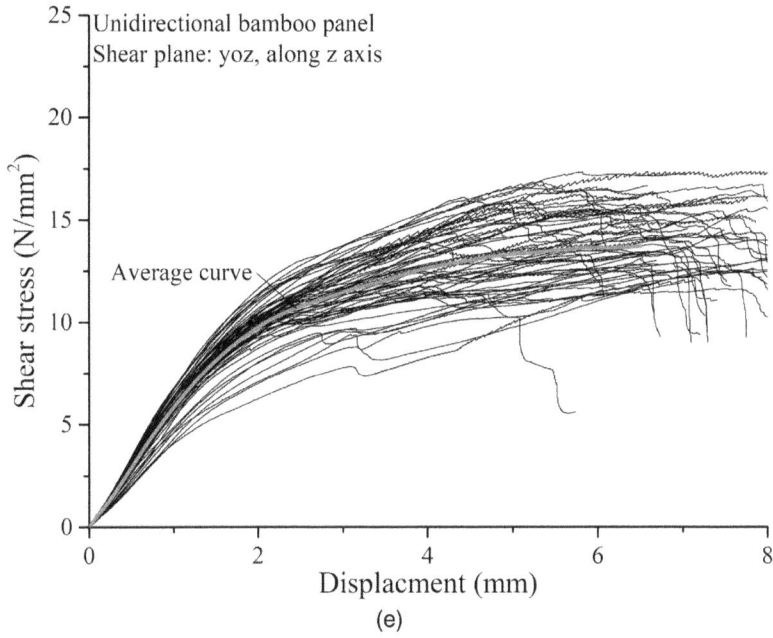

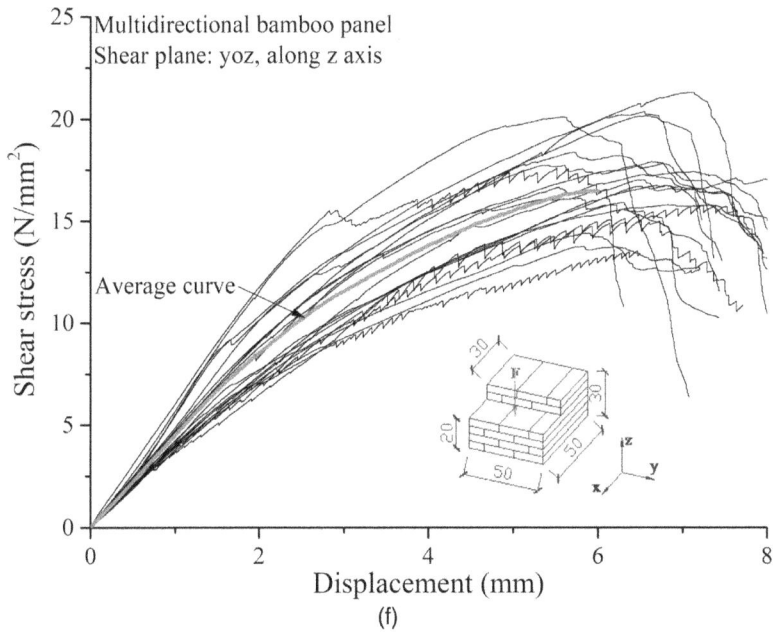

Figure 3.24 Continued

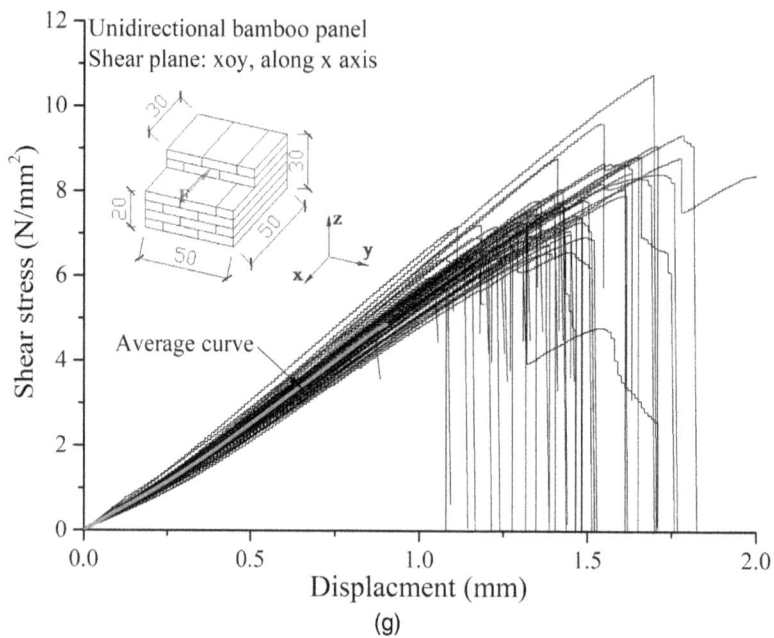

(g)

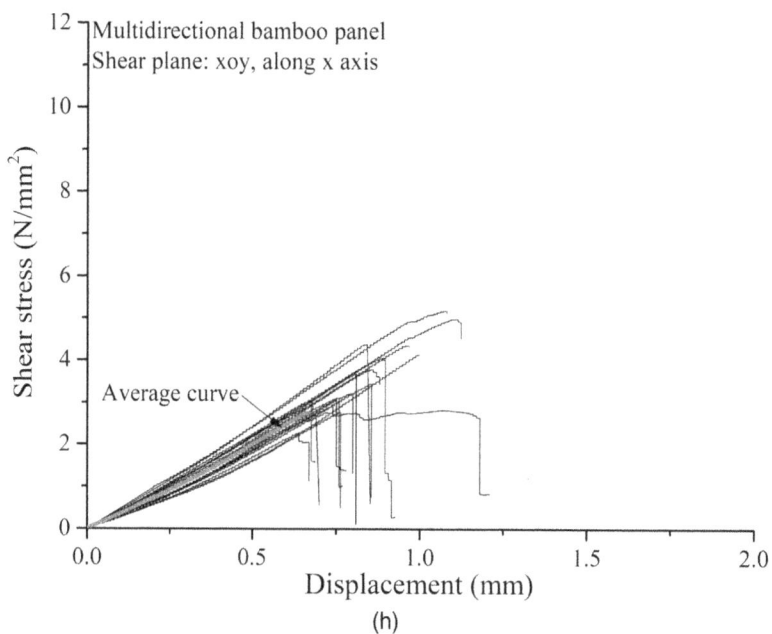

(h)

Figure 3.24 Continued

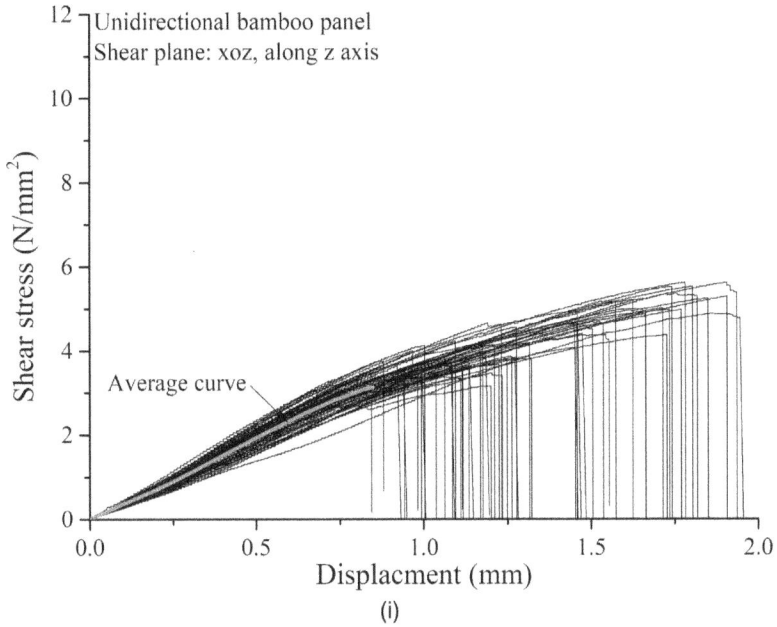

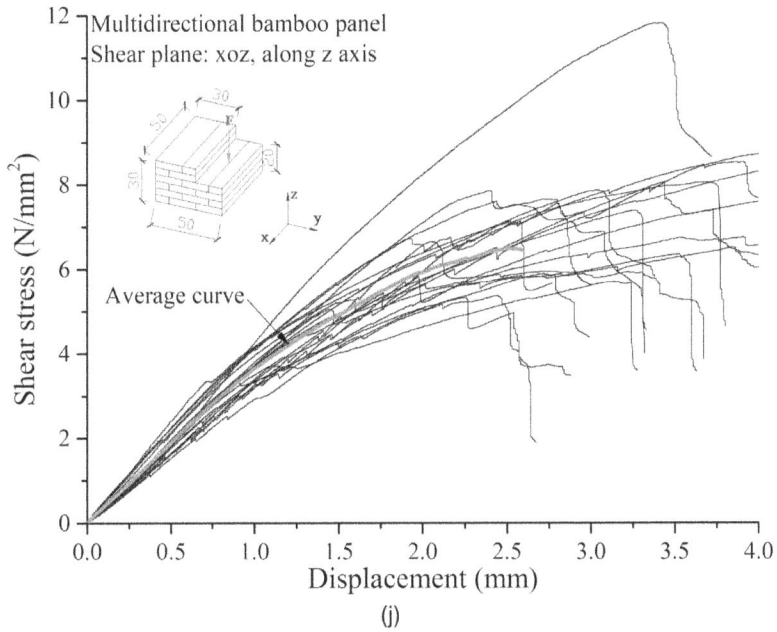

Figure 3.24 Continued

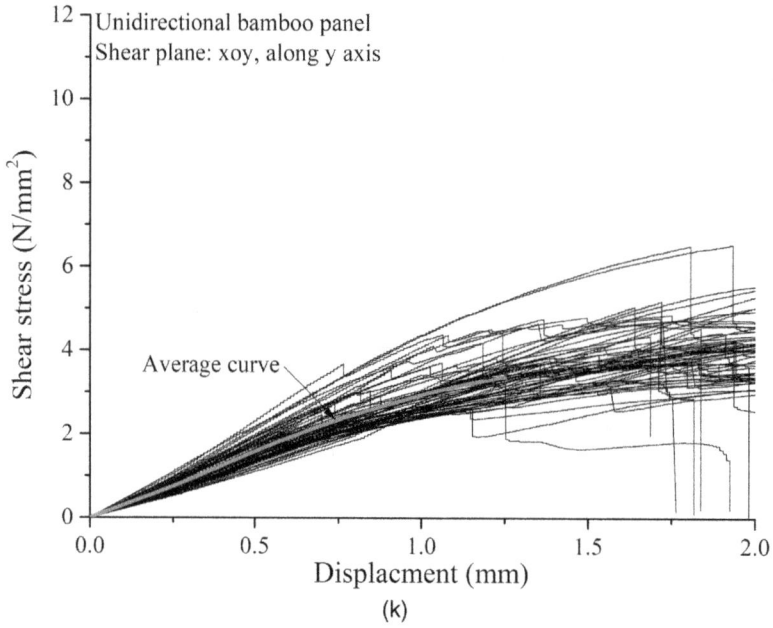

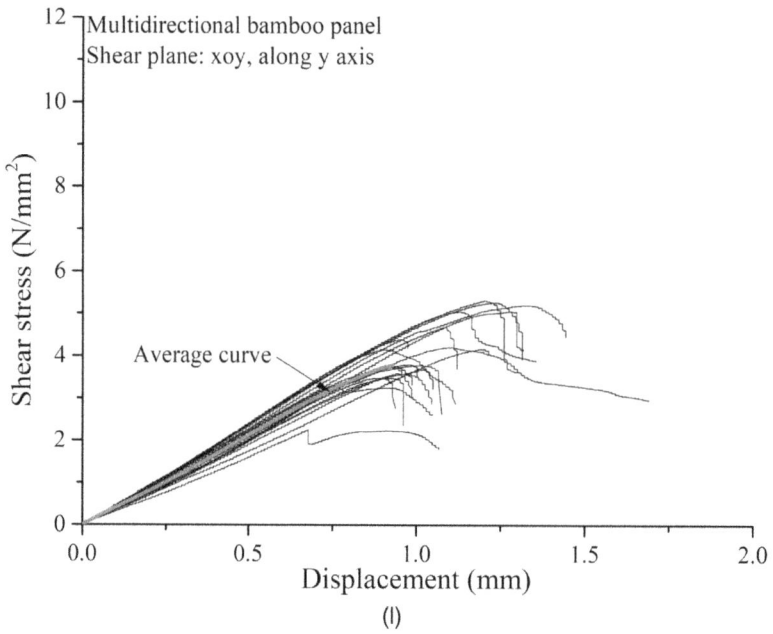

Figure 3.24 Continued

was measured for the relative displacement between the loading platens. Nonetheless, the differences between the two types of glubam are well demonstrated in Figure 3.24, showing the effects of bidirectional bamboo fiber configuration in the thin-strip glubam. The shear strength of thin-strip bamboo panels is about 1.8 times that of thick-strip bamboo panels. Thus, cross lamination of bamboo to create a type of CLB could improve the shear strength and behavior. From the observation of failure patterns shown in Figure 3.22 and Figure 3.23, as well as the shear stress and deformation behaviors shown in Figure 3.24, the shear resistance of glubam when subjected to shear along the cross section perpendicular to the thickness (z-direction) is not dependable, due to the lack of bamboo fiber across the shear plane. Thus, such loading conditions should be avoided in design, unless special measures are taken, such as a special retrofit.

3.3.6 Torsional Behaviors of Glubam

A torsion test can provide pure shear stresses in the section of a specimen, allowing measurement of the pure shear modulus and strength.[30–32] However, studies on torsional behaviors of timber are relatively rare compared with studies on loading behaviors such as bending. Vafai and Pincus (1973)[33] obtained the shear strength and failure modes of timber beams using torsion tests. The relationships between shear strength and deformation were obtained from torsional loading and strain gauge measurements. Lekhnitskii (1963)[34] calculated the shear strength through an orthotropic approach with strain gauge both on the longitudinal-tangential (LT) plane and the longitudinal-radial plane (LR). Yoshihara and Ohta (1997)[35] found the relationship between shear stress and shear strain of rectangular wood bars, and a 5% to 7% difference of shear stresses in the LT and LR planes. Riyanto and Gupta (1998)[30] evaluated the shear strength of structural size timber elements. Most of the specimens failed with cracks starting from the middle span of the longitudinal-radial plane of the beam. Khokhar (2011)[36] investigated the variation in shear modulus along the length of joists and the influence of knots on shear modulus and suggested four general failure modes on torsion.

To the author's knowledge, there has been no study to date on the torsional behavior of bamboo and engineered bamboo. Recently, doctoral student Wu, assisted by undergraduate students Mao and Li from the author's group took the initiative on studying the torsional behaviors of engineered bamboo and timber. The glubam specimens are made of two types of strips, with different degrees of carbonization. The shapes of specimens include square (15 × 15 mm) and circular sections (diameter = 15 mm).

The torsion testing device, as shown in Figure 3.25, was used with manually applied torque. The overall rotation of the specimen was measured by the built-in goniometer. This angular deformation may include the twisting deformation of the portions included in the grips. In order to monitor the purer torsional deformation of the specimen, the Digital Image Correlation (DIC) technique was adopted. Using DIC for the central portion of the specimen, the twisting angle measurement was first calibrated by

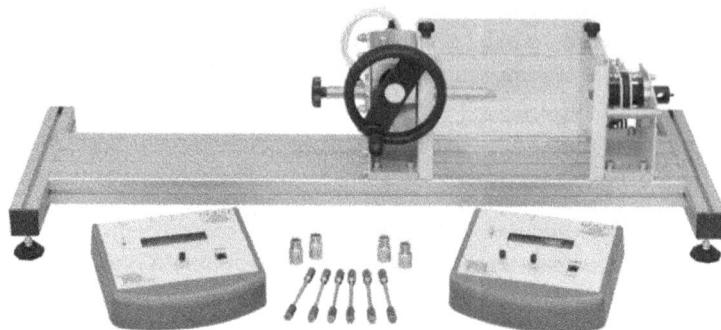

Figure 3.25 Torsional testing device.

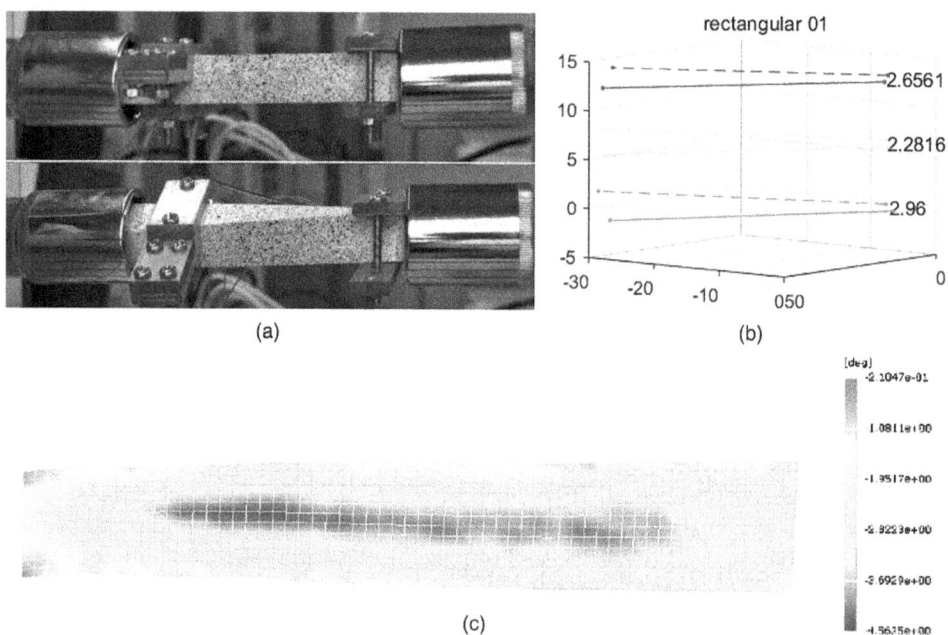

Figure 3.26 DIC measurement: (a) specimen with sprays; (b) track analysis of selected points; (c) shear strain nephograms.

comparing it with the angular deformation detected using a custom-made goniometer consisting of a strain gauge – instrumented deflectometer.

Figure 3.26 (a) shows the specimen with random dots sprayed on its surface during testing, and Figure 3.26 (b) conceptually shows the tracking analysis of the movements of the selected point. The strain nephograms of a sample specimen obtained using the DIC measurement is exhibited in Figure 3.26 (c). The dark colors along the center line of the specimen indicate the region with larger shear strain, which agrees with the theory of elasticity solution that the largest shear stress is located along the center line on each face of a rectangular prism under torsion. That essentially supports the numerical findings of finite element (FE) analysis of composite glulam.[37]

Based on the DIC data, the maximum shear strain can be obtained. The shear stress can be calculated from the recorded torque values, T, using the theoretical equations for maximum shear stresses in elastic materials.

$$\text{For square section}: \tau_{max} = \frac{4.81T}{a^3} \tag{3.8a}$$

$$\text{For circular section}: \tau_{max} = \frac{2T}{\pi r^3} \tag{3.8b}$$

where, a is the side length of the square section; and r is the radius of the circular section. Note that the maximum shear stresses for a rectangular section is at the middle of each side.

Figure 3.27 (a) and (b) shows the maximum shear stress-strain diagrams for both unidirectional thick-strip glubam and the bidirectional thin-strip glubam, respectively. It is shown that the thick-strip glubam specimens have higher strength compared with the thin-strip glubam specimens; however, the latter seem to have better deformability.

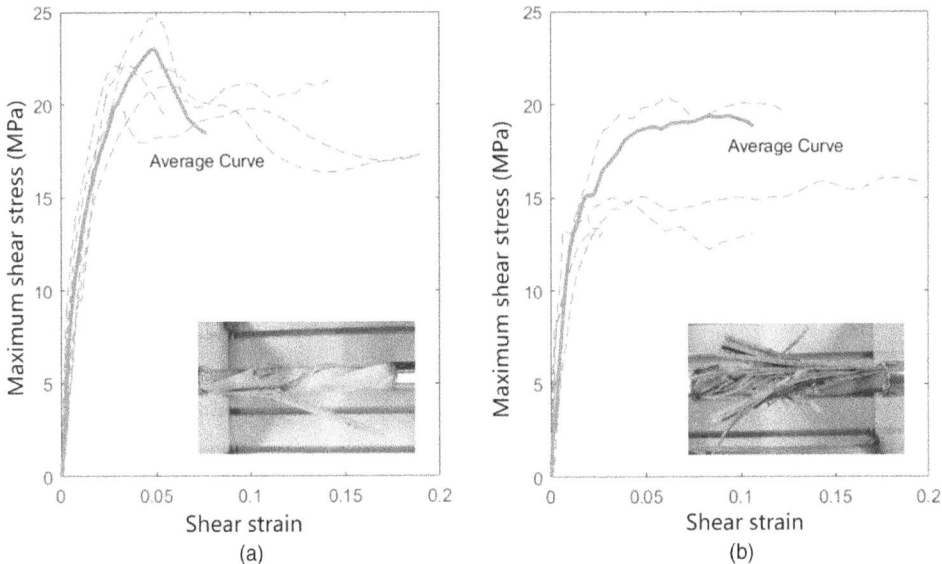

Figure 3.27 Shear stress-strain relationships for rectangular glubam specimens of (a) thick-strip; (b) thin-strip.

3.3.7 Comparison of Basic Mechanical Properties of Engineered Bamboo

The basic mechanical properties, including tensile and compressive strengths, flexural strengths, and shear strengths of glubam obtained recently by the author's research team[2,5,8,38] and other researchers[3,4,6,9,39–41] are summarized in Table 3.1 (a), (b) and (c), respectively. Since the torsional study used steel is still at its pilot phase, the results are not compared. The mechanical property values are shown as the average, with the standard derivation values shown in the parentheses. It should be noted that studies by Colombian researchers[6,9] are based on engineered bamboo made with Guadua bamboo, which is known to be stronger than the Moso bamboo used by other researchers, particularly those from China and North America.

As shown in Table 3.1 (a) and 3.1 (b), the material modulus of thick-strip Guadua bamboo-based materials is generally higher than those of Moso-based materials. For the thick-strip engineered (Moso) bamboo, the strength values achieved by the author's team and the other researchers are seen to have differences; however, they are approximately in the same range. For the thin-strip glubam, the specimens used in earlier studies[1] were made with bamboo strips that were chord-cut, whereas the recent testing specimens were from glubam made with radial-cut strips from bamboo culms. Despite this difference in manufacturing methods, the strength values are reasonably close.

3.4 Stress-strain Models for Engineered Bamboo

Similar to wood, bamboo is a bio-based material with severely non-homogeneous properties. Theoretical modeling of constitutive relationships for such material is significantly complex. For engineered bamboo, particularly, the thin-strip based glubam, it is possible to establish constitutive models similar to those for fiber- reinforced composites. Such an effort is underway for the bi-axial behavior of the planar glubam board. In this book, two simple design models are introduced for axial stress-strain behavior of glubam in the main bamboo fiber direction.

Based on the basic mechanical testing data, a simple linear model has been established for the two types of glubam.[5] As illustrated in Figure 3.28, nine parameters are used to describe the model,

$$\text{Tension:} \quad 0 \le \varepsilon \le \varepsilon_{tu}, \quad \sigma = E_t \varepsilon \tag{3.9a}$$

$$\text{Compression:} \ \varepsilon_{ce} \le \varepsilon \le 0, \quad \sigma = E_c \varepsilon \tag{3.9b}$$

$$\varepsilon_{co} \le \varepsilon \le \varepsilon_{ce}, \quad \sigma = E_c \varepsilon_{ce} + (\varepsilon - \varepsilon_{ce}) \frac{f_c - E_c \varepsilon_{ce}}{\varepsilon_{co} - \varepsilon_{ce}} \tag{3.9c}$$

$$\varepsilon_{cu} \le \varepsilon \le \varepsilon_{co}, \quad \sigma = f_c \tag{3.9d}$$

where tension is treated as positive for stresses and strains; values of f_t, E_t, f_c and E_c are given in Table 3.1 (a) according to the mechanical test results; ε_{tu} is the ultimate tension strain, given by $\varepsilon_{tu} = f_t / E_t$, which is about 0.012 for thick-strip bamboo and 0.0078 for thin-strip bamboo; ε_{ce} is the strain of proportional limit, which is 0.00322

Table 3.1a Mechanical properties of engineered bamboo in tension and compression

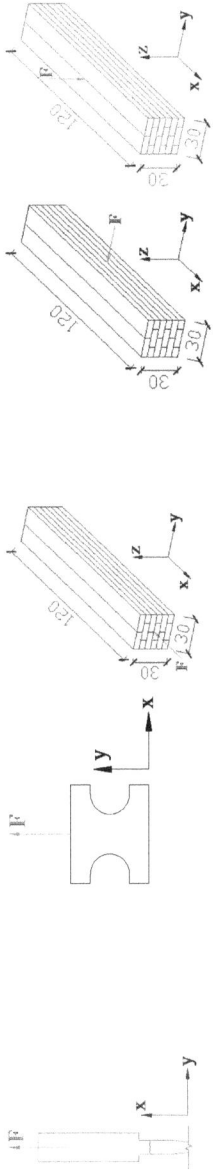

Properties	Tension			Compression			
	f_{tx}	E_{tx}	f_{ty}	f_{cx}	E_{cx}	f_{cy}	f_{cz}
Thick-strip Li et al. (2020)[5]	119.3 (22.0)	10,508 (1025)	5.9 (1.1)	57.7 (4.7)	9777 (769)	14.7 (1.1)	30.2 (5.2)
Huang et al. (2013)[39]	138 (24.47)	13,680 (740)		61.76 (4.48)	11,450 (440)		
Sinha et al. (2014)[40]	61 (-)	13,410 (-)		60.77 (-)			9.59 (-)
Sharma et al. (2015)[4]	90 (23.4)		2 (0.26)	77 (3.85)			22 (1.54)
Li et al. (2018)[3]	84.53 (9.59)	7007 (549.2)		68.8 (2.49)	9393 (435.3)		
Wu & Xiao (2018)[38]	106.7 (20.3)			71.2 (4.5)	11,000 (-)		
Correal et al. (2014)[6]	143.1 (31.8)	18,345 (3559)	3.2 (0.9)	62 (1.9)	32,271 (4098)	5.3 (1.6)[*1]	3.5 (0.8)[*1]
Chen et al. (2019)[41]	107.7 (10.7)	11,143 (-)		56.3 (7.2)	11,022 (-)		
Thin-strip Li et al. (2020)[5]	80.5 (17.3)	10,746 (761)	13.3 (2.1)	39.5 (7.0)	9810 (794)	29.1 (3.4)	12.3 (1.2)
Xiao et al. (2013)[1]	82.9(16.4)	10,400(1976)	16.9 (4.09)[*2]	58.0 (2.6)	25.3 (2.94)[*2]		

Note: *1, data are based on 0.05 strain; *2, Yang et al. (2015).[27]

Table 3.1b Mechanical properties of glubam in bending

Properties		$f_{m,xz}$	$E_{m,xz}$	$f_{m,xy}$	$E_{m,xy}$	f_{myz}	f_{myx}
Thick-strip	Li et al. (2020)[5]	104.6 (7.5)	8682 (764)	104.9 (7.6)	9052 (599)	10.2 (2.8)	9.3 (1.5)
	Huang et al. (2013)[39]						
	Sharma et al. (2015)[4]	77~83 (4.6~6.7)	11~3GPa (0.55~0.78)				
	Wu & Xiao (2018)[38]	111.2 (-)		114.2 (-)			
	Correal (2014)[6]	103 (10.9)	12,720 (649)	122.4 (6.0)	13,260 (53)		
	Sinha et al. (2014)[40]		12,190 (-)		22,300 (-)		
Thin-strip	Li et al. (2020)[5]	88.5 (10.1)	10,291 (1723)	99.4 (10.0)	14,708 (1702)	39.3 (7.3)	26.7 (2.6)
	Xiao et al. (2013)[1]			99 (11)	9400 (927)		

Table 3.1c Mechanical properties of glubam in shear

Properties		τ_{yx}	τ_{xy}	τ_{xz}	τ_{zx}	τ_{yz}	τ_{xy}
Thick-strip	Li et al. (2020)[5]	8.3 (1.4)	8.4 (0.9)	9.8 (1.1)	7.4 (1.1)	4.2 (0.7)	4.4 (0.8)
	Sharma et al. (2015)[4]	16 (0.8)					
	Takeuchi et al. (2018)[9]	6.0 (0.58)			4.34 (0.56)	2.99(0.30)	2.16 (0.36)
					4.61 (0.92)[*1]		2.16 (0.56)
	Wu & Xiao (2018)[38]	7.5			6.6		
	Correal (2014)[6]	9.5 (1.4)					
	Sinha et al. (2014)[40]	15.67 (-)					
Thin-strip	Li et al. (2020)[5]	14.9 (1.9)	18.8 (2.0)	8.5 (1.6)	3.5 (0.8)	7.4 (1.5)	4.1 (0.8)
	Xiao et al. (2013)[1]	7.4 (1.54)	7.2 (0.86)				
	Xiao et al. (2017)[8]	14.7 (1.89)	16.00 (2.16)		4.61 (1.39)		3.08 (1.07)

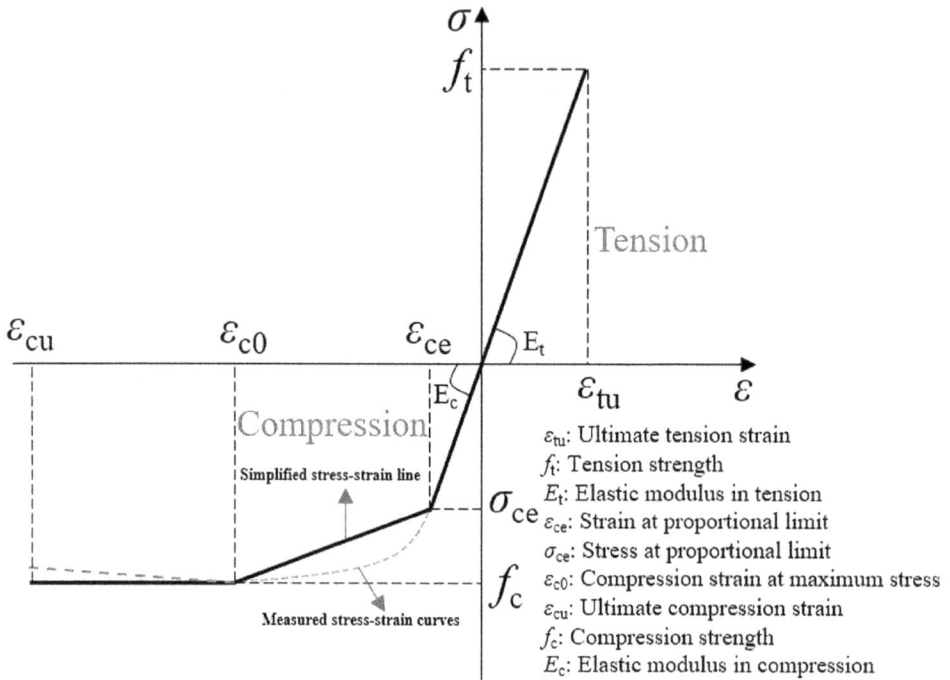

Figure 3.28 Simplified stress-strain model.
Source: Yang (2013).[27]

and 0.00275 for thick- and thin-strip bamboo, respectively; ε_{co} is the compression strain at maximum stress, which is 0.0233 for thick-strip bamboo and 0.01 for thin-strip bamboo; ε_{cu} is the ultimate compression strain, which is about -0.035--0.03 for thick- and thin-strip bamboo panels. It should be mentioned that the values selected in the model are based on the recent testing data and should be used as reference only. Modifications might be needed; if the manufacturing method for a particular engineered bamboo is changed dramatically, the values need to be re-evaluated. Figure 3.29 exhibits the simplified models for the glubam with thick-strip and thin-strip, respectively.

Huang et al. proposed a combined linear line and curve-based model for parallel strand bamboo (PSB).[42] Shen et al. also suggested using a polynomial function to represent the compressive curve of engineered bamboo.[43] In this book, a more sophisticated model is suggested by using the so-called Popovics' equation[44] for the compressive behavior,

Tension: $0 \le \varepsilon \le \varepsilon_{tu}, \quad \sigma = E_t \varepsilon$ (3.10a)

Compression: $\varepsilon_{cu} \le \varepsilon \le 0, \quad \sigma = f_c \dfrac{xr}{r-1+x^r}$ (3.10b)

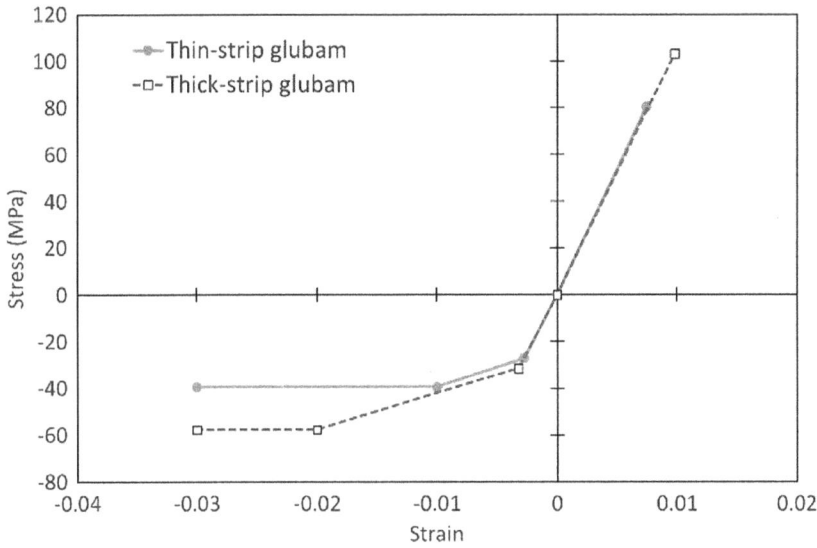

Figure 3.29 Segmental linear axial stress-strain relationships for two types of glubam.

where, x is the relative compressive strain, and $x = \varepsilon / \varepsilon_{co}$; r is a coefficient, and $r = E_c / (E_c - f_c / \varepsilon_{co})$; ε_{tu} is the ultimate tension strain, given by $\varepsilon_{tu} = f_t / E_t$, which is about 0.012 for thick-strip bamboo and 0.0078 for thin-strip bamboo; ε_{co} is the compression strain at maximum stress, which is 0.0233 for thick-strip bamboo and 0.01 for thin-strip bamboo; ε_{cu} is the ultimate compression strain, which is about -0.035--0.03 for thick- and thin-strip bamboo panels. The four basic mechanical parameters of f_t, E_t, f_c and E_c are given in Table 3.1 (a). This refined model uses six material properties and is shown in Figure 3.30 for the thick-strip and thin-strip glubam.

The suggested axial stress-strain model is essentially design-oriented. A more sophisticated theoretical constitutive model should be developed in future, based on the theory of composite materials. More detailed micro and macro examinations of bamboo and the engineered materials are necessary. Recently, the relationship between the volume of bamboo fiber and corresponding strength based on the image analysis method was studied by Penellum et al.,[45] and Akinbade et al.[46] The author believes more attention needs to be directed to such a fundamental research area.

3.5 Long-term Creep Properties of Glubam

Research on the long-term creep behaviors of bamboo and glubam is very limited. The author's research team carried out a pilot testing program on the tensile and compressive creep deformation of thin-strip glubam at different stress levels in an indoor environment.[47,48] Since this research was conducted several years ago, the bamboo strips were chord-cut from the culms, rather than using the radial cut for glubam produced in

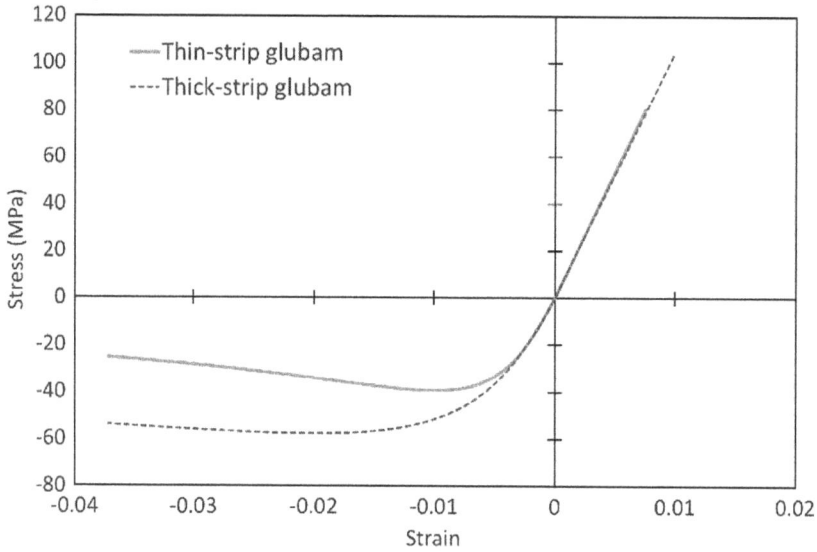

Figure 3.30 Refined stress-strain model for glubam with curved compressive behavior.

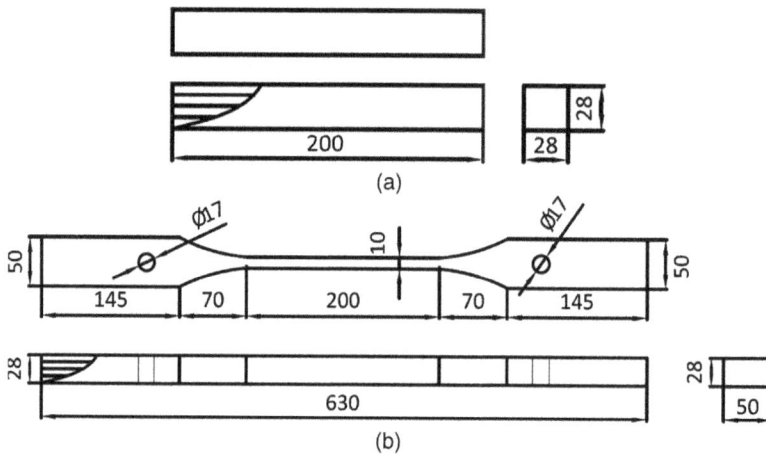

Figure 3.31 Long-term creep test specimens (mm): (a) compression; (b) tension.

recent years. Nonetheless, the research provides the tensile and compressive creep properties and behaviors of glubam useful for analysis of glubam components.

The long-term creep test specimens were designed according to ASTM D143,[22] and referencing the relevant provisions of GB/T 50329.[49] Details of the compressive and tensile specimens are shown in Figure 3.31 (a) and (b), respectively.

The long-term creep test was carried out on a special creep testing device, as shown in Figure 3.32. In the tensile creep device, three specimens are connected to two steel loading plates in parallel to withstand a constant force, which is applied using an oil jack through four loading plates and high-strength steel bars. The applied tensile force is measured (as compressive force) using a dial gauge positioned in between two loading plates. In the compression creep device, three test specimens are connected in series, and each end of the specimen is provided with a unidirectional hinge. The hinges are perpendicular to each other for a glubam specimen and are positioned at the center of a thick steel diaphragm which can slide along the vertical bars. The compressive force is applied using an oil jack and measured using a dial gauge. In both the tensile and compressive loading devices, springs are provided between the loading plates, to maintain the constant loading. Due to the creep deformation of the specimens and due to the pressure decay in the oil jack, the intended applied force may decrease despite the use of the springs; therefore additional jacking force is added occasionally during the course of the creep testing.

The creep deformation of each test specimen was measured by dial gauges attached to the specimen using a specially manufactured jig. As shown in Figure 3.32, two dial gauges are installed on the jig to measure the deformation on both sides of each specimen, and the obtained deformation values are averaged to eliminate the effects of possible bending deformation. The dial gauge has a gauge length of 150 mm and a resolution of 0.0067 for strain measurement. During the loading, the readings of each device were recorded every four hours from 8:00 to 23:00 in the first three days, followed by every half day in the next month, and then recorded every three days for the next two

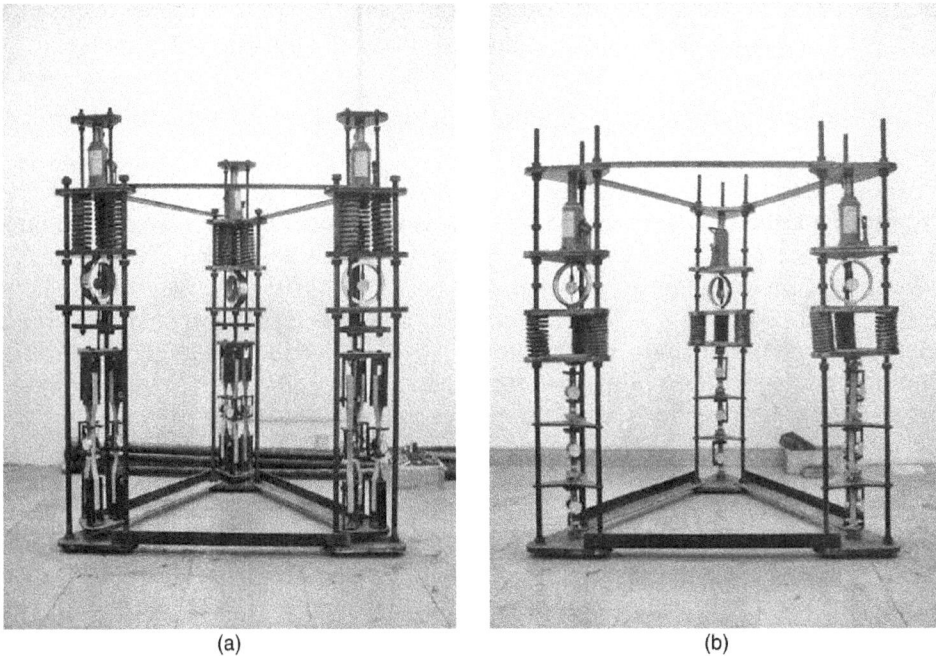

(a)   (b)

Figure 3.32 Long-term creep testing apparatus

(a) Tensile creep test device (b) Compression creep test device.

Table 3.2 Creep test load level (based on the average value of compressive strength)

Compression creep test (section size 28mm × 28mm, 3 series)			Tensile creep test (section size 28mm × 28mm, 3 parallel)		
Stress ratio	Maximum stress (MPa)	Applied load (kN)	Stress ratio	Maximum stress (MPa)	Applied load (kN)
0.6	35.076	27.500	0.6	35.076	29.464
0.4	23.384	18.333	0.4	23.384	19.643
0.2	11.692	9.167	0.2	11.692	9.821

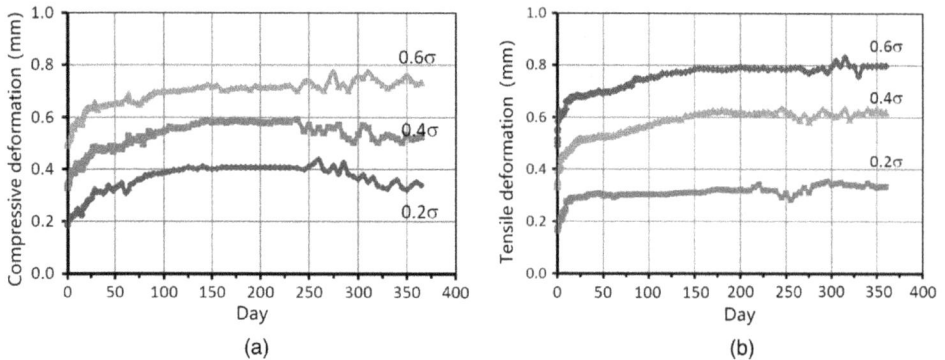

Figure 3.33 Average creep deformation-time curves under uniaxial compression (a); tension (b).

months. Thereafter, the readings were taken and recorded every five days until the end of the creep test, which lasted for a year.

According to the average strength value of the materials obtained from the short-term mechanical tests, the stress levels of the long-term creep tests were divided into three levels; i.e., 0.2, 0.4, and 0.6 times the average strength, as shown in Table 3.2.

The long-term creep tests of glubam began on April 24, 2011 and lasted for a whole year until April 24, 2012. The tests experienced a complete four-season cycle of spring, summer, fall (autumn), and winter, although in an in-room condition. The one-year-long creep testing results of glubam are shown in Figure 3.33 (a) and (b) for compression and tension, respectively. Each curve represents the average deformation of three specimens corresponding to time.

As can be seen from Figure 3.33, the creep deformation trends at different stress levels are essentially similar. In the first 50 days, due to the spring and summer seasons, the relative humidity fluctuations in the Changsha area were relatively severe, and the fluctuations of the creep deformation were large. During the fall (autumn) and winter seasons, the creep deformations tend to be smoother. Toward the end of the testing, the fluctuation in the creep deformations became severe due to the rainfall spring season. Despite the in-room condition, the obtained testing results of creep deformation of glubam do not rule out the influence of temperature and humidity changes.[48]

It can also be seen from the figure that in the initial stage of loading, the increase of creep deformation of the specimens is larger and faster. The rate of change of creep then gradually slows down, decreases to a minimum, and then remains essentially unchanged. And at the same stress level, the deformation of the tensile creep and the compression creep are substantially the same.

The creep data obtained by the author's research team is verified according to the provisions of the American ASTM specification. The mid-span creep of wood components according to ASTM D6815[50] should be:

$$D30 - Di > D60 - D30 > D90 - D60 \tag{3.11}$$

and,

$$FD90 = D90 / Di < 2.0 \tag{3.12}$$

In the equations, Di is initial elastic deformation; $D30$, $D60$, and $D90$ are deformations at day 30, day 60, and day 90 after loading. Applying these rules against the data indicates that the creep testing results meet the requirements of the specification, and the ratio of the total deformation of the test specimen to the initial deformation under different stress conditions is between 1.5 and 1.9, less than 2.0. This indicates that the creep test results are reliable, that the creep of glubam has the general material creep characteristics, and the creep deformation is stable under the conditions of the test.

3.6 Aging Behavior of Glubam

3.6.1 Accelerated Aging Tests

The weathering resistance of engineered bamboo is extremely important; however, only very limited research is available. To study aging effects on mechanical behaviors of glubam, Shan et al.[51] carried out a testing program with artificial climatic environment simulation of the influence of solar light and rainfall on glubam, using the method of ultraviolet accelerated aging.

The accelerated aging test is based on the climate of Changsha, China. In Changsha, there are 1300 to 1800 hours of sunshine per year, and the average annual precipitation is 1200–1700 mm. The equipment used in the test was an ultraviolet weathering test chamber, as shown in Figure 3.34. The test chamber has the functions of irradiation, condensation, and automatic spray circulation, simulating outdoor conditions such as sunlight and rainfall. In the test, the intensity of the ultraviolet light source was about 4.3 times higher than that of natural sunlight, and the water spray amount was about 300 mm per hour. The cycle of the simulation test was first to irradiate for 8 hours to simulate the degradation of glubam sheets by sunlight in a climatic environment; then to condense and spray water for 1 hour each to simulate the dew action and rainfall process in a natural climate. Therefore, it can complete 2.4 cycles of a single test per day. Table 3.3 shows the basic information of the artificial accelerated aging test and the equivalent aging time. The test selects the aging time according to the total amount of irradiation.

(a)   (b)

Figure 3.34 Ultraviolet resistance to weather aging test box: (a) external; and (b) internal views.

Table 3.3 Basic information of artificial accelerated aging test

Test box time	Equivalent aging time	
	According to the amount of radiation	*According to the amount of water spray*
12 d	240 d	600 d
24 d	480 d	1200 d
48 d	960 d	2400 d

Table 3.4 Aging testing matrix

Specimen group	Treatment
Group 1	Exposed cutting surface
Group 2	The cutting surface is sealed with asphalt paint
Group 3	The cutting surface is sealed with GFRP

3.6.2 Aging Test Specimens and Treatment

The aging test specimens were randomly selected from a batch of glubam boards and cut into samples (length 400mm, width 200mm, and thickness 28mm). The test samples were divided into three groups, as shown in Table 3.4. Each group of samples included three glubam plates and were placed in the test box for aging cycles.

In accordance with Chinese standard JG/T 199, the sample specimens were treated with a water content adjustment before and after the aging cycles, by storing them in an environmental box with a temperature of 20°C±2°C, relative humidity of 65%±5% for two weeks.[52] After the water content treatment, the sample dimensions were measured and recorded. Glubam exhibits significant deformation under dry and wet cycles, especially in the thickness direction of the sheet. The increment of this deformation is directly related to the aging performance; therefore after the sample aging cycles, the deformation needs to be measured. The measurement point is shown in Figure 3.35. After this

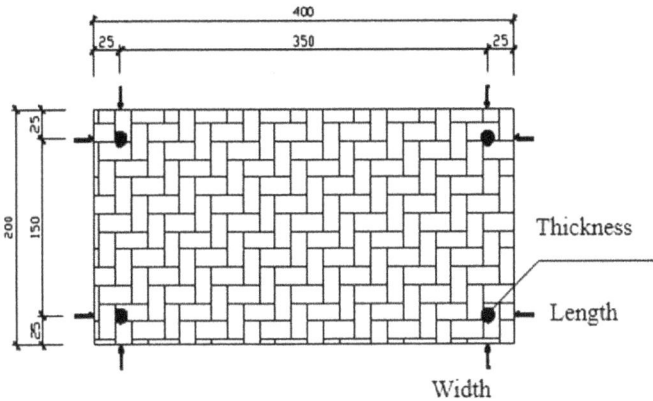

Figure 3.35 Measured point layout of deformation measurement.

Table 3.5 Thickness deformation measurement results

Accelerated aging time	Exposed cutting face	Asphalt paint edge banding	GFRP edge banding
12 d	14.2%	11.7%	5.19%
24 d	18.6%	17.2%	10.3%
48 d	23.6%	20.5%	14.1%

process, mechanical testing specimens were cut from the aged glubam plates, and their basic mechanical properties were studied, including tensile strength and modulus of elasticity, compressive strength, flexural strength and elastic modulus, as well as internal bond strength. The basic aging properties of bamboo are obtained by comparing the test results before and after aging.

3.6.3 Test Results and Analysis

After the aging testing cycles, the dimensional expansion rate of each group of specimens is measured in three directions. The results show that the changes in the planar dimensions are relatively small and essentially can be neglected. However, as shown in Table 3.5, the glubam specimens exposed at the cut edges show a significant increase in deformation in the thickness direction, and this increased with increasing aging time. However, if the edges are sealed with asphalt or glass fiber reinforced polymer (GFRP), the thickness deformation slowed down; in particular, the GFRP edge sealing produced a better effect. The method was also used by the author's research team for actual glubam structures.

Each group of glubam samples subjected to the accelerated aging test was used to manufacture specimens for basic mechanical testing, including tensile strength and tensile modulus, compressive and bending strength, and the results are shown in Figure 3.36 (a) to (f). All the mechanical properties exhibit a downward trend with the increase of

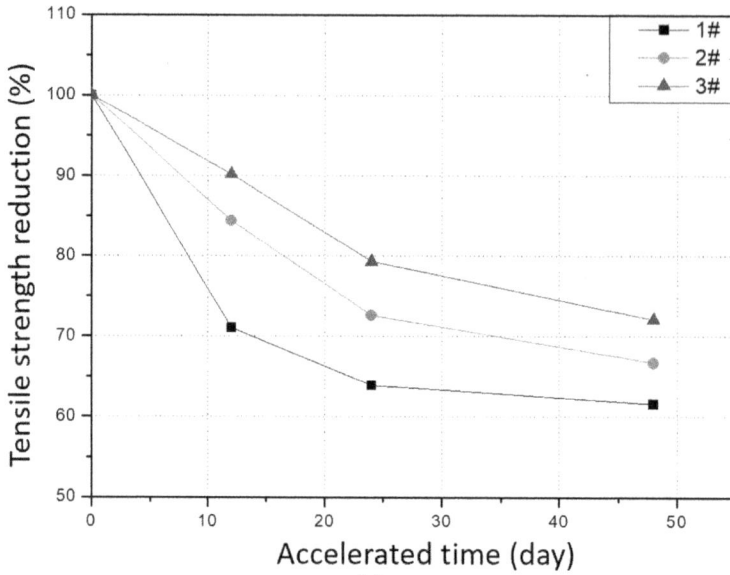

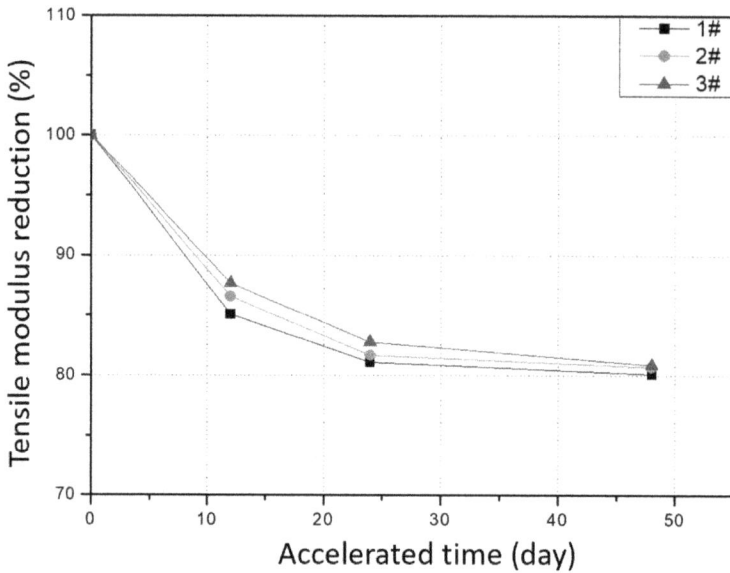

Figure 3.36 Property reduction after accelerated aging: tensile strength (a) and tensile modulus (b); compressive strength (c); internal bond strength (d); Properties after accelerated aging: bending strength (e); and flexural modulus (f).

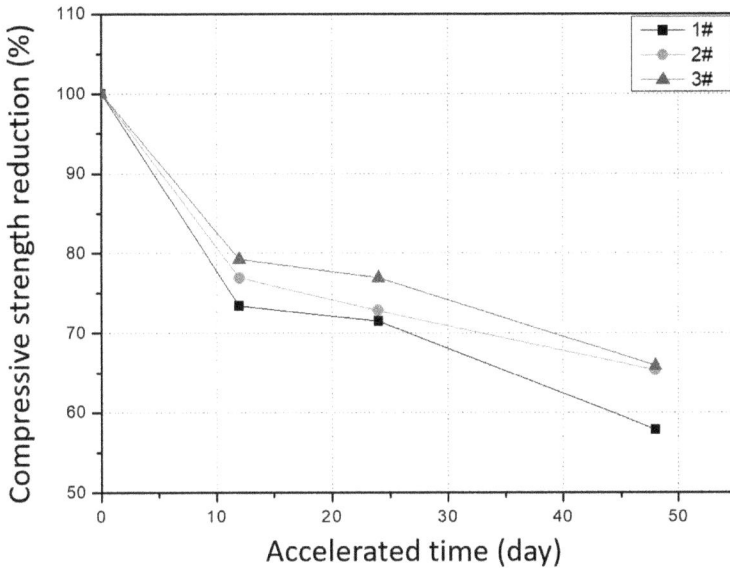

(c)

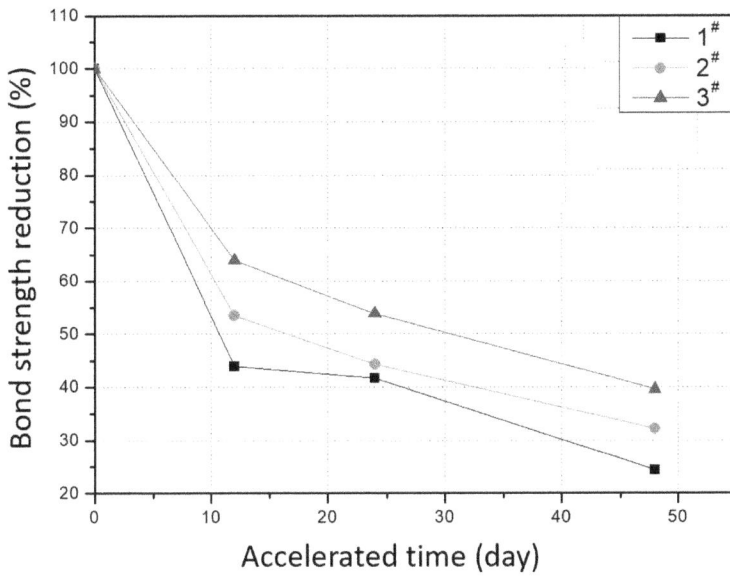

(d)

Figure 3.36 Continued

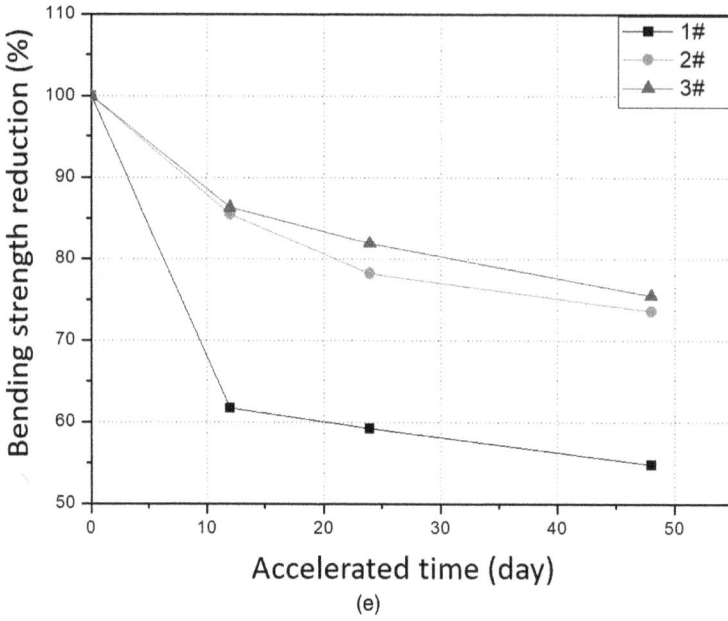

(e)

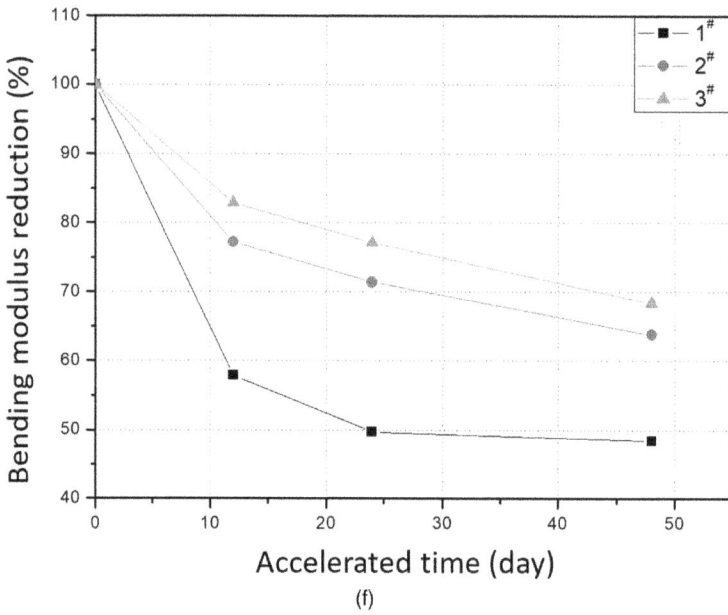

(f)

Figure 3.36 Continued

aging time; however, the sealing of the cutting edges is seen to be effective in reducing the aging effects on mechanical properties.[51]

The change in thickness and mechanical properties, denoted as P_t, corresponding to aging exposure time, can be normalized against the initial properties, P_o, and the aging factor β can be obtained from the following equation,

$$\beta = P_t / P_0 \tag{3.13}$$

The relationship between the aging factor β and the aging time t is shown in Figure 3.37 for all the tested properties. It can be seen from the figure that all the mechanical strength results tend to decrease monotonously with the aging time. Moreover, in all mechanical properties, the internal bond strength is most sensitive to the aging time, while other strengths are closely related to it. Therefore, the internal bond strength aging factor $\beta_{b,t}$ and the other intensity aging factor β have an internal bond strength correlation coefficient η, which is defined by the following equation,

$$\eta = \beta_{s,t} / \beta_{b,t} \tag{3.14}$$

where $\beta_{s,t}$ is the aging factor of a mechanical property after the aging time t; and $\beta_{b,t}$ is the aging factor of the bond strength after the aging time t.

The correlation coefficient, η, can be seen in Figure 3.38 for the mechanical properties. As shown in Figure 3.38, as most of the coefficients are close to each other, a common linear regression relationship is suggested, as follows,

$$\eta = 1.0917 + 0.0013t \tag{3.15}$$

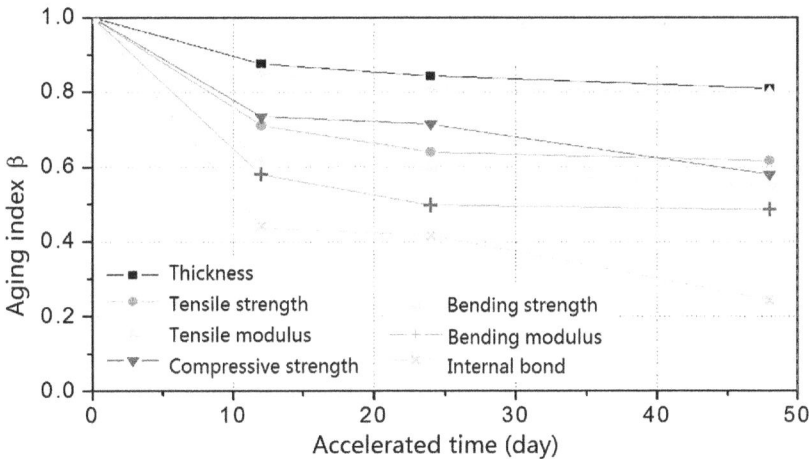

Figure 3.37 Relationship between aging factor β and accelerated aging time.

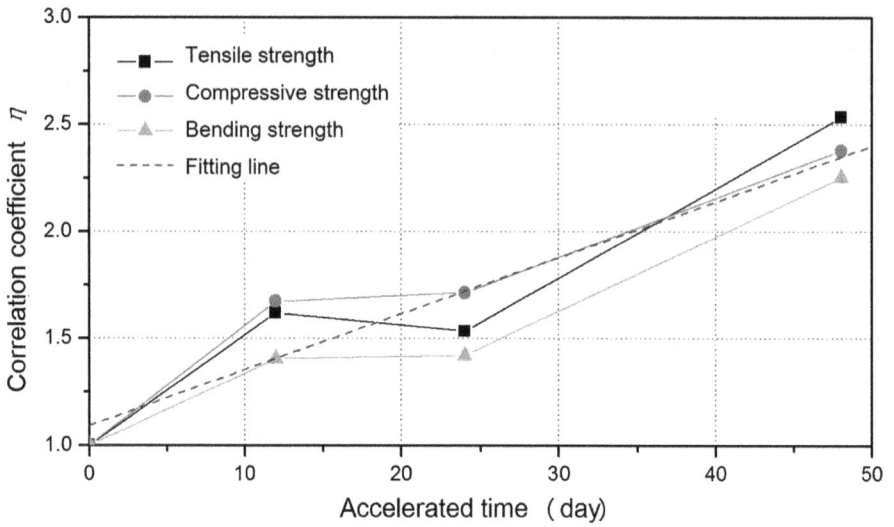

Figure 3.38 Fit curve of correlation coefficient η.

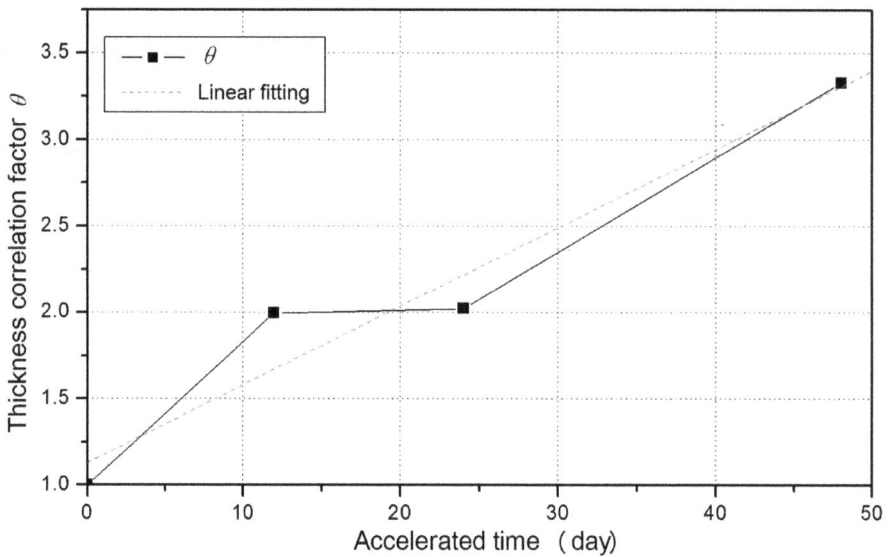

Figure 3.39 Fit curve of correlation coefficient θ.

On the other hand, the change of thickness is a geometric parameter, which can be directly measured as t_o and t_t, for the thicknesses prior to and after experiencing the aging cycles. The thickness after aging, t_t, can be calculated based on the initial thickness, t_o, multiplying a coefficient θ, which is obtained using regression analysis, and shown in Figure 3.39.

$$\theta = 1.1275 + 0.0023t \tag{3.16}$$

The pilot research program conducted by the author's research team shows that the aging effects on the mechanical behaviors of glubam are related to the internal bonding strength of the bamboo bundles/strips and a deterioration relationship with time can be established. Based on the limited research, it is recommended that direct exposure of the glubam material to rainfall should be avoided for buildings expected to have a long duration of usage.

3.7 Behavior of Engineered Bamboo under High-strain Rate Loading

Studies on the dynamic behavior of engineered bamboo in comparison with timber have recently been carried out by the author's research group.[53] The research was focused on the compressive behavior of glubam using a Split Hopkinson Pressure Bar (SHPB) device, as well as dynamic bending using a pendulum impact machine.

3.7.1 Pendulum Impact Behavior

As shown in Figure 3.40 (a), a pendulum-type impact testing device with a maximum impact energy of 85 J (started at an upward angle of 60°) was employed to study the dynamic bending behaviors of two type of glubams: the thick-strip glubam (G1) and the thin-strip glubam (G2). All the specimens had a length of 350 mm and a clear span of 300 mm. The cross sections for the thick-strip and the thin-strip glubams were 20 × 20 mm and 30 × 30 mm, respectively, as shown in Figure 3.40 (b). The thick-strip

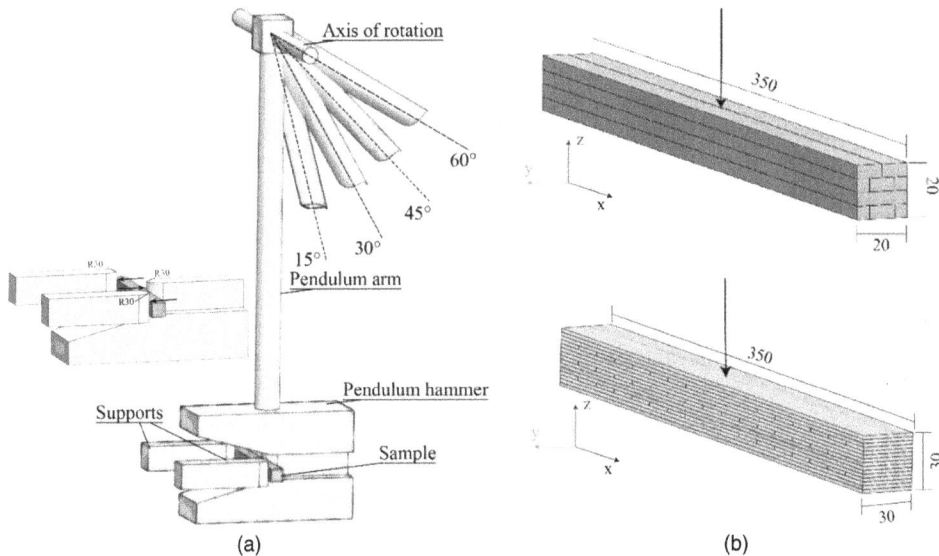

Figure 3.40 Test apparatus and specimens: (a) Schematic diagram of pendulum impact machine; (b) Schematic diagrams of thick- and thin-strip glubam specimens, tested by Chen.

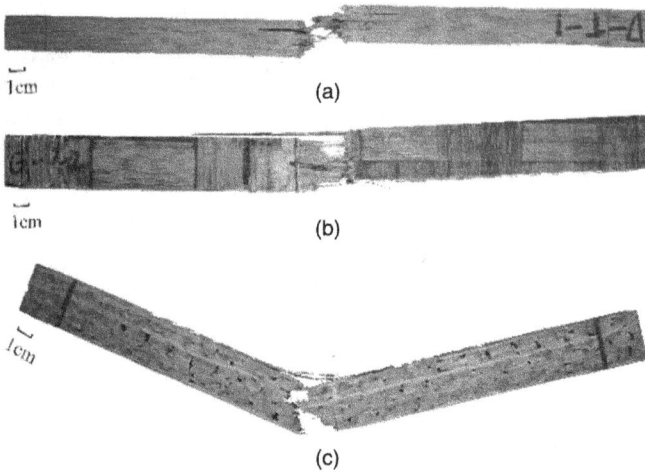

Figure 3.41 Typical failure modes for each type of glubam, tested by Chen.

glubam specimens included various conditions of carbonization (the subscripts of n, m, and d represent no carbonization, medium carbonization, and deep carbonization, respectively). The medium-carbonized glubam was heated in the 125°C-chamber for 80 minutes, and the deep-carbonized glubam was in the 140°C-chamber for 110 minutes. The thick-strip glubam specimen series is designated as $G1_{n,yz}$, $G1_{m,yz}$, $G1_{d,yz}$, in which the second subscript character indicates the direction of the bending moment vector and the third subscript character represents the direction of the impact force at the onset of the impact. The thin-strip glubam specimens are $G2_{yz}$, $G2_{xz}$, in which the first subscript character indicates the direction of the bending moment vector and the second subscript character represents the direction of the impact force. For the thick-strip glubam, the impact force was perpendicular to the bamboo fiber direction, whereas the tests on thin-strip glubam were conducted for impact in the direction perpendicular to both the longitudinal direction (main fiber) and the transverse direction (less fiber). In total, four start angles were designed, corresponding to four levels of potential impact energy.

As shown in Figure 3.41, all specimens failed, with rupture of bamboo fibers due to bending at onset of impact. Some specimens of thick-strip glubam exhibited horizontal shear cracking, particularly due to their lack of bamboo fibers in the transverse direction.

As an important characteristic parameter, the impact toughness is quantized by Eq.3.17.

$$A_w = 1000Q / bh \qquad (3.17)$$

where, A_w (kJ/m^2) is the toughness, Q (in J) is the absorbed energy, b (mm) is the width of the sample, h (mm) is the thickness of the sample. Hence, the impact toughness can be compared between the specimens in Figure 3.42. The overall observation shows that the impact toughness is related to the type of glubam and the loading directions.

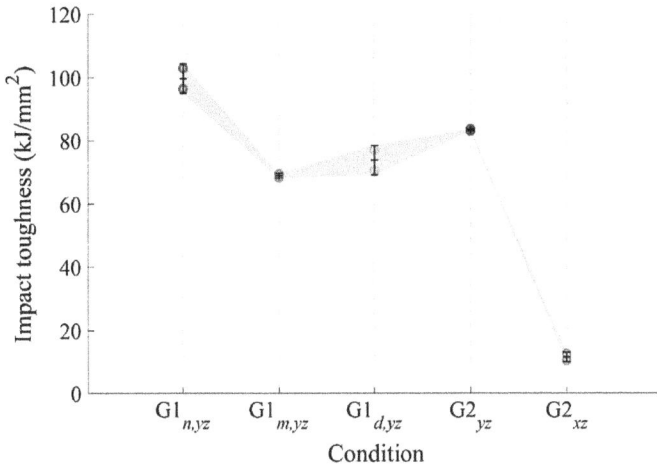

Figure 3.42 Toughness of all types of specimens.

For thick-strip glubam, the average performances are similar for different degrees of carbonization; however, the impact toughness of specimens without carbonization is highest when impacted in the flatwise direction. The impact behavior of the thin-strip glubam is similar to the thick-strip glubam if the bending is against the main bamboo fiber direction (longitudinal bending) with flatwise impact (bending moment vector in transverse direction, or y-direction). For the thin-strip glubam subjected to transverse bending, the toughness is very low since the amount of bamboo fiber for bending resistance is minimal.

3.7.2 High-strain Rate Compressive Behavior

The two types of glubam (thick-strip and thin-strip) and Douglas fir cylindrical specimens were tested[53] to characterize their quasi-static and high-strain rate compressive characteristics covering a large range of strain rates from $2.2\times10^{-3}s^{-1}$ to around $1\times10^3 s^{-1}$. Quasi-static tests using a universal testing machine and dynamic tests using an expressly-designed aluminum Split Hopkinson Pressure Bar (SHPB) apparatus (Figure 3.43) were performed in two loading directions: longitudinal and transverse. Different air gun pressures were adopted to achieve a strain rate of around 300, 600, and 900 s^{-1}. In the following, the cylinder specimens are identified using the following label made up of two parts: (i) material where "T" refers to Douglas fir, "G1" to thick-strip glubam, and "G2" to thin-strip glubam; (ii) direction where "L" designates longitudinal and "T" designates transverse.

The typical failure mode of Douglas fir and the two types of glubam under high-strain rate compression is splitting into pieces, with higher air pressure leading to smaller pieces of debris (as shown in Figure 3.44). Figure 3.45 summarizes the compressive stress-strain curves of quasi-static and SHPB tests. The asterisk markers indicate the critical points. The results indicate that the strength and stiffness in the transverse direction are always less than the corresponding quantities in the longitudinal direction.

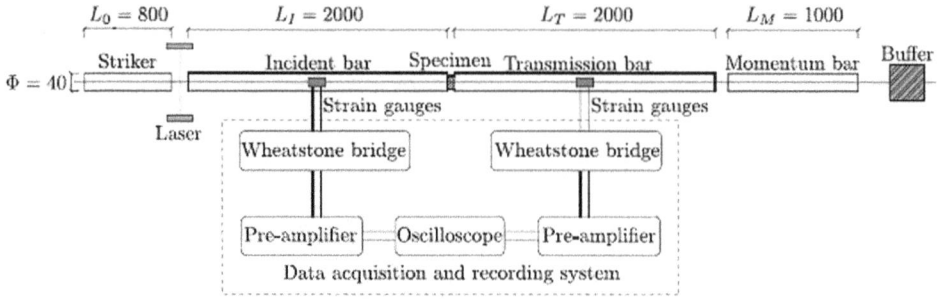

(a)

(b)

Figure 3.43 The aluminum Split Hopkinson Pressure Bar (SHPB) apparatus: (a) Schematic design; (b) photo of the apparatus.

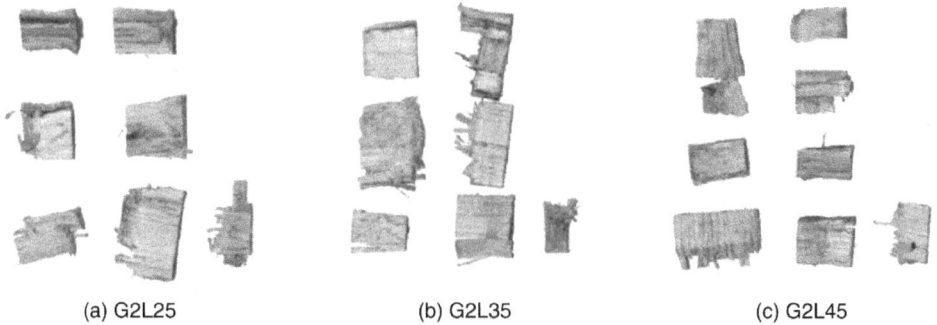

(a) G2L25   (b) G2L35   (c) G2L45

Figure 3.44 Typical failure modes of specimens under high-strain rate compression with different loading rates (the number which forms the last element of the label represents the different loading rates (air pressure); e.g., G2L25 means the specimens made of thin-strip glubam, loading in the longitudinal direction with an air pressure of 0.25 MPa).

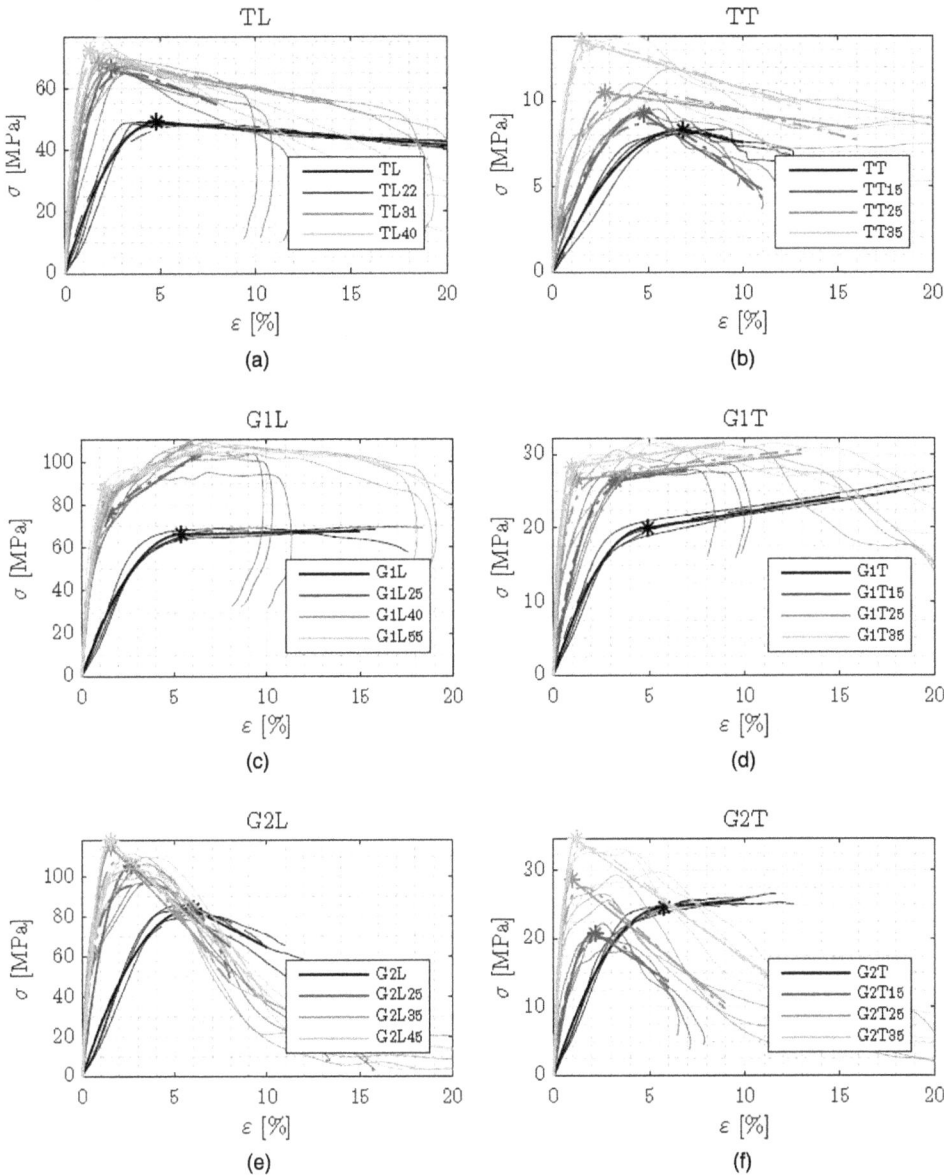

Figure 3.45 Summary of stress-strain relationship of quasi-static and SHPB tests and fitted curves with critical point of first model: (a) TL; (b) TT; (c) G1L; (d) G1T; (e) G2L; (f) G2T.

The stiffness of all the materials investigated also increases following the increase of the strain rate in both longitudinal and transverse directions. The results of this investigation showed that Douglas fir and glubam are strain-rate sensitive, and the degree of sensitivity depends on the loading direction.

3.8 Thermal and Fire Behaviors of Engineered Bamboo

Despite the increasing numbers of research studies related to the mechanical behavior of bamboo and engineered bamboo in recent years, our understanding related to the thermal and fire-resistant qualities of bamboo or engineered bamboo is still very limited. This section provides a summary of recent studies by the author's research team and other scholars on these important subjects.

3.8.1 Thermal Effects

Thermal conductivities and resistance are basic physical properties of construction materials. This information is attracting more attention with today's trend toward energy-efficient buildings. The thermal performance of raw bamboo and engineered bamboo composite materials has also been investigated.[54–57] Kiran et al.[54] show that the thermal conductivity of bamboo increases with an increase in density. The thermal conductivity and thermal diffusivity of raw bamboo is not constant along the radial direction.[55] Huang et al. reveal the increase of the specific heat capacity of raw bamboo in line with the temperature.[55, 56] The study by Shah et al.[57] demonstrates that the thermal conductivity of engineered bamboo composites increases with density, exhibiting anisotropic behavior, and engineered bamboo composites have the same or lower thermal conductivity in comparison with wood at the same density.

To provide the thermal performance characteristics of lightweight frame walls using different combinations of timber, oriented strand board (OSB), and glubam, the author's research group[58] conducted comprehensive thermal studies with Guarded Hot Plate (GHP) tests on a total of 11 materials, including glubam, timber, plywood, ply-bamboo, OSB, etc., for different configurations of lightweight frame walls. Figure 3.46 shows the details of the test setup, which is designed according to ISO Standard 8302.[59]

During the test, the specimen of size 300 × 300 mm is placed between the cold and the hot plates, as shown in Figure 3.46 (a). The temperature of the hot surface, T_1, and

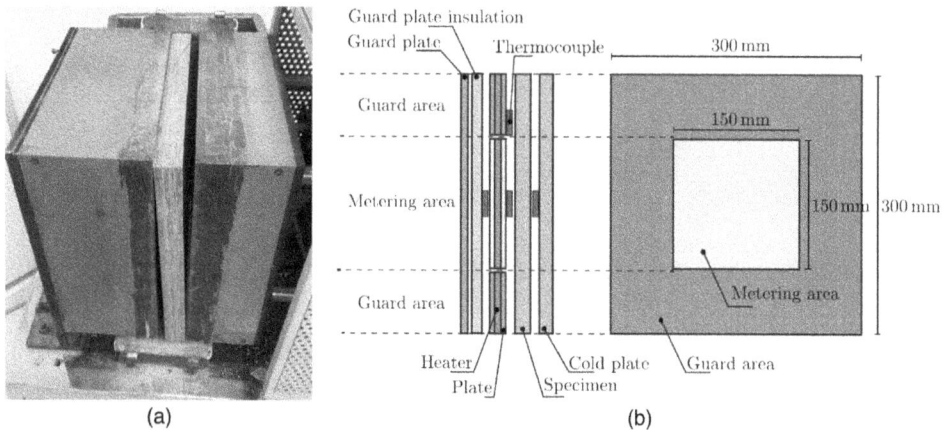

Figure 3.46 Photo (a) and schematic (b) of GHP apparatus.

the temperature of the cold surface, T_2, are set. The hot plate is separated into two parts: measurement area (150 × 150 mm) and guard area. These are heated by two different heating circuits. Different thermocouples are installed on both the cold and the warm side in order to check that steady-state conditions are reached. When steady-state conditions are reached, measurements are carried out for 1 hour. The thermal conductivity, λ, is evaluated as:

$$\lambda = \Phi d / (T1 - T2) A \tag{3.18}$$

where Φ is the heat flow equal to the average electric power under steady-state conditions in the metering area heating circuit; d is the specimen thickness; and A is the metering area. At the end of this procedure, for each material, two specimens were tested. The effect of temperature on thermal conductivity was studied by varying the temperatures, leaving the same gradient ($T_1 - T_2 = 10°C$). In particular, the following conditions were tested ($T_1 - T_2$): 10°C–20°C, 20°C–30°C, 30°C–40°C, and 40°C–50°C. Test results are shown in Table 3.6 and Figure 3.47.

Table 3.6 Characteristics and thermal conductivity of the different materials tested

Material	d (mm)	V (cm³)	ρ (kg/m³)	T_2-T_1 (°C)	λ (W/mK)	$\bar{\lambda}$ (W/mK)	C_v (%)
Glubam orthogonal	28.15 (28.36)	25.58 (28.52)	855.3 (824.0)	10–20	0.154 (0.152)	0.150 (0.146)	4.1 (4.6)
				20–30	0.154 (0.152)		
				30–40	0.140 (0.136)		
				40–50	0.154 (0.145)		
Glubam parallel	31.17 (31.71)	28.16 (28.52)	855.6 (852.8)	10–20	0.242 (0.244)	0.231 (0.234)	3.3 (5.1)
				20–30	0.233 (0.247)		
				30–40	0.222 (0.217)		
				40–50	0.226 (0.228)		
Plybamboo orthogonal	17.80 (16.63)	16.16 (15.13)	709.7 (728.0)	10–20	0.125 (0.134)	0124 (0.124))	0.8 (6.8)
				20–30	0.125 (0.129)		
				30–40	0.125 (0.113)		
				40–50	0.123 (0.118)		
Plybamboo parallel	20.81 (20.41)	18.69 (18.35)	689.9 (682.5)	10–20	0.188 (0.198)	0.174 (0.185)	7.6 (8.2)
				20–30	0.183 (0.197)		
				30–40	0.154 (0.160)		
				40–50	0.170 (0.186)		
SPF orthogonal	29.53 (28.00)	26.78 (26.16)	461.4 (479.3)	10–20	0.089 (0.093)	0.091 (0.090)	3.7 (4.3)
				20–30	0.089 (0.084)		
				30–40	0.088 (0.089)		
				40–50	0.096 (0.094)		
SPF parallel	31.30 (31.14)	28.13 (27.74)	537.7 (533.2)	10–20	0.198 (0.174)	0.194 (0.181)	4.4 (4.9)
				20–30	0.188 (0.171)		
				30–40	0.184 (0.185)		
				40–50	0.205 (0.193)		
OSB orthogonal	19.18 (19.00)	17.33 (17.62)	687.8 (692.3)	10–20	0.103 (0.096)	0.103 (0.094)	1.3 (2.6)
				20–30	0.105 (0.094)		
				30–40	0.102 (0.095)		
				40–50	0.102 (0.090)		

(continued)

Table 3.6 Cont.

Material	d (mm)	V (cm³)	ρ (kg/m³)	T_2-T_1 (°C)	λ (W/mK)	$\bar{\lambda}$ (W/mK)	C_v (%)
OSB parallel	22.00 (21.87)	19.88 (19.09)	685.0 (655.4)	10–20	0.226 (0.204)	0.210 (0.192)	6.6 (8.9)
				20–30	0.219 (0.211)		
				30–40	0.190 (0.167)		
				40–50	0.204 (0.185)		
Ferro-cement jacket	22.42 (19.19)	18.43 (17.28)	2031.7 (2078.3)	10–20	0.442 (0.348)	0.452 (0.366)	2.6 (3.7)
				20–30	0.440 (0.384)		
				30–40	0.460 (0.359)		
				40–50	0.467 (0.372)		
Rock wool	35.24 (35.33)	50.56 (46.15)	52.2 (74.5)	10–20	0.038 (0.035)	0.038 (0.037)	0.7 (4.4)
				20–30	0.038 (0.036)		
				30–40	0.038 (0.038)		
				40–50	0.038 (0.039)		
Gypsum board	11.69 (11.65)	10.62 (10.57)	683.8 (682.1)	10–20	0.171 (0.171)	0.148 (0.142)	12.5 (119.9)
				20–30	0.162 (0.139)		
				30–40	0.131 (0.132)		
				40–50	0.129 (0.128)		

Source: Wang et al. (2018).[58]

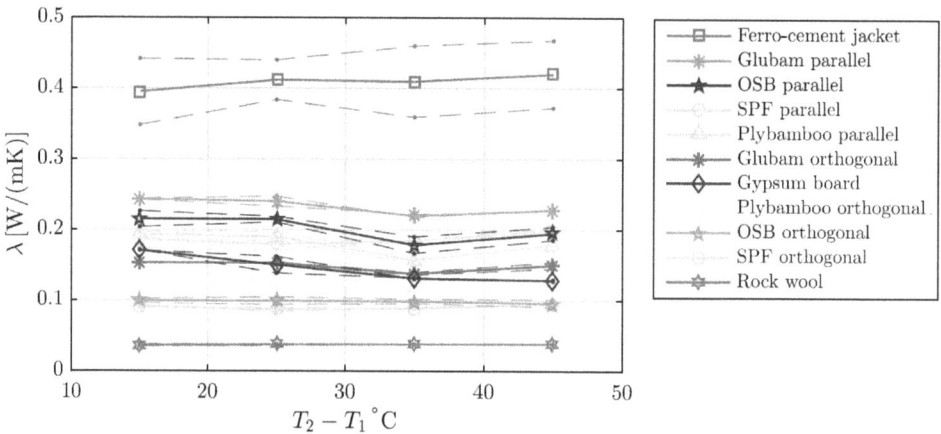

Figure 3.47 Thermal conductivity for the different materials related to two-side temperature differences $T_2 - T_1$. For each material, dashed lines indicate the results of the two specimens and a solid line indicates the mean value.

The thermal conductivity of the bamboo-based materials (glubam and plybamboo) is slightly higher compared with that of the wood-based materials (SPF and OSB), with a difference equal to approximately +5%. Moreover, it is shown that the thermal conductivity of these materials exhibits an anisotropic behavior. In fact, in the direction parallel to the fibers or the layers, these materials are characterized by a thermal conductivity rate approximately 40% higher. On the other hand, the effect of the temperature (in the range from 10°C to 50°C) was found to be negligible, as long as the two-side temperature difference is the same (in this case, 10°C).

3.8.2 Flammability and Fire Behavior of Glubam

Two series of fire simulation tests by the author's research team demonstrated the satisfactory fire performance of glubam houses.[60,61] The fire endurance of the unprotected prefabricated bamboo house was longer than 35 minutes, while the full-scale room unit, in which the glubam members were protected by gypsum boards, stood more than an hour of fire. Mena et al.[62] presented an experimental study on fire combustion and resistance of Guadua a.k. bamboo. The experimental results illustrate that the original and glue-laminated Guadua a.k. bamboo showed better fire behavior and provided better results than plywood. Tests on two types of engineered bamboo products were conducted by Xu et al.[63] to examine their combustion properties. The test results show that the laminated bamboo behaved in a similar way to a softwood. However, the bamboo scrimber has a better resistance to fire. Furthermore, Solarte et al. also presented some tested results of a flammability analysis of a thick-strip laminated bamboo product.[64] Some experiments were conducted by Pope et al. to measure the in-depth heating and charring rate of the laminated bamboo board with thick bamboo strips.[65]

A study led by Huo[66] provided an evaluation of the combustion performance of glubam board according to ISO 5660-1 using a cone calorimeter.[67] The thin-strip laminated bamboo board for glubam with a bamboo fiber ratio of 1:1 and a 10 mm thickness was tested in Huo et al.'s study. Fir plywood samples were also tested to provide the comparative benchmark combustion behavior. The plywood consisted of three ply, 2.8 mm each. Both glubam and plywood samples had similar dimensions of 100 mm × 100 mm × 8 mm (1.0 mm tolerance). The materials were conditioned at 23±2°C and 50±5% relative humidity until reaching a stable state in weight. The test indicated that the densities of glubam and plywood were 811 kg/m³ and 529 kg/m³, respectively.

The tests were carried out by placing the glubam and plywood specimens into a horizontally positioned sample holder in the cone calorimeter apparatus. A piece of aluminum foil was used to insulate the samples from heat losses to the holder. The combustion tests were carried out at five levels of heat flux; i.e., 15, 25, 35, 50, and 70 kW/m². The test of each sample (both glubam and plywood) at one heat flux level was repeated three times. All the data were recorded every five seconds.

The experimental parameters included time to ignition (TI), heat release rate (HRR), peak heat release rate ($pHRR$), smoke production rate (SPR), carbon monoxide yield (YCO) and carbon dioxide yield (YCO_2). The measured results of glubam were compared with those of a typical plywood.

Table 3.7 illustrates the measured peak heat release rates ($pkHRR$) at different heat fluxes, which indicate that the $pkHRR$ values of glubam are lower than those of plywood, and the time corresponding to the $pkHRR$ (t_p) of glubam is later than that of plywood except at 15 and 25 kW/m². The $pkHRR$ of glubam is significantly higher than that of plywood at 15 kW/m², and the t_p of glubam is earlier than that of plywood. However, the $pkHRR$ of glubam is a little higher than that of plywood at 25 kW/m², while the t_p of glubam is later than that of plywood at this temperature. Figure 3.48 shows the time duration (t_p) corresponding to $pkHRR$ versus heat flux relations for glubam and plywood. It can be shown from Figure 3.48 that the heat flux has no obvious effect on the t_p versus I relationships and there is little difference between glubam and plywood. It indicates that glubam does not have a higher HRR than plywood.

Table 3.7 Summary of main test results for glubam and plywood

Heat flux		15 kW/m²	25 kW/m²	35 kW/m²	50 kW/m²	70 kW/m²
t_{ig} (s)	glubam (GLG*)	989 (769)	139 (133)	71 (44)	44 (25)	17
	plywood	1197	62	30	15	7
pkHRR (kW/m²)	glubam	241.6	253.9	185.5	263.1	304.8
	plywood	192.1	235.7	277.6	291.1	378.4
t_p (s)	glubam	1065	430	375	270	215
	plywood	1280	315	290	235	180
av.YCO (kg/kg)	glubam	0.032	0.027	0.029	0.026	0.027
	plywood	0.040	0.039	0.036	0.035	0.033
Av.YCO₂ (kg/kg)	glubam	0.965	0.860	0.830	0.832	0.848
	plywood	1.028	0.886	0.888	0.870	0.895

Note: *Meno et al. (2012).[62]

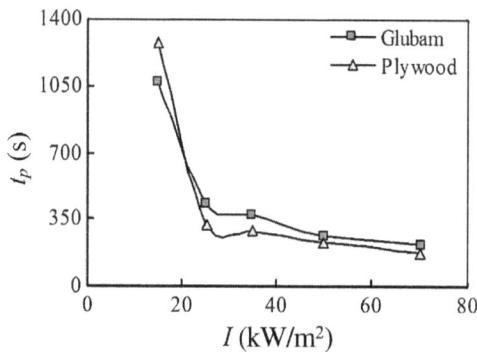

Figure 3.48 Time duration (t_p) corresponding to peak heat release rate (*pkHRR*) versus heat flux relations for glubam and plywood.

Huo et al.[64] also found that the burning behavior of glubam is similar to that of the glue-laminated Guadua bamboo (GLG) boards and bamboo scrimber reported in the literature.[63,64] The thin-strip glubam is generally less flammable than the laminated bamboo (LBL, or thick bamboo strip-based glubam) boards and bamboo scrimber in the perpendicular direction.[64] Four common constructional wood species are more flammable than glubam and the laminated Guadua (GLG). The ignition time mainly depends on the anatomical structure of bamboo and the wood species. The effect of the manufacturing method and the addition of glue on the fire behavior is considered secondary.

Researchers at the University of Queensland are carrying out extensive research into the fire behavior of engineered bamboo with thick-strip lamination. Particularly, Gonzalez et al. used novel test methods including the environmental chamber and the heating blanket to test bamboo samples at steady-state conditions.[68] They found that the round and laminated bamboo experiences a significant reduction in its mechanical properties at elevated temperatures, with results similar to other ligno-cellulosic materials such as timber. They indicate that a reduction in the modulus of elasticity can lead to a shift of failure mode of the engineered bamboo column under compression, from crushing to buckling.

It should be pointed out that the number of studies on thermal and fire behaviors of engineered bamboo is still very limited. There is still an urgent need for both fundamental studies on combustion mechanisms, and structural behaviors under elevated temperature, as well as fire protection measures.

3.9 Acoustic Properties of Engineered Bamboo

Nowadays, noise pollution, which has a negative influence on human health, has become one of the major environmental pollutions due to the development of industrialization and the traffic system.[69] More and more engineers and researchers are working on the development of strategies and materials to control noise. Different types of walls, such as autoclaved aerated concrete walls, double skin walls, green walls, etc., are applied to meet the requirement of sound insulation of a building.[70,71] In terms of materials, the conventional acoustic insulation materials used in buildings are glass wool and mineral-based materials such as rock wool or glass fiber.[72,73] With the growing concern about the health risks caused by these industrial or mineral fibers, as well as the increasing emphasis on energy conservation, researchers have conducted a number of studies on organic natural fibers and green materials for their acoustic characteristics.[74,75] For instance, Zulkifh et al.[76] investigated the sound absorption coefficient and transmission loss index of a multi-layer coir fiber panel and found it has good acoustic properties that make it a potential replacement for synthetic-based products. Cucharero et al. measured the acoustic absorption properties of hardwood and softwood pulp fiber foams and concluded that further processing and smaller dimensions of these fibers contribute to better acoustic absorption performance.[77] The sound absorption coefficients of bamboo fibers and bamboo fiberboard were measured by Koizumi et al.,[78] and the results show that bamboo fibers have better acoustic performance than plywood with the same density. However, studies on the sound insulation properties of wood-based and bamboo-based materials are limited.

In order to gain a full understanding of as many physical properties of glubam as possible, the author's research group also studied its acoustic behaviors. Essentially, until now, there has been no study of the acoustic properties of engineered bamboo, and we know little about those of engineered wood.

In a recent study by Dr. Wen and Kong of the author's group in ZJUI, a four-microphone impedance tube (see Figure 3.49) was used to measure the sound transmission loss indexes of glubam and SPF. The principle of this acoustic test can be briefly summarized: A sound wave emanating from the sound source generates a plane sound wave in the source tube, and it is separated into three parts once it reaches the sample surface; i.e., absorbed by the sample, reflected at the surface, and penetrating through the sample. The part of the sound wave penetrating through the sample generates a plane sound wave in the receiving tube, and it is absorbed or reflected at the absorbent ending.[79] Two microphones are located at each side of the sample, and the standing wave separation method is applied to separate incident and reflected waves. According to the data from the four microphones, the sound reduction index TL can be calculated using,

$$tp = \frac{\sin(k \cdot s_1) \cdot p_3 \cdot e^{jks_2} - p_4}{\sin(k \cdot s_2) \cdot p_1 - p_2 e^{-jks_1}} e^{jk(L_1 + L_2)} \tag{3.19}$$

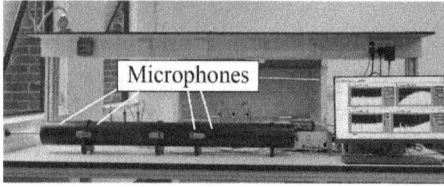

Figure 3.49 A four-microphone impedance tube for sound transmission loss measurement.

and,

$$TL = -20log_{10}|tp|(\text{dB})$$

(3.20)

where, p_1, p_2, p_3 and p_4 are the sound pressures recorded by the four microphones, s and L are the distances between the sample and microphones, and k is the wave number.

The samples of thin-strip glubam, thick-strip glubam, and SPF were all cut into circular disk shapes with diameters of 100 mm or 29 mm for the acoustic testing at a low frequency of 50–1000 Hz and a high frequency of 500–6300 Hz, respectively. For each material, three samples were prepared and tested to obtain an average value. The results of the sound reduction indexes of two types of glubam and SPF are plotted in Figure 3.50. It can be noted that the bamboo-based material glubam behaves better in terms of sound insulation than SPF. Between the two types of glubam, thick-strip glubam has a larger sound reduction index than the thin-strip one. This is due to the fact that the thick strips are finely shaped and laminated with more accurate alignment and reduction of gaps among the strips.

3.10 Future Research Needs and Research Update

Despite increased interest and the growing number of research studies, the full characterization of various types of engineered bamboo is still at its early stage. Significant attention and research efforts are urgently needed, particularly relating to durability, temperature effects (both high temperature and low temperature), corrosion, etc.

Furthermore, a composite mechanics approach should be adopted in establishing a theoretical platform for analyzing engineered bamboo materials and structures. It is notable that the fiber component in bamboo strips plays the dominant role in the mechanical behaviors of bamboo lamina or glubam.[80,81] The study by Yu et al.[81] has shown that the average longitudinal tensile modulus and tensile strength of Moso bamboo fibers ranges from 32 to 34.6 GPa and 1.43 to 1.69 GPa, respectively, significantly higher than nearly all the published data for wood fibers. Despite its scientific importance, research on single fibers is extremely difficult and may not be fully useful to quantitatively establish the connection between the mechanical properties of a single bamboo fiber and that of the laminated bamboo, at least at the current stage. The behavior of bamboo fiber bundles may be relatively convenient to test and the results can be integrated into the composite mechanics approach for bamboo and its lamina. Figure 3.51 shows the

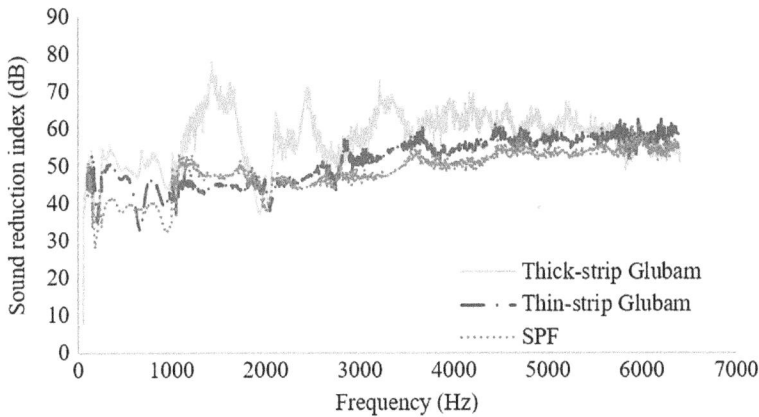

Figure 3.50 Sound reduction index of glubam and SPF by Wen and Kong.

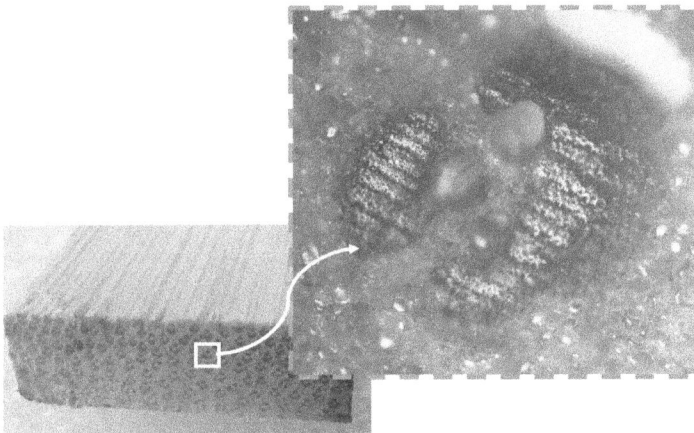

Figure 3.51 Bamboo fiber bundles.
Source: Photo courtesy of Ms. C.Q. Chen, ZJUI.

optical microphotograph of the bamboo fiber bundles. Such bundles form dark dots in the bamboo section, which we refer to as "monkey face."

Research studies have undertaken several approaches to extract fiber bundles by means of chemical or physical tools.[82] During chemical extraction, it is difficult to confirm the section areas of fiber bundles, due to the uncertainty in controlling the amount of materials to be removed in the extraction process. Hence, an entirely physical extraction method may be more reliable, leading to regular-shaped fiber bundles and experimental uniformity. Recently, doctoral candidate C.Q. Chen employed a diamond wire- cutting machine to accurately cut the bamboo strip from a sectional mesh, as shown in Figure 3.52. The bamboo fiber bundle specimen is shaped with a 0.45 mm square section and a length of 70 mm. And 40 mm was conserved for testing. The

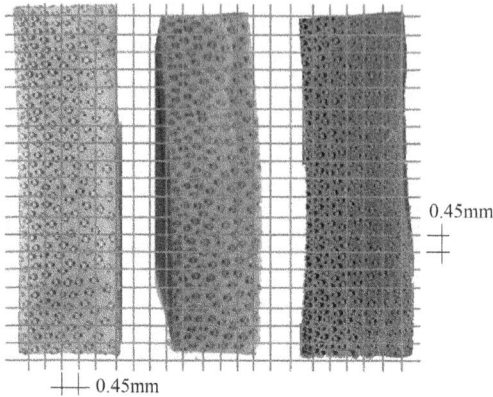

Figure 3.52 Meshing of sections of bamboo strips: (a) without carbonization; (b) with medium carbonization; (c) with deep carbonization.

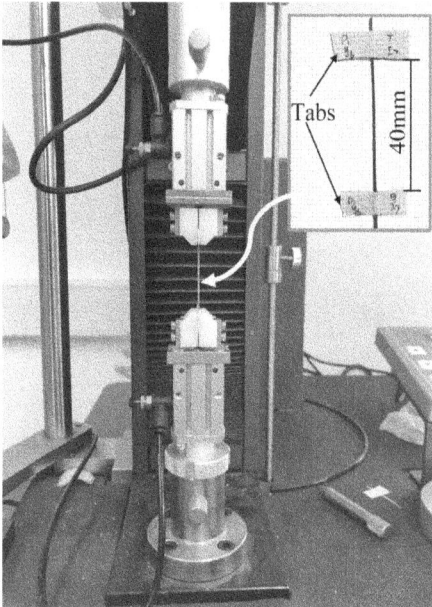

Figure 3.53 Universal testing machine (50 N).

number of bamboo fiber bundles and the number of fibers within the 0.45 mm square section can be relatively easily determined using microscopic photographs.

A 50 N capacity universal testing device (Figure 3.53) was used by Chen to apply tension force to the bamboo bundle specimen at a loading rate of 0.5 mm/min. To decrease the influence of local loads, canvas was used as the tab reinforcement at both ends of the bundle. The carbonized effects on bamboo strips and bamboo fiber strips were assessed as well.

There are many variations of the locations of breaks in bamboo fiber bundles because the bamboo fiber bundle extracted by a physical method retains all the ingredients of raw bamboo. In other words, the bamboo fiber bundle is more analogous to a composite material, combined of a base material and fibers.[81] Hence, the crack might happen in any place longitudinally except the reinforced ends. Figure 3.54 (a) exhibits an SEM (scanning electron microscope) image of the broken bundle, and Figure 3.54 (b) shows the tensile force and deformation relationships for bamboo bundles at different locations within the strip depth.

As shown in Figure 3.54, it is clear that the bamboo fiber strengths of the bundles are different at a different depth within the wall thickness. The bundles close to the bark

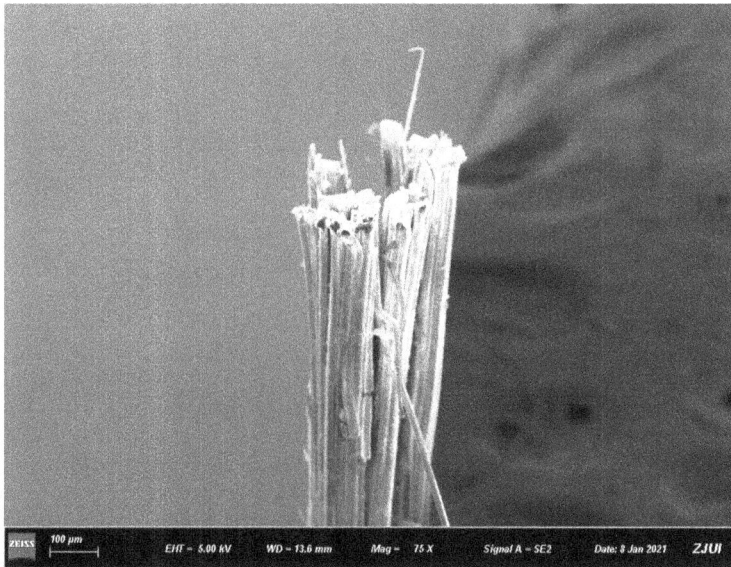

(a)

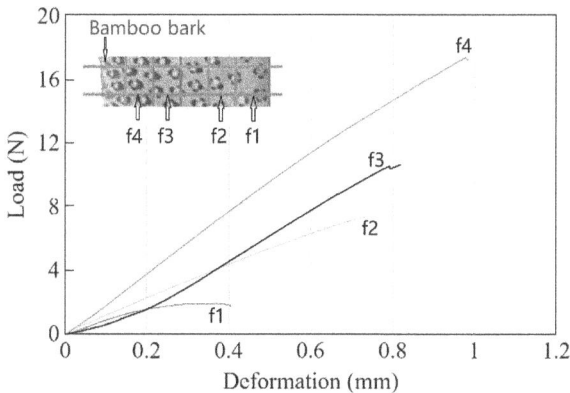

(b)

Figure 3.54 Tensile test results of bamboo bundles: (a) SEM image; (b) load-deformation relationships.

or skin have a higher strength and stiffness compared with those closer to the bamboo yellow (inner skin).

Efforts have recently been made to study the fracture behaviors of engineered bamboo, though mostly on bamboo scrimbers.[84–86] In analyzing the transverse fracture of bamboo composites, Liu et al.[84] adopted an interesting approach by considering the bamboo fiber bundles as the "aggregates," analogically comparing them with concrete. They explained the failure process and final fracture of bamboo fiber bundles caused by the propagation of cracking in the matrix among the fiber bundles.

It is to be hoped that with more testing data on micro and macro behaviors of bamboo and engineered bamboo materials, a more sophisticated method for the analysis and design of engineered bamboo can be established in future.

References

[1] Xiao, Y., Yang, R.Z., & Shan, B. (2013). Production, environmental impact and mechanical properties of glubam. *Journal of Construction and Building Materials*, 44, 765–773.

[2] Li, Z., Yang, G.S., Zhou, Q., Shan, B., & Xiao, Y. (2019). Bending performance of glubam beams made with different processes. *Advances in Structural Engineering*, 22(2), 535–546.

[3] Li, H., Wu, G., Zhang, Q., Deeks, A.J., & Su, J. (2018). Ultimate bending capacity evaluation of laminated bamboo lumber beams. *Construction and Building Materials*, 160, 365–375.

[4] Sharma, B., Gatóo, A., Bock, M., & Ramage, M. (2015). Engineered bamboo for structural applications. *Construction and Building Materials*, 81, 66–73.

[5] Li, Z., He, X.Z., Cai, Z.M., Wang, R., & Xiao, Y. (2021). Mechanical properties of engineered bamboo boards for glubam fabrication. *ASCE Journal of Materials in Civil Engineering*, 33(5).

[6] Correal, J.F., Echeverry, J.S., Ramírez, F., & Yamín, L.E. (2014). Experimental evaluation of physical and mechanical properties of Glued Laminated Guadua angustifolia Kunth. *Construction and Building Materials*, 73, 105–112.

[7] Xiao, Y., Zhou, Q., & Shan, B. (2010). Design and construction of modern bamboo bridges. *Journal of Bridge Engineering*, 15(5), 533–541.

[8] Xiao, Y., Wu, Y., Li, J., & Yang, R.Z. (2017). An experimental study on shear strength of glubam. *Construction and Building Materials*, 150, 490–500.

[9] Takeuchi, C.P., Estrada, M., & Linero, D.L. (2018). Experimental and numerical modeling of shear behavior of laminated Guadua bamboo for different fiber directions. *Construction and Building Materials*, 177, 23–32.

[10] Yang, R.Z., Xiao, Y., & Lam, F. (2014). Failure analysis of typical glubam with bidirectional fibers by off-axis tension tests. *Construction and Building Materials*, 58, 9–15.

[11] Liu, B., Chen, Z.Y., Yin, Y.F., Fan, C.M., Jiang, X.M., & Guo, Q.R. (2008). Comparison of two standards for evaluating physical and mechanical properties of bamboo materials. *China Wood Industry*, 22(4).

[12] National Technical Committee 263 on Bamboo and Rattan of Standardization Administration of China. (1995). GB 15780-1995: Testing methods for physical and mechanical properties of bamboos (in Chinese). China Building Industry Press, Beijing.

[13] National Urban–Rural Development Administration. (2007). JG/T 199-2007: Testing methods for physical and mechanical properties of bamboo used in building (in Chinese). National Urban–Rural Development Administration, Beijing.

[14] Gao, L., Zheng, W., & Liang, C. (2008). Properties and utilization research of structural bamboo-based composites (in Chinese). *World Bamboo and Rattan*, 5.

[15] Japanese Ministry of Agriculture, Forestry and Fisheries. (2003). Notice (JAS) No. 237: Structural veneer lumber. Japanese Ministry of Agriculture, Forestry and Fisheries, Tokyo.

[16] European Committee for Standardization (CEN). (1999). EN1194-1999: Wood structure. Glulam – Determination of performance grading and eigenvalues. European Committee for Standardization, Brussels.

[17] Zhang, Y.T., & He, L.P. (2007). Comparison of mechanical properties for glued laminated bamboo wood and common structural timbers (in Chinese). *Journal of Zhejiang Forestry College*, 24(1), 100–104.

[18] Ye, L.M., Jiang, Z.H., Ye, J.H., & Meng, J.X. (1991). Study on board of reconsolidated bamboo (in Chinese). *Journal of Zhejiang Forestry College*, 8(2), 133–140.

[19] Wang, Z., & Guo, W.J. (2003). Status on new architecture materials and its development(in Chinese). *World Bamboo and Rattan*, 3.

[20] Zhang, X.D., Li, J., Wang, Q.Z., & Zhu, Y.X. (2005). Mechanical property prediction of laminated wood-bamboo composite and analysis (in Chinese). *Journal of Nanjing Forestry University* (Natural Sciences Edition), 29(6), 103–105.

[21] Jiang, S.X., Cheng, D.L., Zhang, X.C., & Cui, H.Y. (2008). Pilot study on process and properties of high temperature heat-treated reconstituted bamboo lumber (in Chinese). *China Forestry Science and Technology*, 6(22), 80–82.

[22] ASTM International. (2014). ASTM D143-14: Standard test methods for small clear specimens of timber. ASTM International, West Conshohocken, PA, www.astm.org

[23] The State Bureau of Quality and Technical Supervision. (1991). GB 1927-1943-91: Wood physical and mechanical properties test method. Standards Press of China, Beijing.

[24] National Technical Committee 41 on Timber of Standardization Administrator of China. (2009). GB/T 1928-2009: General requirements for physical and mechanical tests of wood (in Chinese). China Building Industry Press, Beijing.

[25] Ministry of Housing and Urban–Rural Development of the People's Republic of China. (2007). GB 50005-2007: Code for design of timber structures (in Chinese). China Building Industry Press, Beijing.

[26] Kwan, S.H., Shin, F.G., & Yipp, M.W. (1987). Consideration of bamboo as a natural composite material (in Chinese). *Acta Materiae Compositae Sinica*, 4(4), 79–83.

[27] Yang, R.Z. (2013). Research on material properties of Glubam and its application (in Chinese). Dr. of Engineering thesis supervised by Y. Xiao, Hunan University.

[28] International Organization for Standardization. (2004). ISO 13061-2004: Physical and mechanical properties of wood. International Organization for Standardization, Geneva, Switzerland, www.iso.org

[29] Yoshihara, H., & Fukuda, A. (1998). Influence of loading point on the static bending test of wood. *Journal of Wood Science*, 44(6), 473–481.

[30] Riyanto, D.S., & Gupta, R. (1998). A comparison of test methods for evaluating shear strength of structural lumber. *Forest Product Journal*, 48 (02), 83–92.

[31] Davalos, J.F. (2002). Shear moduli of structural composites from torsion tests. *Journal of Composite Materials*, 36(10).

[32] Zhang, H.X., Mohamed, A., Smith, I., & Xiao, Z. (2012). Evaluation of shear constant of timber glulam composite with photogrammetric approach. In B.H.V. Topping & Y. Tsompanakis (Eds.), *Proceedings of the 14th International Conference on Computing in Civil and Building Engineering, Moscow, Russia, 27–29 June*, London.

[33] Vafai, A., & Pincus, G. (1973). Torsional and bending behaviour of wood beams. *ASCE Journal of Structural Engineering*, 99(ST6), 1205–1221.

[34] Lekhnitskii, S.G. (1963). *Theory of elasticity of an anisotropic body*. MIR Publishers, Moscow.

[35] Yoshihara, H., & Ohta, M. (1997). Shear stress/shear strain relationship in wood obtained by torsion test. *Mokuzai Gakkashi*, 43(6), 457–463.

[36] Khokhar, A.M. (2011). *The evaluation of shear properties of timber beams using torsion test method*. Edinburgh Napier University, Edinburgh.

[37] Gupta, R., & Siller, T. (2005). Stress distribution in structural composite lumber under torsion. *Forest Products Journal*, 55.

[38] Wu, Y., & Xiao, Y. (2018). Steel and glubam hybrid space truss. *Engineering Structures*, 171, 140–153.

[39] Huang, D., Zhou, A., & Bian, Y. (2013). Experimental and analytical study on the non-linear bending of parallel strand bamboo beams. *Construction and Building Materials*, 44, 585–592.

[40] Sinha, A., Way, D., & Mlasko, S. (2014). Structural performance of glued laminated bamboo beams. *Journal of Structural Engineering*, 140(1), 04013021.

[41] Chen, G., Yu, Y., Li, X., & He, B., (2019). Mechanical behavior of laminated bamboo lumber for structural application: An experimental investigation. *European Journal of Wood and Wood Products*, https://doi.org/10.1007/s00107-019-01486-9

[42] Huang, D.S., Bian, Y.L., Huang, D.M., Zhou, A.P., & Sheng, B.L. (2015). An ultimate-state-based-model for inelastic analysis of intermediate slenderness PSB columns under eccentric-ally compressive load. *Construction and Building Materials*, 94, 306–314.

[43] Shen, Y.R., Huang, D.S., Zhou, A.P., & Hui, D. (2016). An inelastic model for ultimate state analysis of CFRP reinforced PSB beams. *Composites Part B Engineering*, 115(April), 266–274.

[44] Popovics, S. (1973). A numerical approach to the complete stress-strain curves for concrete. *Cement and Concrete Research*, 3(5), 583–599.

[45] Penellum, M., Sharma, B., Shah, D.U., Foster, R.M., & Ramage, M.H. (2018). Relationship of structure and stiffness in laminated bamboo composites. *Construction and Building Materials*, 165, 241–246.

[46] Akinbade, Y., Harries, K.A., Sharma, B., & Ramage, M.H. (2020). Variation of through-culm wall morphology in P. edulis bamboo strips used in glue-laminated bamboo beams. *Construction and Building Materials*, 232, 117248.

[47] Xiao, Y., Li, L., & Yang, R.Z. (2014). Long-term loading behavior of a full-scale glubam bridge model. *Journal of Bridge Engineering*, 19(9), 04014027.

[48] Li, L., & Xiao, Y. (2016). Creep behavior of glubam and CFRP-enhanced glubam beams. *ASCE Journal of Composites for Construction*, 20(1).

[49] Ministry of Housing and Urban–Rural Development and General Administration of Quality Supervision, Inspection and Quarantine of the People's Republic of China. (2012). GB/T 50329-2012: Standard for methods testing of timber structures (in Chinese). China Building Industry Press, Beijing.

[50] ASTM International. (2009). ASTM D6815-09: Standard specification for evaluation of duration of load and creep effects of wood and wood-based products. ASTM International, West Conshohocken, PA.

[51] Shan, B., Chen, J., & Xiao, Y. (2012). Mechanical properties of glubam sheets after artificial accelerated aging: Novel and non-conventional materials and technologies. In Y. Xiao et al. (Eds.), *Proceedings of NOCMAT-13-2011 Conference, Key Engineering Materials*, Vol. 517. Trans Tech Publications, Durnten, Zurich, Switzerland.

[52] Ministry of Construction, China.. (2007). JG/T-199-2007: Testing methods for physical and mechanical properties of bamboo used in building (in Chinese). Standards Press of China, Beijing.

[53] Zhou, S.C., Demartino, C., & Xiao, Y. (2020). High-strain rate compressive behavior of Douglas fir and glubam. *Construction and Building Materials*, 258, 119466.

[54] Kiran, M., Nandanwar, A., Naidu, M.V., & Rajulu, K.C.V. (2012). Effect of density on thermal conductivity of bamboo mat board. *International Journal of Agriculture and Forestry*, 2(5), 257–261.

[55] Huang, P., Chang, W.S., Shea, A., Ansell, M.P., & Lawrence, M. (2014). Non-homogeneous thermal properties of bamboo. In S. Aicher, H.W. Reinhardt, & H. Garrecht (Eds.), Materials

and joints in timber structures: *Recent developments of technology*. RILEM Bookseries Vol. 9, Springer, Dordrecht, The Netherlands, pp. 657–664.

[56] Huang, P., Zeidler, A., Chang, W.S., Ansell, M.P., Chew, Y.J., & Shea, A. (2016). Specific heat capacity measurement of Phyllostachys edulis (Moso bamboo) by differential scanning calorimetry. *Construction and Building Materials*, 125, 821–831.

[57] Shah, D.U., Bock, M.C., Mulligan, H., & Ramage, M.H. (2016). Thermal conductivity of engineered bamboo composites. *Journal of Materials Science*, 51(6), 2991–3002.

[58] Wang, J.S., Demartino, C., Xiao, Y., & Li, Y.Y. (2018). Thermal insulation performance of bamboo- and wood-based shear walls in light-frame buildings. *Energy and Buildings*, 168, 167–179.

[59] International Organization for Standardization. (1991). ISO 8302: Thermal insulation – Determination of steady-state thermal resistance and related properties – Guarded hot plate apparatus. International Organization for Standardization, Geneva, Switzerland, www.iso. org

[60] Zhou, Q., She, L.Y., Xiao, Y., Shan, B., Huo, J.S., Ma, J., & Yang, R.Z. (2011). Fire-resistance simulation and test of prefabricated bamboo house (in Chinese). *Journal of Building Structures*, 32(7), 60–65.

[61] Xiao, Y., & Ma, J. (2012). Fire simulation test and analysis of laminated bamboo frame building. *Construction and Building Materials*, 34, 257–266.

[62] Mena, J., Vera, S., Correal, J.F., & Lopez, M. (2012). Assessment of fire reaction and fire resistance of Guadua angustifolia kunth bamboo. *Construction and Building Materials*, 27(1), 60–65.

[63] Xu, Q., Chen, L., Harries, K.A., & Li, X. (2017). Combustion performance of engineered bamboo from cone calorimeter tests. *European Journal of Wood and Wood Products*, 75(2), 161–173.

[64] Solarte, A., Hidalgo, J.P., & Torero, J.L. (2018). Flammability studies for the design of fire-safe bamboo structures. In *WCTE 2018 – World Conference on Timber Engineering, Seoul, South Korea*. WCTE.

[65] Pope, I., Hidalgo, J.P., Osorio, A., Maluk, C., & Torero, J.L. (2019). Thermal behaviour of laminated bamboo structures under fire conditions. *Fire and Materials*, 45(3).

[66] Huo, J.S., Ma, J.F., & Xiao, Y. (2020). Flammability assessment of glubam with cone calorimeter tests. *ASCE Journal of Materials in Civil Engineering*, 33(5), DOI: 10.1061/(ASCE) MT.1943-5533.0003670.

[67] International Organization for Standardization. (2002). ISO 5660-1: Reaction-to-fire tests – Heat release, smoke production and mass loss rate – Part 1: Heat release rate (cone calorimeter method). International Organization for Standardization, Geneva, Switzerland, www. iso.org

[68] Gonzalez, M.G., Madden, J., & Maluk, C. (2018). Experimental study on compressive and tensile strength of bamboo at elevated temperatures. In *WCTE 2018 – World Conference on Timber Engineering, Seoul, South Korea*. WCTE.

[69] Goines, L., & Hagler, L. (2007). Noise pollution: A modern plague. *Southern Medical Journal*, 100(3), 287–294.

[70] Azkorra, Z., Pérez, G., Coma, J., Cabeza, L.F., Bures, S., Ílvaro, J.E. et al. (2015). Evaluation of green walls as a passive acoustic insulation system for buildings. *Applied Acoustics*, 89, 46–56.

[71] Omrany, H., Ghaffarianhoseini, Ali, Ghaffarianhoseini, Amirhosein, Raahemifar, K., & Tookey, J. (2016). Application of passive wall systems for improving the energy efficiency in buildings: A comprehensive review. *Renewable and Sustainable Energy Reviews*, 62, 1252–1269.

[72] Papadopoulos, A.M. (2005). State of the art in thermal insulation materials and aims for future developments. *Energy and Buildings*, 37, 77–86.

[73] Li, T.T., Chuang, Y.C., Huang, C.H., Lou, C., & Lin, J. (2015). Applying vermiculite and perlite fillers to sound absorbing/thermal-insulating resilient PU foam composites. *Fibers and Polymers*, 16, 691–698.

[74] Khedari, J., Charoenvai, S., & Hirunlabh, J. (2003). New insulating particleboards from durian peel and coconut coir. *Building and Environment*, 38(3), 435–441.

[75] Khedari, J., Charoenvai, S., & Hirunlabh, J. (2004). New low-cost insulation particleboards from mixture of durian peel and coconut coir. *Building and Environment*, 39(1), 59–65.

[76] Zulkifh, R., Mohd Nor, M.J., Tahir, M.F. Mat, Ismail, A.R., & Nuawi, M.Z. (2008). Acoustic properties of multi-layer coir fibres sound absorption panel. *Journal of Applied Sciences*, 8(20), 3709–3714.

[77] Cucharero, J., Ceccherini, S., Maloney, T., Lokki, T., & Hanninen T. (2021). Sound absorption properties of wood-based pulp fibre foams. *Cellulose*, 28, 4267–4279.

[78] Koizumi, T., Tsujiuchi, N., & Adachi, A. (2002). The development of sound absorbing materials using natural bamboo fibers and their acoustic properties. In C.A. Brebbia, & W.P. DeWilde (Eds.), *Inter-Noise and Noise-Con Congress and Conference Proceedings.*Institute of Noise Control Engineering, Dearborn, MI, pp. 713–718.

[79] Zhu, Y.J., Wei, M., & Xiong, W.B. (2014). Experimental study of acoustic impedance tube method based on measurements. *Technical Acoustics*, 33(6), 201–204.

[80] Phong, N.T., Fujii, T., Chuong, B., & Okubo, K. (2012). Study on how to effectively extract bamboo fibers from raw bamboo and wastewater treatment. *Journal of Materials Science Research*, 1(1), 144.

[81] Yu, Y., Tian, G.L., Wang, H.K., Fei, B.H., & Wang, G. (2020). Mechanical characterization of single bamboo fibers with nanoindentation and microtensile technique. *Holzforschung*, 65, 113–119, DOI: 10.1515/HF.2011.009.

[82] Li, H., & Shen, S. (2011). The mechanical properties of bamboo and vascular bundles. *Journal of Materials Research*, 26(21), 2749–2756.

[83] Osorio, L., Trujillo, E., Van Vuure, A.W., & Verpoest, I. (2011). Morphological aspects and mechanical properties of single bamboo fibers and flexural characterization of bamboo/epoxy composites. *Journal of Reinforced Plastics and Composites*, 30(5), 396–408.

[84] Liu, W., Yu, Y., Hu, X., Han, X., & Peng, X. (2019). Quasi-brittle fracture criterion of bamboo-based fiber composites in transverse direction based on boundary effect model. *Composite Structures*, 220, 347–354.

[83] Xie, P., Liu, W., Hu, Y.C., Meng, X.M., & Huang, J.K. (2020). Size effect research of tensile strength of bamboo scrimber based on boundary effect model. *Engineering Fracture Mechanics*, 239, 107319.

[84] Liu, Y.Y., Huang, D.S., & Zhu, J.J. (2021). Experimental investigation of mixed-mode I/II fracture behavior of parallel strand bamboo. *Construction and Building Materials*, 288, 123127.

Chapter 4

Design Strength of Glubam

Because engineered bamboo is a new material, the elements of structural design using engineered bamboo have not been fully established. In this chapter, the author attempts to lay out the framework for the determination of the design strengths of glubam, based on both allowable stress method and the probability-based ultimate limit state design philosophy, referencing the existing design specifications for timber structures.

4.1 Specified and Characteristic Material Properties

4.1.1 Quality Assurance

The engineered bamboo, glubam, is made from a two-step manufacturing process. For reliable design of glubam structures, the quality of glubam materials and components needs to be assured.

As discussed in Chapter 2, the manufacturing process of engineered bamboo boards under elevated temperature and pressure is well established, primarily in China. Despite the fact that some of these engineered bamboo boards are not originally intended for making glubam structural elements, the existing guidelines discussed in Chapter 1 (re-listed in Table 4.1) for manufacture, quality control, inspection, and certification can be adopted. For the specified material strength, the average mechanical values shown in Chapter 3 can generally be adopted. However, the specific values of a certain batch of engineered bamboo boards to be used in the design and construction of a glubam structure should be evaluated to make sure that the actual values exceed those used in design. The author suggests taking the following two measures to certify the specified strength of engineered bamboo.

4.1.1.1 Factory Testing Record

Similar to obtaining the mill certificates for rebars in reinforced concrete structure design, the factory output testing record is useful. For established manufacturers, material tests are required by third-party or certified engineers to obtain the strength and modulus for a certain number of batches of engineered bamboo boards produced. As an example, Table 4.2 shows the strength record of certification testing for a batch of engineered bamboo boards. The samples were cut from a board randomly chosen from the batch. Based on the tensile test results of six specimens in the transverse

DOI: 10.1201/9781003204497-4

Table 4.1 Adopted standards for glubam base materials.

Ref. No.	Standards	Objectives
[1]	GB 20240 Bamboo flooring	Bamboo flooring
[2]	GB 21179 Decorative bamboo veneered panel	Decorative bamboo panels
[3]	LY 1072 Laminated bamboo strip lumber	Bamboo panels, tests
[4]	LY 1073 Experimental methodologies of physical mechanical properties of laminated bamboo strip lumber	Bamboo strips, tests
[5]	LY 1575 Strip plybamboo for bottom boards of trucks and buses	Plybamboo boards
[6]	GB 13123 Bamboo-mat plywood	Bamboo mat
[7]	GB 13124 Methods of testing bamboo-mat plywood (partially replaced by GB 13123)	Bamboo mat, tests
[8]	JB/T 6564 Plybamboo for export machine packing boxes	Plybamboo, packing boxes
[9]	JG 3059 Plybamboo form with steel frame	Plybamboo-steel
[10]	LY 1574 Plybamboo for concrete-form	Plybamboo-concrete

Table 4.2 Example of testing records for engineered bamboo boards (density 0.87)

No.	Trans./Long.	Thickness	Width	Max. force (N)	MOR (MPa)	MOE (MPa)	Loading rate
1	Trans.	28.5	50.1	1795	37.44	4215	20mm/min.
2	Trans.	28.5	50.1	1865	37.69	4485	20mm/min.
3	Trans.	28.7	50.2	1905	39.56	4869	20 mm/min.
4	Trans.	28.5	50.0	1820	38.02	4315	20 mm/min.
5	Trans.	28.6	50.1	1900	38.56	4623	20 mm/min,
6	Trans.	28.5	50.2	1732	35.36	4053	20 mm/min.
1	Long.	28.7	50.2	4455	98.72	9652	50 mm/min.
2	Long.	28.5	50.2	4320	96.96	9427	50 mm/min.
3	Long.	28.7	50.1	4640	100.90	9982	50 mm/min.
4	Long.	28.5	50.2	4624	99.00	10,084	50 mm/min.
5	Long.	28.5	50.0	4235	96.42	9567	50 mm/min.
6	Long.	28.5	50.1	4429	98.51	9838	50 mm/min.

and the longitudinal directions of the engineered boards, the average strengths can be determined.

4.1.1.2 Evaluation Tests

Besides the manufacturer certification reports, it is recommended that the fabricator or contractor involved in a specific glubam structure project conducts an evaluation testing. The evaluation testing results should be certified by the structural engineer who has performed the design. For each required mechanical property, it is recommended to use no more than five specimens randomly sampled from no less than three randomly chosen boards, if the batch is produced by a manufacturer with a good tracking record.

For the second manufacturing process with cold-pressing lamination, the key issues are to ensure the good quality of the glued interface, including cleanness and smoothness before applying the adhesive, an adequately spread amount of adhesive, adequate pressure (normally 1.0 MPa to 1.5 MPa), evenness of pressure, adequate room temperature (normally not below 20°C), etc. It has been shown that the glued interface strength can be enhanced to assure that the failure of the glued interface does not occur prior to the failure of the base materials.[11] Due to the fact that evaluation procedures and guidelines for the quality of cold-pressing lamination have not been fully established, it is the practice of the author's research team to apply a reduction to the material strengths or to rely on full-scale component testing to assess the quality of the glubam structural components.

4.1.2 Characteristic Properties

To develop a future design code, it is important to establish the characteristic strength values of glubam materials. Based on experimental results, the characteristic strength values of glubam can be estimated. Following the estimation method suggested by Appendix F in Chinese standard GB 50005 for timber structures,[12] the characteristic strength values, with 95% probability of exceedance, can be estimated as:

$$f_k = m - kS$$

(4.1)

where m is the average value of each set of test results, S is the standard deviation of strength values, k is the characteristic factor with 75% confidence level, which is given in reference sources[12–14] and is shown in Table 4.3 for the test numbers. Because laminated bamboo is an industrialized product, using a factor of 1.645 for the materials produced from a certified manufacturer is suggested. Therefore, the characteristic values of the tensile strengths of the material data shown in Table 4.2 can be calculated as follows:

Longitudinal direction: f_k = 98.42-1.645 × 1.59 = 95.79 MPa

Transverse direction: f_k = 37.77-1.645 × 1.40 = 35.47 MPa

At present, the full certification process for the manufacturer of laminated bamboo may still need to be established.

In this book, based on the recent test results[15] given in Table 3.1, the characteristic strength values of the two types of glubam boards are calculated and shown in Table 4.4. The characteristic strength values of engineered wood products along with the average density values are also shown in Table 4.4 for comparison.

4.2 Allowable Stress Design (ASD)

4.2.1 General Concept of ASD

The concept of allowable (also referred to as permissible) stress design (ASD) is that the working stress under the service load (so-called unfactored) determined using elastic structural analysis shall not exceed the permissible stress. The permissible stress is set by a code; that is, reduced strength value equal to the average failure strength of a material

Table 4.3 Values of characteristic factor k, corresponding to the number of tests n

n	10	11	12	13	14	15	16	17	18	19	20	21	22
k	2.104	2.074	2.048	2.026	2.008	1.991	1.977	1.964	1.952	1.942	1.932	1.924	1.916
n	23	24	25	26	27	28	29	30	31	32	33	34	35
k	1.908	1.901	1.895	1.889	1.883	1.878	1.873	1.869	1.864	1.860	1.856	1.853	1.849
n	36	37	38	39	40	41	42	43	44	45	46	47	48
k	1.846	1.842	1.839	1.836	1.834	1.831	1.828	1.826	1.824	1.822	1.819	1.817	1.815
n	49	50	55	60	65	70	80	90	100	120	140	160	180
k	1.813	1.811	1.802	1.795	1.788	1.783	1.773	1.765	1.758	1.747	1.739	1.733	-
n	200	250	300	350	400	450	500	600	700	800	900	≥ 1000	-
k	1.723	1.714	1.708	1.703	1.699	1.696	1.693	1.689	1.686	1.683	1.681	1.645	-

Table 4.4 Characteristic strength values of bamboo boards and structural timber (in MPa)

1-Engineered bamboo boards, thick-strip $\bar{\rho} = 631$ kg/m³ (left); thin-strip 862 kg/m³ (right)

$f_{t,x,k}$		$f_{t,y,k}$		$f_{c,x,k}$	
78.6	45.7	3.7	9.3	47.7	25.9
$f_{c,y,k}$		$f_{c,z,k}$		$f_{b,xz,k}$	
12.5	22.6	9.7	19.9	89.8	68.2
$f_{b,xy,k}$		$f_{b,yz,k}$		$f_{b,yx,k}$	
90.1	79.5	4.8	24.9	6.2	21.5
$\tau_{yx,k}$		$\tau_{xy,k}$		$\tau_{xz,k}$	
5.8	11.2	6.8	15.0	7.8	5.4
$\tau_{zx,k}$		$\tau_{yz,k}$		$\tau_{zy,k}$	
5.5	1.9	3.0	4.5	3.0	2.6

2-Structural timber from EU marked as C40, $\bar{\rho} = 450$kg/m³

Bending strength	Compression strength	Tension Strength
38.6	22.4	24

3-Machine stress-rated dimension lumber from North America marked as 2850Fb-2.3E, $\bar{\rho} = 500$kg/m³

41.3	28.2	33.3

4-Structural timber from New Zealand marked as SG15, $\bar{\rho} = 530$kg/m³

41.0	35.0	23.0

1 Characteristic strength values of wood products given according to parallel-to-grain directions. 2 Only wood products of the highest strength grade are given in this table.

divided by a safety factor, typically ranging from 1.5 to 2.0. The concept of ASD is a traditional design method, but not necessarily obsolete and is still being used in practice. The main reason why the ASD method is still viable might be because with wood structures, the material is essentially elasto-brittle and is expected to work within the elastic range. Considering the similarity between glubam and glulam and timbers, and the fact that ASD is still utilized in the United States and elsewhere, it is of value to establish the allowable stresses for glubam.

The mechanical properties of glubam obtained in the tests described in Chapter 3 are different from the actual stress state of the glubam member used in actual design. It is not adequate to apply the material strength directly to the design of a structure of glubam. Based on the ASD method, the material strength should be reduced and used as the allowable stress which can be reliably larger than the possible working stresses under load effects. The general equation can be expressed as,

$$f \leq F_a = \frac{f_s}{k} \tag{4.2}$$

where f is the working stress corresponding to service loading conditions; F_a is the allowable stress which is equal to the specified strength f_s divided by a safety factor k. The specified strength was originally taken as the specified proportional limit strength, and for linear elasto-brittle material, taken as the average failure strength. However, the value is now typically taken as the characteristic strength with certain probability of assurance.

In wood structure design, the difference between the strength of the clear material samples and the design strength of the actual wood structural members mainly comprises variations in wood strength, wood defects, load durability, unpredictable overload, stress concentration, and possible errors in design and construction, etc. In order to ensure the safety of wood structure design, the influence of these factors on the strength of the wood must be considered comprehensively, and the strength value of the clear material sample should be reasonably reduced.

4.2.2 ASD-based Previous Chinese Code

There are differences in the calculation methods of allowable stress for wood in various countries, but the basic principles are essentially the same. In China, the allowable stress design (ASD)[16] of timber structures is no longer used in practice. However, to compare it with the U.S. ASD,[17,18] it is introduced herein. The allowable stress of wood F_a is calculated as follows:

$$F_a = \sigma_{12} K_1 K_2 K_3 K_4 K_5 / (K_6 K_7) \tag{4.3a}$$

$$\text{or, } F_a = [\sigma]_{min} K_2 K_3 K_4 K_5 (K_6 K_7) \tag{4.3b}$$

where, σ_{12} is the average value of the strength at a moisture content of 12%; $[\sigma]_{min}$ is the minimum value of the strength of the test measurement after considering the influence of the variability; K_1 is the strength variation coefficient for wood; K_2 is the long-term load coefficient; K_3 is the wood defect coefficient; K_4 is the drying defect coefficient; K_5 is the stress concentration factor; K_6 is the overload coefficient; and K_7 is the structural deviation coefficient.

Despite the fact that the earlier Chinese GB code[16] for allowable strength design of timber structures was mainly established for structures with wood logs and saw-cut woods, the concept and the basic parameters are referenced, in the attempt to develop an ASD framework for glubam. The basic material of glubam is bamboo, and its physical and mechanical properties have many similarities with wood. Thus, it is reasonable to refer to the allowable stress calculation method for wood in developing the allowable stresses of glubam. Since glubam is a composite material man-made through well-established industrialized manufacturing processes with reasonable quality control, the defects are well dispersed and the variability of the material properties should be significantly improved, when compared with wood. Taking into account various influencing factors, according to the calculation method of allowable stress for wood, the author's research team suggested the reduction factors of glubam materials, as shown in Table 4.5. The reduction factors of wood are also indicated in Table 4.5.

In Table 4.5, the main considerations in determining the various reduction factors are as follows:

Table 4.5 Reduction factors for mechanical properties of glubam

Component type	K_1' (K_1)	K_2	K_3' (K_3)	K_4' (K_4)	K_5	K_6	K_7	Total reduction factor K' (K)
Static tensile	0.50 (0.50)	0.72	0.46 (0.38)	0.94 (0.85)	0.90	1.20	1.10	0.099 (0.074)
Static compression	0.76 (0.72)	0.72	0.80 (0.67)	1.00 (1.00)	—	1.20	1.10	0.309 (0.245)
Bending strength	0.72 (0.70)	0.72	0.62 (0.52)	0.88 (0.80)	—	1.20	1.10	0.200 (0.148)
Shear stress	0.69 (0.66)	0.72	0.96 (0.80)	0.83 (0.75)	—	1.20	1.10	0.279 (0.201)

Note: The reduction factors K_i' are the suggested values for glubam and reduction factors K_i represent the original values for timber.

1 The reduction factor of glubam is based on the corresponding reduction factors for wood design. Considering the difference of material properties, the main adjustment reduction factors include strength variation coefficient K_1, material defect coefficient K_3, and drying coefficient K_4.

2 For the strength variation coefficient K_1, it is suggested that the value for wood be increased by 0.01 for every 10% reduction in the strength variation coefficient K_1 of glubam compared with timber. Therefore, the strength variation coefficient for wood is multiplied by an adjustment factor of 1.00, 1.06, 1.03, and 1.04 respectively, for the tensile, flexural, and shear strength variation coefficient of glubam. However, if the characteristic value is used for $[\sigma]_{min}$, then there is no need to apply the strength variation coefficient K_1.

3 For the material defect coefficient K_3, since glubam has undergone a strict material selection and industrial production, the degree of influence of defects in the raw materials is greatly reduced, and multiplication by the factor of 1.2 is selected as the adjustment of this coefficient. For similar reasons, the dry defect coefficient K_4 is adjusted by multiplying by a factor of 1.1.

4 Since the other coefficients K_2, K_5, K_6, K_7, etc., are less affected by the types of materials, no adjustment is made and it is suggested that the factors for wood are used for glubam.

4.2.3 U.S. ASD Design

In the United States, wood structure design is mainly based on the National Design Specifications (NDS),[17,18] in which the allowable stress is calculated as,

$$Fa = F_{NDS-S}\Pi C_i \tag{4.4}$$

where F_{NDS-S} is the reference design value, i.e., the basic ASD value for a specific strength, such as bending, etc.; and ΠC_i is the product of all applicable adjustment C factors. The reference design value F_{NDS-S} is given in tables in the NDS Supplement[18] for species of wood, grades, and structural applications. The general issues involved in the determination of the reference design value F_{NDS-S},[19,20] and considerations for obtaining an equivalent F_{NDS-S} of glubam are discussed herein:

1   For a given species of wood, clear specimens are tested following the procedures of ASTM D143.[21] The results are assumed to reasonably fit a normal distribution curve, so that the characteristic value or the so-called 5% exclusion limit can be determined by subtracting 1.645 times the standard deviation. For glubam, the testing of various strengths of the basic engineered bamboo board based on ASTM D143[21] can be considered an identical procedure to predicting the strength of clear wood specimens. The characteristic strength values for glubam are shown in Table 4.4. Note that compression perpendicular to grain and modulus of elasticity are based on average values.

2   The ASTM D143[21] tests are conducted with an expected time-to-failure of 10 minutes. Strength determined on the basis of such test results must be modified to adjust the strength values to allowable stresses suitable for normal load duration, defined as 10 years. Adjustment factors are given in Table 8 of ASTM D245 for bending, tension and compression parallel to grain, and horizontal shear.[20] It is suggested that the strength values of glubam be divided by the adjustment factors shown in Table 4.6 (Table 8 of ASTM D245), if attempting to follow the NDS standard. Since in many cases of design, the structural member sizes are determined on the basis of limiting deformation, it is recommended that the modulus of elasticity is not to be increased based on Table 4.6 (meaning to be divided by 1.0 instead of 0.94), for further safety.

3   ASTM D245 allows the adjustment of the characteristic values to account for the usual increases of wood strength and modulus for moisture content below the fiber saturation point.[20] As discussed in Chapter 3, it is suggested that glubam is used in a relatively dry environment and due to the current insufficient information, this adjustment should not be made.

4   Further reduction is needed to actual ASD strengths of wood, which are determined based on applying the most severe strength ratios to bending, tension, and compression parallel to the grain, to account for the effect of various defects, such as slope of grain and knots. A separate strength ratio is applied to shear parallel to the grain to account for checks and splits that may occur. The adjustment is made with correlation of grading of the lumbers. Because glubam is a man-made material manufactured through established quality control procedures and there are no dramatic defects such as wood knots or disorientation of grain, etc., these adjustments are different. With glubam, defects may most likely arise during the second-step lamination process under room temperature. Previous tests have shown that some reductions of compressive strength can result in the specimens with glueline interfaces.[22,23] Considering such reduction and the possible unknown reduction

Table 4.6 Adjustment factors to be applied to the clear wood properties

	Bending strength	Modulus of elasticity in bending	Tensile strength parallel to grain	Compressive strength parallel to grain	Horizontal shear strength	Proportional limit and stress at deformation in compression perpendicular to grain
Softwoods	2.1	0.94	2.1	1.9	2.1	1.67
Hardwoods	2.3	0.94	2.3	2.1	2.3	1.67

due to long-term loading effects, a defect reduction factor of 70% is conservatively recommended for glubam until enough data are accumulated.

The allowable stresses for tension, compression, and bending along the main bamboo fiber direction, and for shear parallel to the main bamboo direction, but across the transverse direction (in the direction with less or no fiber) are shown in Table 4.7, based on the ASTM D5457 approach[24] and the Chinese GB code used previously (values are shown in parentheses). The suggested ASD values for glubam are also compared with the values for softwood and hardwood glulams mainly subjected to bending.[19, 23]

Following the U.S. National Design Specifications (NDS),[17,18] other adjustments are needed for ASD design values, based on the product of all the adjustment factors shown in Table 4.8 and Table 4.9. At the current stage, if one attempts to follow the NDS ASD for the design of glubam structures, the adjustment C factor values for glulam should likely be adopted.

The safety factor for glubam can also be calculated as the reciprocal of the total reduction factor, which is the ratio of the average strength σ to the allowable stress $[\sigma]$, which can be expressed by A:

$$A = 1/K = \sigma/[\sigma]$$ (4.5)

For a wood structure, the safety factor is typically set higher than that of a concrete structure or a steel structure because the wood structure is not uniform and the strength

Table 4.7 Suggested allowable stresses of glubam based on NDS 2018[18] and GB J5-73[16] (in parentheses)

	Glubam-thick	Glubam-thin	Softwood glulam*	Hardwood glulam*
MOE (MPa)	10,510	10,750	8963–14480	8274–11722
Tension, f_x (MPa)	26.2 (11.8)	15.2 (8.0)	4.65–8.62	4.14–6.90
Compression, f_c (MPa)	17.6 (17.8)	9.5 (12.3)	6.38–12.07	5.52–9.65
Bending, f_{by} (MPa)	29.9 (20.9)	22.8 (17.7)	11.03–20.69	8.27–13.79
Shear, f_{vyx} (MPa)	1.93 (2.1)	3.71 (3.12)	1.34–2.07	0.86–1.38

Note: *Reference design values for glulam mainly used in bending from NDS-S (American Wood Council, 2018b).[18]

Table 4.8 Adjustment C factors mainly related to material properties

Adjustment factor	Description	Value
C_D	Load duration factor (applies to ASD only)	0.9–2.0
C_M	Wet service or moisture factor	0.67–1.0 (0.53–0.875 for glulam)
C_F	Size factor	0.4–1.5 (1.0 for glulam, etc.)
C_{fu}	Flat-use factor	0.86–1.2 (1.01–1.19 for glulam)
C_t	Temperature factor	0.5–1.0
C_r	Repetitive member factor	1.0–1.15 (1.0 for timbers, glulam)
C_i	Incising factor	0.8–1.0

Table 4.9 Adjustment C factors based on loading conditions

C_P	Column stability factor
C_L	Beam stability factor
C_V	Volume factor (applies to *glulam* only)
C_b	Bearing area factor
C_T	Buckling stiffness factor
C_c	Curvature factor
C_I	Stress interaction factor (for *glulam* only)
C_{vr}	Shear reduction factor (for notched *glulam*)

Note: Factors are based on loading conditions.

is easily affected by many factors. According to the previous version (GB J5-73) of the Chinese timber structure design specifications,[16] the safety factor of wood structures is generally 3.5–6.0. The suggested safety factor for glubam is basically in the same order as that for wood.

In the suggested allowable stress design, the aging effects due to wet and dry cycles as well as ultraviolet radiation, as discussed in Chapter 3, are not considered and included. Based on limited research until now, it is suggested that thin-strip glubam should not be used for structural components directly exposed to the weathering environment. Additional aging effect studies are also needed for thick-strip glubam. The use of the suggested allowable design should also be based on an adequate manufacturing quality of the engineered bamboo boards and glubam components.

4.3 Limit State Design Recommendations for Glubam Structures

The current Chinese timber structure design code[12] is based on the probabilistic limit state design method. In North America and Europe, the mainstream design standards are also based on limit state design, except that in the United States, the limit state design for wood structures is provided in parallel with the ASD method. In this section, similar to those for ASD, design guidelines based on limit state design are proposed for engineered bamboo structures.

4.3.1 Recommendation Following Chinese GB Code

The Chinese timber structure design code[12] adopts the limit state design method based on probability theory. The reliability index β is used to measure the reliability of structural members, and the design expression of the partial coefficient is adopted. This design method is a combination of deterministic design and semi-empirical semi-probability coefficients.

The general requirement of limit state design is that the structure or component shall meet the critical condition, not exceeding the limit for certain functional requirements of the design. If it does exceed such a state, the structure or component cannot meet the design requirements. In the limit state design method, the partial load and material strength are determined by probability method, while the load and resistance coefficients are essentially semi-empirical.

There are mainly two limit states: the ultimate limit state for safety; and the serviceability limit state for functional requirement.

4.3.1.1 Ultimate Limit State

The Chinese code[12] adopts the following general equation for the ultimate strength design:

$$\gamma_0 S \leq R \tag{4.6}$$

where, γ_0 is the importance index; S is the general expression of the effect from a load or load combination (from a group of different loads); and R is the design capacity, which can be expressed as,

$$R = A_k f \tag{4.7}$$

where, A_k is the general expression of a certain geometrical parameter – for example, the section modulus for bending design, or section area for tension and compression, etc.; f is the design ultimate strength. The design strength f can be expressed by the following equation,

$$f = \left(\frac{K_{DOL} f_K}{\gamma_{R0}} \right) K_d \tag{4.8}$$

where, K_{DOL} is the coefficient of sustained load, taken as 0.72; f_K is the characteristic strength; γ_{R0} is the so-called baseline resistance factor; and K_d is other adjustment coefficients. Note that K_d is added in Eq.4.8 for convenience in discussing the adjustment needs following the Chinese GB code.[12]

The resistance factor is determined based on probabilistic analysis, and for Class II buildings (generally the majority, such as residential buildings of less than 18 stories) with 50 years of service life, the baseline resistance factor γ_{R0} is given in Table 4.10.

The adjustment coefficient K_d mainly considers the type of wood, geometry, moisture contents, building life, and loading combinations, etc. The considerations for K_d and other factors are summarized below:

1   For sawn lumber and round timber, if the shorter dimension of the section is larger than 150 mm, K_d can be taken as 1.10. If the moisture content of the structural

Table 4.10 Baseline resistance factor for Class II buildings

Variable coefficient (Vf)	≤0.1	0.15	0.20	0.25	0.30	0.35	0.40	0.45	0.50	Reliability index
Ductile failure	1.07	1.07	1.09	1.13	1.19	1.25	1.32	1.39	1.48	3.2
Brittle failure	1.18	1.20	1.24	1.31	1.40	1.50	1.62	1.74	1.88	3.7

lumber exceeds 25%, the transverse compressive strength, modulus of elasticity of any species, and bending strength of larch lumber need to be multiplied by an adjustment coefficient of $K_d = 0.9$.

2 For members subjected to compression, the adjustment coefficient needs to be calculated based on the angle between the loading direction and the orientation of grain.

When $\alpha < 10°$,

$$f_{c\alpha} = f_C \tag{4.9}$$

When $10° < \alpha < 90°$,

$$f_{c\alpha} = \left[\frac{f_C}{1 + \left(\dfrac{f_c}{f_{c,90}} - 1 \right) \dfrac{\alpha - 10°}{80°} \sin \alpha} \right] \tag{4.10}$$

where, $f_{c\alpha}$ is the design compressive strength at an angle to the grain; α is the angle between the force and grain directions; f_c is the design compressive strength parallel to the grain; $f_{c,90}$ is the design compressive strength perpendicular to the grain.

3 For different conditions of building usage, the adjustment coefficient K_{adj} for design strength and elastic modulus should be taken following Table 4.11.

4 The design strength and modulus of elasticity (MOE) need to be adjusted based on the duration of the building life, given in Table 4.12.

5 For dimension lumber (not for sawn timber and round timber), the design strength and modulus can be increased based on the code for bending in a flatwise condition and for visually graded lumbers.

Table 4.11 Adjustment coefficient for structures under different conditions

Condition	Adjustment coefficient	
	Compressive strength	MOE
Exposed environment	0.9	0.85
Long-term productive high temperature environment (40°C–50°C on surface)	0.8	0.8
Constant load	0.8	0.8

Condition	Adjustment coefficient	
	Compressive strength	MOE
Aiming at wood structures	0.9	1.0
Temporary construction or maintenance	1.2	1.0

Table 4.12 Adjustment coefficient for timber structures of different life

Life span (years)	Adjustment coefficient	
	Strength	MOE
5	1.10	1.10
25	1.05	1.05
50	1.00	1.00
≥100	0.90	0.90

Table 4.13 Strength and modulus adjustment coefficients for snow and wind load design condition

	Adjustment coefficient	
	Strength	MOE
Snow load	0.83	1.0
Wind load	0.91	1.0

Source: Reproduced from Table 4.3.10 of GB 50005.[12]

6   If the ratio of floor live load and dead load ρ (= Q_k/G_k) <1.0, the adjustment coefficient should be calculated as follows,

$$K_d = 0.83 + 0.17\rho \le 1.0 \tag{4.11}$$

When designing for snow and wind loads, the adjustment coefficients for strength and elastic modulus should be made according to Table 4.13.

The Chinese design code also provides an appendix (Appendix F in GB 50005) for factory manufactured timber products which are not included in the code.[12] For engineered bamboo, the characteristic strengths, f_K shown in Table 4.4, are calculated following Appendix F of the GB 50005 code, and the design strength without considering adjustment coefficient K_d is determined based on $f = \left(\dfrac{K_{DOL} f_K}{\gamma_{R0}} \right)$, where K_{DOL} is taken as 0.72 following the GB code, and the resistance factor γ_{R0} is based on the reliability analysis, following the recommendations provided in the GB code.[12]

The adjustment factor K_d for engineered bamboo is recommended to follow the requirements of the GB code[12] as discussed above; however, the adjustment for the angle between the direction of load application and grain orientation can be neglected, except in the special case where there is actually an angle between the two directions.

4.3.1.2 Serviceability Limit State

Another limit that design must take into account is the serviceability limit state to assure the functionality of a building. In the Chinese GB 50005 code,[12] the design must ensure

Table 4.14 Deflection limits for bending members

No.	Component types			Deflection limit
1	Purlin	$l \leq 3.3m$		l/200
		$l > 3.3m$		l/250
2	Rafters			l/150
3	Bending components in ceiling			l/250
4	Beam floors and joists			l/250
5	Studs	Rigid veneers		l/360
		Ductile veneers		l/250
6	Roof trusses	Industrial buildings		l/120
		Civil buildings	No painted ceilings	l/180
			Painted ceilings	l/240

Note: *l* is the length of the calculation span.

that the effects of certain service load combinations do not exceed the required limit values; this is expressed as,

$$S_d \leq C \tag{4.12}$$

where S_d is the loading effects under the service load combination; and C is the code-specified limit, such as deflection, etc. In most situations of designing wood structures, the serviceability limit check is for the deflection of members or floors. The Chinese GB code provides the limit values of deflection as shown in Table 4.14.[12]

For engineered bamboo structures, it is recommended that the requirements of Table 4.14 are followed if attempting to achieve the same level of requirements as for timber as set down in Chinese code GB 50005.[12]

4.3.2 Load and Resistance Factor Design (LRFD)

North America has taken a leadership role on research in limit state design methodologies; however, the implementation of limit state design into design codes is relatively slow in the United States. The trend in the late 1990s started to replace the existing allowable stress design (ASD) methodologies with load and resistance factor design (LRFD). The American wood industry responded to this trend by publishing the limit-state-based *Load and Resistance Factor Design (LRFD) Manual for Engineered Wood Construction* (American Forest & Paper Association.[25] Currently, ASD and LRFD approaches are both adopted in practice. The LRFD method was first incorporated into the 2005 version of the National Design Specifications. In this book, the latest (2018) version of NDS,[17] the LRFD for wood structure design is introduced and recommendations are attempted for engineered bamboo following the NDS LRFD.

The general goal of the LRFD approach is to assure a capacity that exceeds the demand, with certain acceptable reliability, similar to most of the limit-state-based design approaches, such as the current Chinese GB 50005 code. In the U.S. NDS,[17] the LRFD design is based on a stress check, probably because of the attempt to align with the parallel ASD approach.[17] However, in this book, the LRFD is introduced mainly using the load and resistance capacity, similar to the design of other structures, such as steel and reinforced concrete structures. In general, the LRFD design can be expressed as,

$$\sum \gamma_i Q_i \leq \phi R_n \qquad (4.13)$$

where Q_i is the nominal load due to a specific loading condition; γ_i is load factors, ranging from 1.2 to 1.7, based on ASCE-7;[26] ϕ is resistance factor, taken as 0.75–0.9, for different conditions; R_n is nominal or ideal resistance against a particular limit (yield or fraction, etc.). The factored capacity ϕR_n is also called dependable resistance. The above Eq.4.13 is a general format for LRFD design, and for wood structures, a time effect factor λ (Table 4.15) is added to the resistance side,

$$\sum \gamma_i Q_i \leq \lambda \phi R_n \qquad (4.14)$$

The time effect factor λ takes into account the duration of the loading on a wood member. As shown in Table 4.15, the factor is dependent on the combination of loads acting on the structural member. These factors were derived using a probabilistic-based approach to achieve a consistent probability of failure for different load durations. The time effect factor does not apply to compression stress perpendicular to the grain, the modulus of elasticity for deflection calculations, or the buckling modulus of elasticity.

The resistance factor ϕ depends on the usage of a member, and the general values for LRFD design of wood structures are shown in Table 4.16.

Table 4.15 Time effect factor, λ (LRFD only)

Load combination	λ
1.4D	0.6
1.2D + 1.6L + 0.5 (Lr or S or R)	0.7 when L is from storage; 0.8 when L is from occupancy; 1.25 when L is from impact[a]
1.2D + 1.6 (Lr or S or R) + (L or 0.5W)	0.8
1.2D + 1.0W + L + 0.5 (Lr or S or R)	1.0
1.2D + 1.0E + L + 0.2S	1.0
0.9D + 1.0W	1.0
0.9D + 1.0W	1.0

a Time effect factor for design of connections or structural members pressure-treated with water-borne preservatives or fire-retardant chemicals shall not be greater than 1.0.

Source: American Wood Council (2018b).[18]

Table 4.16 Wood LRFD resistance factor ϕ values

Members and loading conditions	ϕ
Compression	0.90
Flexure	0.85
Tension	0.80
Shear	0.75
Connections	0.65

Table 4.17 Format conversion factor, K_F for LRFD

Application	Property	K_F
Member	F_b	2.54
	F_t	2.70
	F_v, F_{rt}, F_s	2.88
	F_c	2.40
	$F_{c\perp}$	1.67
	E_{min}	1.76
All connections	(All design values)	3.32

Source: Table 2.3.5, American Wood Council (2018b).[18]

In calculating the nominal capacity R_n, specified strength F_r should be used. The U.S. NDS-S[17] provides a format conversion factor K_F as shown in Table 4.17,[18] to convert the reference design value, F_{NDS-S}, i.e., the basic ASD value, into the specified strength F_r for LRFD.

Similar to the ASD, other adjustments are needed using C factors, as follows:

$$Fr = K_F F_{NDS-S} \, \Pi C_i \tag{4.15}$$

where the adjustment factors C_i are given in Table 4.8 and Table 4.9.

For engineered bamboo design based on LRFD, following the NDS procedures is suggested, and simply using the same form conversion factor K_F, in order for engineered bamboo to be possibly considered in design in North America. In future, further calibration of the factors are needed.

4.3.3 Limit State Design Based on European Code EC-5

The Euro-code EC-5[14] specifies the design strength, x_d, of materials based on the following general equation,

$$x_d = k_{mod} X_k \, / \, \gamma_m \tag{4.16}$$

where, X_k represents the characteristic strength values; k_{mod} is the modification factor taking into account the effect on the strength parameters of the load duration, the moisture content in the structure, and service conditions; and γ_m is the partial safety factor for the material property, taken as 1.3 for timber and wood-based materials.

4.3.4 Summary of Design Strength Values

Using the three codes and specifications for timber structure design, the main design values are established for glubam and are summarized and compared in Table 4.18.

4.4 Future Code Development

The suggestions made in this chapter are for the purpose of providing a preliminary framework and suggesting material strength values for the design of glubam structures.

Table 4.18 Limit state design strengths for engineered bamboo

	GB 50005[1]		LRFD[2]		EC-5[3]	
	Glubam-thick	Glubam-thin	Glubam-thick	Glubam-thin	Glubam-thick	Glubam-thin
Tension, f_{tx} (MPa)	32.4	18.4	49.5	28.7	42.3	24.6
Compression, f_{cx} (MPa)	20.4	10.7	29.6	16.0	25.7	13.9
Bending, f_{by} (MPa)	38.4	29.0	53.2	40.5	48.4	36.7
Shear, f_{vyx} (MPa)	2.4	4.7	3.9	7.5	3.1	6.0

1 The GB 50005 strength values are based on $K_{DOL}f_K/\gamma_{R0}$.
2 The LRFD strength values are calculated as $K_F \phi F_{NDS-S}$.
3 The values are given based on service condition 1 (the average moisture content in most softwoods will not exceed 12%), and long-term loading (6 months to 10 years), thus $k_{mod} = 0.7$.

The work may still be premature. The author does believe that it is the right direction to establish design specifications and guidelines for engineered bamboo following the existing established framework for timber structure design. Nonetheless, further accumulation of experimental data is needed for systematic calibration of the probability-based design methods for engineered bamboo structures.

References

[1] National Technical Committee 263 on Bamboo and Rattan of Standardization Administration of China. (2006). GB 20240-2006: Bamboo flooring (in Chinese). China Building Industry Press, Beijing.

[2] National Technical Committee 198 on Wood-based Panels of Standardization Administration of China. (2007). GB 21179-2007: Decorative bamboo veneered panel (in Chinese). China Building Industry Press, Beijing.

[3] National Forestry Administration. (2002). LY 1072-2002: Laminated bamboo strip lumber (in Chinese). National Forestry Administration, Beijing.

[4] National Forestry Administration. (1992). LY 1073-92: Experimental methodologies of physical mechanical properties of laminated bamboo strip lumber (in Chinese). National Forestry Administration, Beijing.

[5] National Forestry Administration. (2000). LY 1575-2000: Strip plybamboo for bottom boards of trucks and buses (in Chinese). National Forestry Administration, Beijing.

[6] General Administration of Quality Supervision, Inspection and Quarantine of the People's Republic of China. (2003). GB 13123-2003: Bamboo-mat plywood (in Chinese). China Building Industry Press, Beijing.

[7] General Administration of Quality Supervision, Inspection and Quarantine of the People's Republic of China. (1991). GB 13124-91: Methods of testing bamboo-mat plywood (in Chinese). China Building Industry Press, Beijing.

[8] National Industry Administration. (1993). JB/T 6564-93: Plybamboo for export machine packing boxes (in Chinese). National Industry Administration, Beijing.

[9] National Urban–Rural Development Administration. (1999). JG 3059-1999: Plybamboo form with steel frame (in Chinese). National Urban–Rural Development Administration, Beijing.

[10] National Forestry Administration. (2000). LY 1574-2000: Plybamboo for concrete-form (in Chinese). National Forestry Administration, Beijing.

[11] Lu, Y. (2016). Experimental study on modification of cold pressed adhesive and ultrasonic detection of glue joint of bamboo structures. MEng thesis, Hunan University.

[12] Ministry of Housing and Urban–Rural Development of the People's Republic of China. (2017). GB 50005-2017: Code for design of timber structures (in Chinese). China Architecture & Building Press, Beijing.

[13] ASTM International. (2010). ASTM D2915-10: Standard practice for sampling and data-analysis for structural wood and wood-based products. ASTM International, West Conshohocken, PA.

[14] European Committee for Standardization (CEN). (2016). EN 14358-2016: Timber structures – calculation and verification of characteristic values. European Committee for Standardization, Brussels.

[15] Li, L., He, X.Z., Cai, Z.M., Wang, R., & Xiao, Y. (2021). Mechanical properties of engineered bamboo boards for glubam fabrication. *ASCE Journal of Materials in Civil Engineering*, 33(5).

[16] Academy of Building Sciences of National Capital Construction Commission. (1973). GB J5-73: Timber structures design standards (in Chinese). China Building Industry Press, Beijing.

[17] American Wood Council. (2018a). *National design specification for wood construction*. American Wood Council, Leesburg, VA.

[18] American Wood Council. (2018b). *NDS supplement: National design specification: Design values for wood construction*. American Wood Council, Leesburg, VA.

[19] ASTM International. (2016). ASTM D2555-16: Standard practice for establishing clear wood strength values. ASTM International, West Conshohocken, PA, www.astm.org

[20] ASTM International. (2019). ASTM D245-06: Standard practice for establishing structural and related allowable properties for visually graded lumber. ASTM International, West Conshohocken, PA, www.astm.org

[21] ASTM International. (2007). ASTM D143-07: Standard test methods for small clear specimens of timber. ASTM International, West Conshohocken, PA, www.astm.org

[22] Yang, R.Z. (2013). Research on material properties of glubam and its application (in Chinese). PhD thesis, Hunan University.

[23] Xiao, Y., & Shan, B. (2013). *Modern bamboo structures* (in Chinese). China Architecture and Building Press, Beijing.

[24] ASTM International. (2015). ASTM D5457-15: Standard specification for computing reference resistance of wood-based materials and structural connections for load and resistance factor design. ASTM International, West Conshohocken, PA, www.astm.org

[25] American Forest & Paper Association. (1996). *Load and resistance factor design (LRFD) manual for engineered wood construction*. American Forest & Paper Association, Washington, DC.

[26] ASCE-7 Committee. (2016). *Minimum design loads and associated criteria for buildings and other structures*. American Society of Civil Engineering, Reston, VA.

Chapter 5

Connections in Glubam Structures

Most modern structures consist of various components, and the connections between the components are essential. Significant research on connections in timber structures has been accumulated in industrialized countries.[1–3] For round bamboo culm structures, the connections typically rely on metal connectors, which are often complicated.[4] As a newly emerging form of structures, the studies on engineered bamboo have just begun, with limited understanding on its connection performance.[5] This chapter attempts to provide a summary of available research studies on the behaviors and design of connections for engineered bamboo structures.

5.1 Brief Review of Connections Used in Timber Structures

In timber structures, connecting elements are typically required for: i) forming joints of structural elements, such as beam to column joints, truss joints; ii) extension of member length; iii) forming composite elements, such as a bolted I-shape section, nail laminated timber (NLT), shear walls with studs and plywood sheathing, etc. The joining methods typically include the traditional mortise and tenon, nailing or bolting, mending plate, gluing, etc.

Traditional Chinese timber structures adopt the mortise and tenon connections which are also called carpentry joints, a kind of joinery work still widely used in timber buildings, particularly heavy timber structures. Figure 5.1 (a) illustrates an example of a haunched beam to column joint with mortise and tenon details. Bolted connection is one of the most important forms of connection in the modern timber structure system. Advantages of bolted connections include: i) bolting construction is simple, convenient, and suitable for on-site operation; ii) bolted connections can fully exert the perform-ance of the material, they have uniform force transmission, high bearing capacity, and high safety and reliability; iii) the bolted connection is easier to maintain than other connection methods. Figure 5.1 (b) presents a bolted beam to column connection, as the counterpart to the mortise and tenon form shown in Figure 5.1 (a).

Mending plates are often used to strengthen joints in timber structures, as shown in Figure 5.2. Bolted joints with a metal and a timber mending plate are shown in Figure 5.2 (a) and (b), respectively. Figure 5.2 (c) exhibits nailed mending plates for joints in a truss, whereas Figure 5.2 (d) shows nail-toothed metal plate connected (MPC) joints.

Bolted or nailed connections also rely on many types of metal brackets (either custom-made or standard commercial products) for reliable performance. Figure 5.3

DOI: 10.1201/9781003204497-5

Figure 5.1 Examples: (a) mortise and tenon connection; (b) bolted connection.

Figure 5.2 Mending plated joints: (a) bolted joint with metal mending plate; (b) nailed joint with timber mending plate; (c) nailed joint with metal mending plate; (d) toothed metal plate joint.

Figure 5.3 Metal brackets for timber structures: (a) bracket for beam end support; (b) post and beam bracket; (c) column foundation anchorage; (d) metal hold-downs.

shows several examples of steel plates and brackets for timber connections. Figure 5.4 shows a testing investigation on a large capacity connector which can be used as a hold-down or lateral restrainer for girders.[6] It can be seen from the force and deformation hysteresis curves from the test results shown in Figure 5.4 (b) that the joint has a good bearing capacity and deformability in the tension direction.

Research on connections for engineered bamboo structures is still premature and limited. The following sections summarize several research studies on connections for glubam structures conducted by the author's research group. Many of the glubam structure joints follow the details originally developed for timber structures; however, necessary modifications have been made.

5.2 Embedment Strength of Glubam

As introduced in the previous section, many types of timber joints rely on bolted or nailed connections with or without metal plates. Such connections rely on the dowel action of the bolts or nails inside the holes in the timber. Since the timber is typically softer than the metal, the behavior of such dowel-type connections is strongly influenced

(a)

Displacement (x10^{-2} in)

(b)

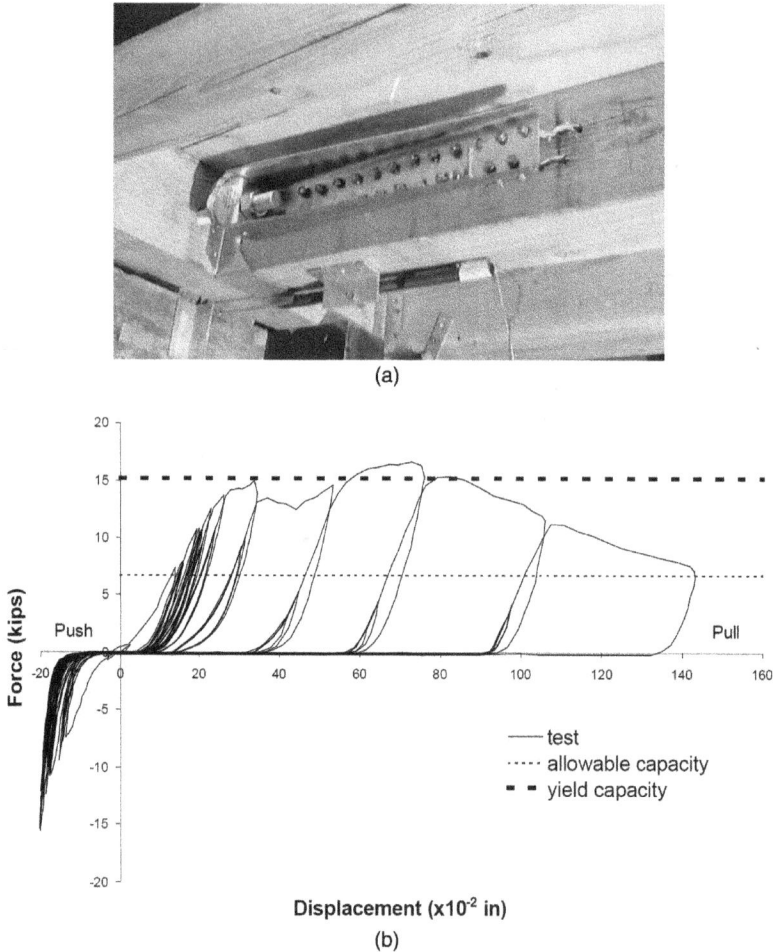

Figure 5.4 Steel connector and load-deformation hysteresis curve.

Source: Xiao & Xie (2003).[6]

by the embedment strength of timber.[7–10] Apparently, for bolted or nailed connections in engineered bamboo structures, studies on embedment strength are also important. This section summarizes the research by the author's research team on the embedment strength of plybamboo used for making glubam.[11–13]

5.2.1 Experimental Programs

Testing of embedment behavior can be conducted by following the guidelines established by ASTM D5764[14] or the European test standard EN 383.[15] The ASTM standard is based on a half-hole test with load applied uniformly along the fastener length to minimize its bending; whereas the EN 383 standard simulates the embedment behavior in a

No.	Type & size of connections
nails	
1	ST nails - 2.00×46.00 mm
2	ST nails - 2.20×56.00 mm
3	High strength nails - 3.40×50.00 mm
4	Common nails - 2.80×60.00 mm
5	High strength nails - 4.00×70.00 mm
6	Common nails - 3.50×80.00 mm
bolts	
7	8.05 mm dia. bolts
8	10.02 mm dia. bolts
9	11.98 mm dia. bolts
10	13.78 mm dia. bolts

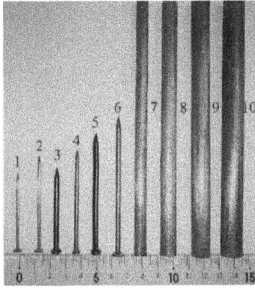

Figure 5.5 Types of fasteners adopted in testing.
Source: Li et al. (2020).[11]

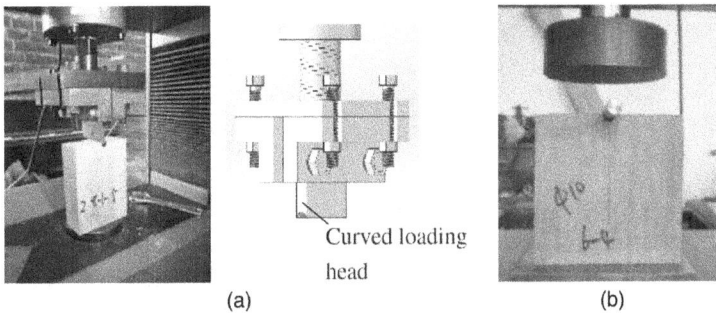

Curved loading
head

(a)   (b)

Figure 5.6 Test loading setup: (a) for nails; and (b) for bolts.
Source: Li et al. (2020).[11]

full-hole test, and the load is applied at the fastener ends. In the research documented in Li et al.[11] and Wang,[12] ASTM D5764[14] is adopted for testing the embedment strength of thick-strip glubam with ten types of fasteners (five types of nails and five types of bolts without threads), as shown in Figure 5.5.

A 20 kN universal testing machine was used, with the load and the displacement of the loading head recorded automatically. For the nail tests, a specially designed loading plate was made, which has a groove at the end to hold the nail during loading, as shown in Figure 5.6 (a). For larger bolts, the load is directly applied through a loading platen, as shown in Figure 5.6 (b). The loading rate was 1.0 mm per min. for all specimens. The test was terminated at an embedment deformation of one-half the fastener diameter or after the maximum load was reached.

As shown in Figure 5.7 (a), due to the nature of the thick-strip configuration in the glubam, three loading directions, each along an axis, are considered. For each loading direction, the fastener is placed in two directions, each aligned with one of the other two axes. Therefore, as shown in Figure 5.7 (b)–(g), for each fastener, there are 3 x 2 = 6 loading cases for embedment behavior and strength, designated as $f_{h,ij}$, here, where $i = x$,

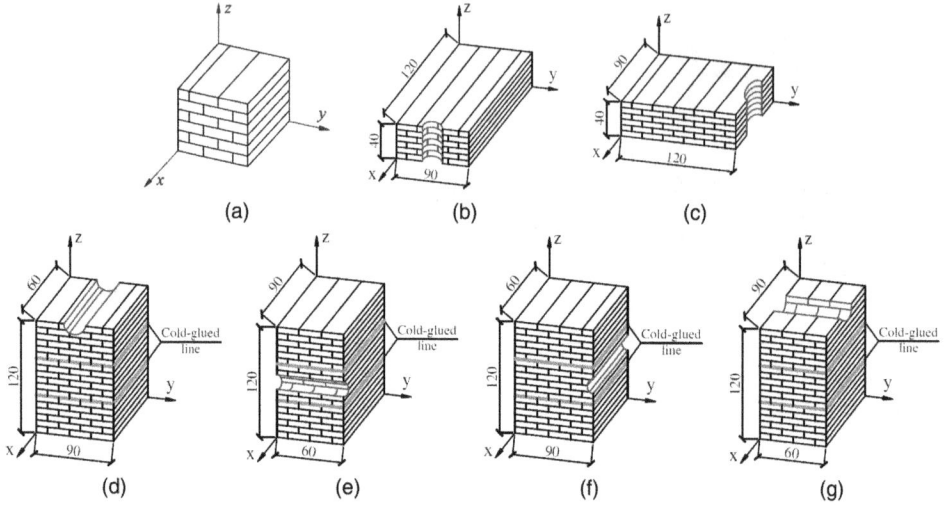

Figure 5.7 Test specimens (in millimeters) for embedment strength: (a) coordinate system; (b) 1-$f_{h;xz}$; (c) 2-$f_{h;yz}$; (d) 3-$f_{h;zx}$; (e) 4-$f_{h;xy}$; (f) 5-$f_{h;yx}$; and (g) 6-$f_{h;zy}$.

Source: Li et al. (2020).[11]

y, z, and $j = x$, y, z, but, $j \neq i$. For each combination of fastener and loading condition, 6 specimens were prepared and tested. In addition, 6 compressive loading tests were also carried out for the specimens in each loading case. Thus, altogether 396 tests were conducted. The embedment stress f_h is calculated by dividing the applied load by the vertical projection of the contact area, or td, and here, t is the thickness of the specimen and d is the fastener diameter.

5.2.2 Embedment Loading Behaviors

The failure mode of the specimens under embedment loading can roughly be divided into two categories, splitting failure and bearing crushing, as observed by Li et al.[11] For loading conditions where the loading or the fastener is along the bamboo fiber, splitting failure occurs; whereas crushing of the bamboo with enlargement along the half hole is typical if the fastener is perpendicular with the bamboo fibers.

The average embedment stress and the displacement relationships for 10 types of fasteners and 6 loading cases are shown in Figure 5.8. The compressive stress and displacement behavior corresponding to each loading condition are also illustrated in Figure 5.8. As shown in Figure 5.8, the stiffness and embedment strength for the cases with smaller diameter nails are generally higher than for those with larger diameter bolts, except for loading cases 4, or $f_{h,xy}$. This probably reflects the local confinement effects. Figure 5.8 also indicates that the compressive stress and displacement curves represent a lower boundary for embedment behaviors, regardless of loading direction and connector size. This implies that the bearing capacity of dowel connections can be conservatively calculated by directly using the compressive strength instead of the embedment strength.[11]

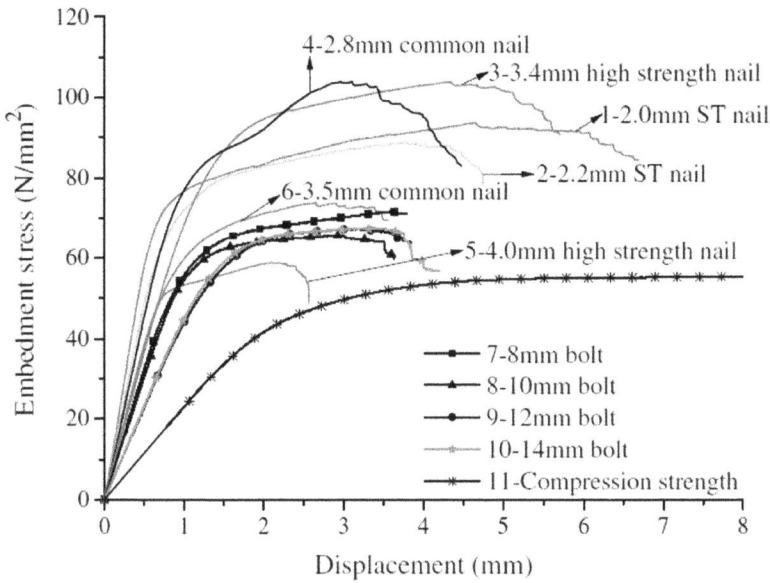

Displacement (mm)

$1\text{-}f_{h,xz}$

(a)

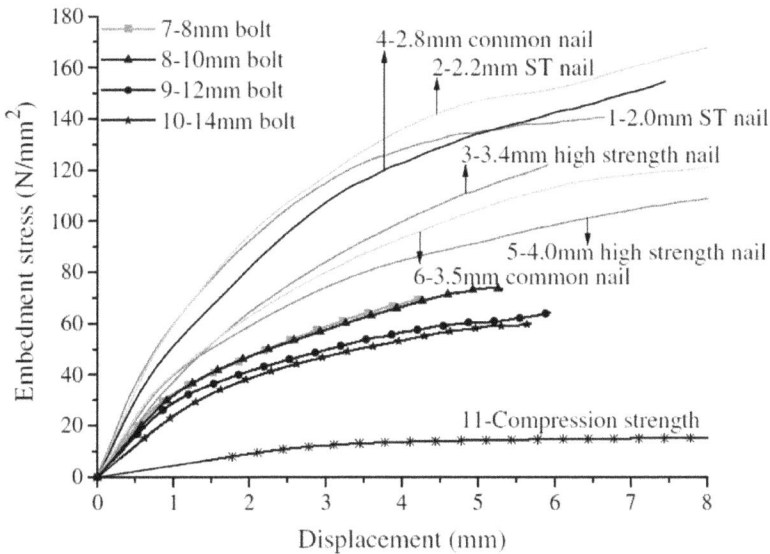

Displacement (mm)

$2\text{-}f_{h,yz}$

(b)

Figure 5.8 Mean embedment curves of glubam under different loading directions with various connection diameters: (a) $1\text{-}f_{h,xz}$; (b) $2\text{-}f_{h,yz}$; (c) $3\text{-}f_{h,zx}$; (d) $4\text{-}f_{h,xy}$; (e) $5\text{-}f_{h,yx}$; and (f) $6\text{-}f_{h,zy}$.

Source: Li et al. (2020).[11]

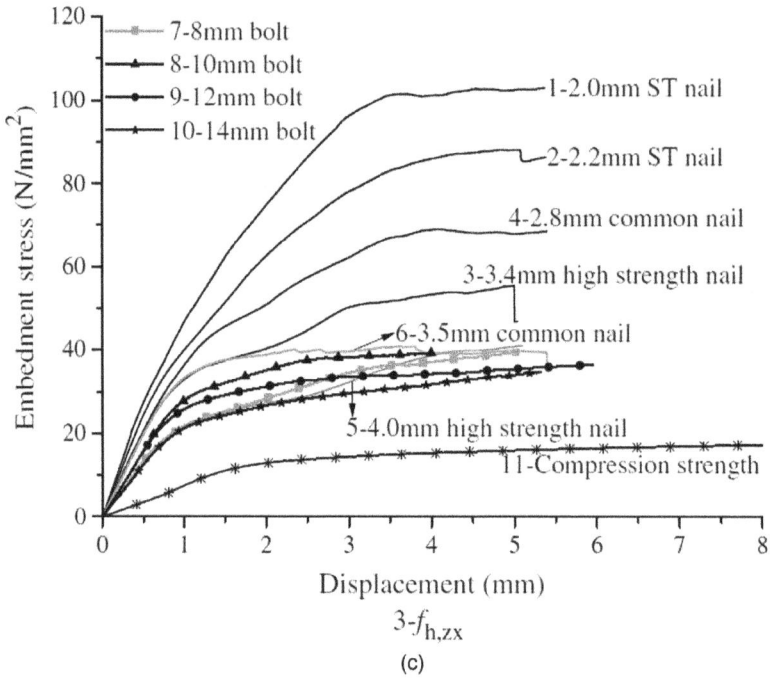

(c)

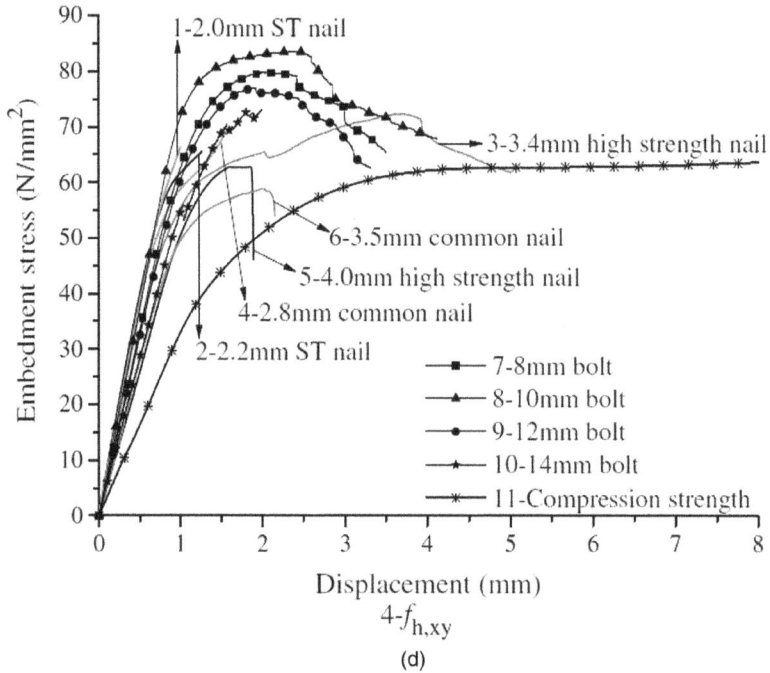

(d)

Figure 5.8 Continued

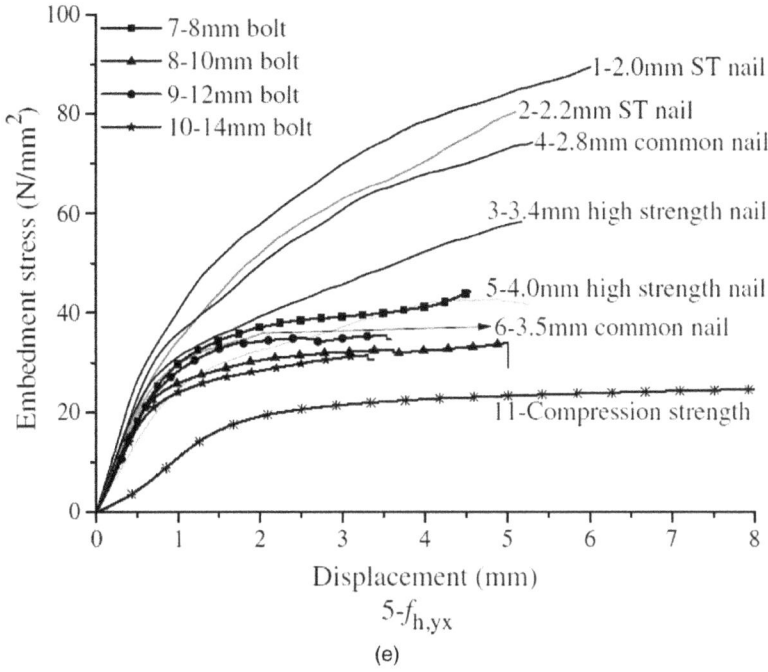

(e)

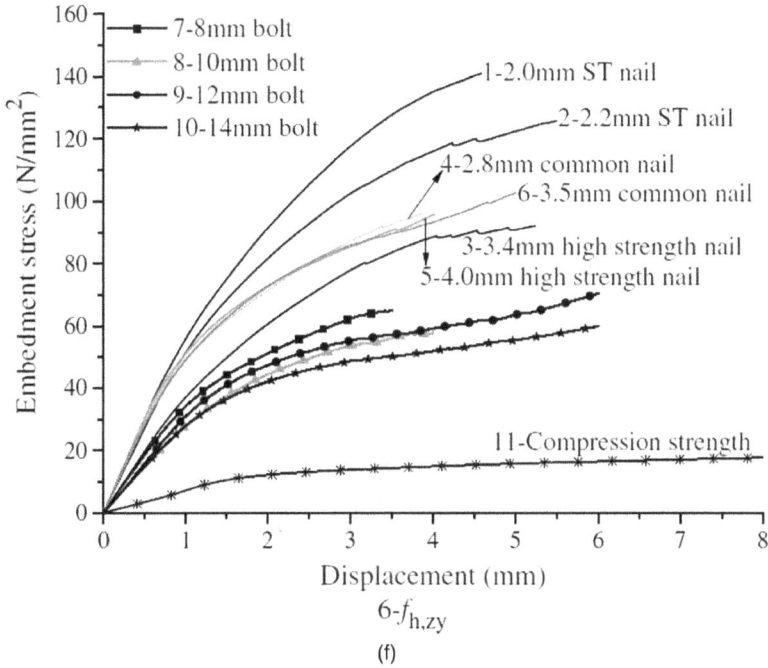

(f)

Figure 5.8 Continued

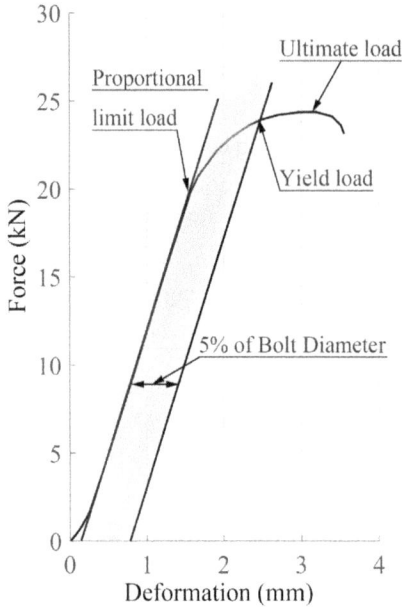

Figure 5.9 ASTM definition of yield load.
Source: ASTM International (2018).[14]

5.2.3 Embedment Strength

As shown in Figure 5.9, the embedment strength is calculated by dividing the bearing-yield load by the fastener diameter and the specimen thickness, according to the definition given in ASTM D5764.[14] The bearing-yield load is determined from the load-deformation curve, based on the so-called 5% offset method. First, a straight line is fitted to the initial portion of the curve and this line is then offset by a deformation equal to 5% of the dowel diameter; the bearing-yield load is found where the offset line intersects the load-deformation curve. However, for the cases where the offset line does not intersect the curve, the maximum load should be used as the bearing-yield load.

Based on the test results, Li et al.[11] calculated the embedment strengths for all the specimens with different loading conditions, and adopted the following equation [16] to provide the trend lines of the average strength.

$$f_h = A\rho^B d^C \tag{5.1}$$

where, f_h = embedment strength; ρ = timber or bamboo density; d = fastener diameter; and A, B, and C are statistical coefficients. The measured coefficient of variation (COV) of the diameter of ST (staple) nails is 0.28%; for common nails, it is 0.19%; and for high-strength nails, it is 0.39%. Table 5.1 shows the regression values for the coefficients A, B, and C, for all the six loading directions.[11] For the same data set of Li et al., Wang provided regression values for the coefficients in Eq.5.1, separately for nails and bolts.[12]

Table 5.1 Linear regression values for coefficients in Eq.5.1

Loading direction	Parameters		
	A	B	C
1-$f_{h,xz}$	10.2	0.3	-0.06
2-$f_{h,yz}$	221	-0.2	-0.16
3-$f_{h,zx}$	45	-0.01	-0.20
4-$f_{h,xy}$	0.5	0.7	0.20
5-$f_{h,yx}$	57	-0.08	-0.08
6-$f_{h,zy}$	4.4	0.4	-0.09

Following the so-called design-by-testing method suggested by Eurocode EN 1990 Annex D,[17] Li et al.[11] estimate the characteristic and design values of the embedment strength, denoted as $f_{h;k}$ and $f_{h;d}$, respectively. As examples, the results for loading along the bamboo fiber direction (1-$f_{h;xz}$) and across the fiber direction (2-$f_{h;yz}$), are shown in Figure 5.10 (a) and (b), respectively.

Using the dataset of Li et al.,[11] another simpler approach is attempted herein. Following the ASTM D143[18] or the Chinese GB 50005 code,[19] the characteristic values for each testing group are estimated, and then the regression analysis for the coefficients in Eq.5.1 is carried out. For the most important directions (directions 1 and 2, in Figure 5.7), Eq.5.1 is rewritten as follows,

$$f_{h;xz} = 121\, p^{-0.101} d^{0.056} \tag{5.2a}$$

$$f_{h;yz} = 780\, p^{-0.462} d^{0.106} \tag{5.2b}$$

For the design embedment strength, it is suggested that the approaches described in Chapter 4 are followed, considering the embedment as compression.

5.2.4 Embedment Strength of Different Glubams

The above discussions are based on the test results of thick-strip plybamboo boards for glubam. The author's group also conducted another pilot testing program on the embedment strength of thin-strip, bidirectional plybamboo boards,[13] with only one type of bolt diameter of 12.0 mm. The plybamboo specimens were 28 mm thick with a longitudinal to transverse bamboo strip (or fiber) ratio of 4:1. The edge and end distances are both 48 mm. The tests were conducted using the tensile testing method. Five tests were executed, each for the main bamboo fiber direction ($f_{h;xz}$) and the transverse direction ($f_{h;yz}$). The results of 5% offset strength are shown in Table 5.2

The average embedment strength and standard deviation are also shown in Table 5.2. Using Eq.4.1 ($f_k = m - kS$, where m is the average value of each test results, S is the standard deviation of strength values, k is the characteristic factor with a 75% confidence level, taken as 2.464, based on ASTM D2915[20]), the characteristic embedment strength values can be calculated and are shown in Table 5.2. By comparing the average and the characteristic values of Table 5.2 with those for the 14 mm diameter bolts

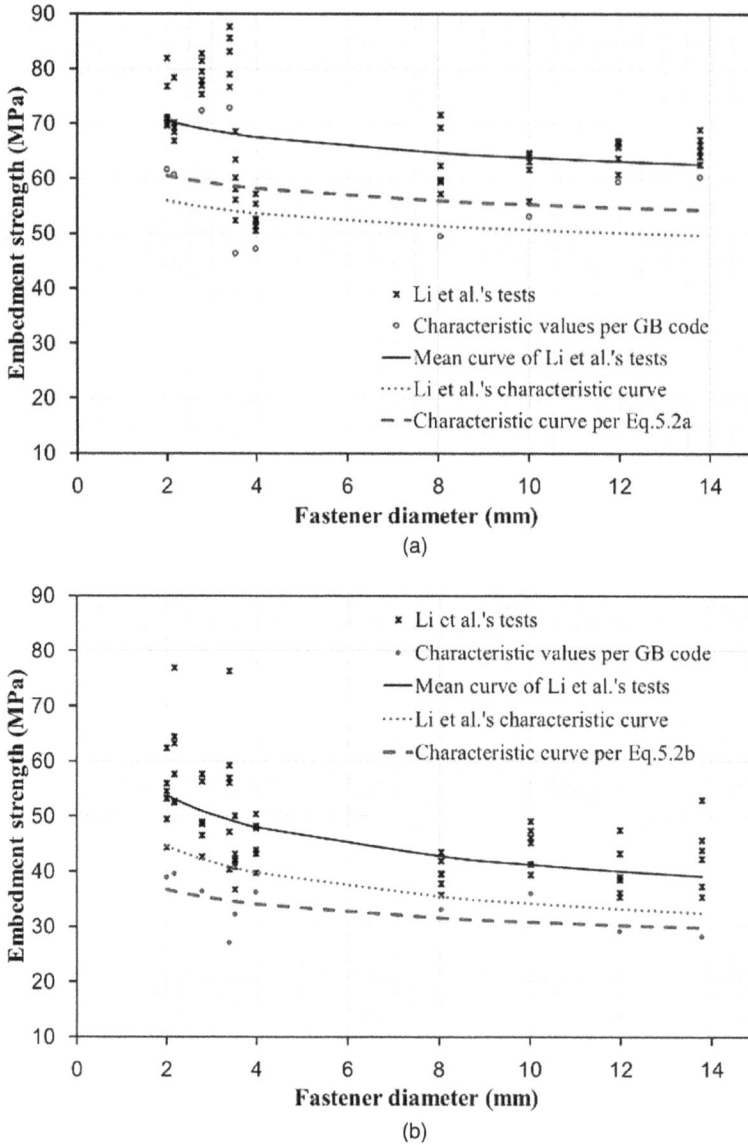

Figure 5.10 Embedment strength: (a) 1-$f_{h;xz}$; (b) 2-$f_{h;yz}$.

shown in Figure 5.10, one can see that the results for the two types of glubam are close. The strength values in the longitudinal direction of the thin-strip glubam are lower than those for thick-strip glubam; however, the opposite is true for the transverse direction.

Using Eq.5.2, and substituting the density of 880 kg/m³ for thin-strip glubam used in the testing, the characteristic embedment strength for the longitudinal and transverse directions can also be calculated as 52.69 MPa, and 25.72 MPa, respectively, which are smaller, but in reasonably close order to those shown in Table 5.2.

Table 5.2 Embedment strength of thin-strip plybamboo with 12.0 mm bolts

Loading direction	Specimen	5% Yield load (kN)	Embedment strength (MPa)	Average embedment strength (MPa)	Standard deviation (MPa)	Characteristic embedment strength (MPa)
$f_{h;xz}$	1-1	22.685	63.01			
	1-2	21.877	60.77			
	1-3	24.602	68.34	64.24	2.91	57.07
	1-4	23.706	65.85			
	1-5	22.769	63.25			
$f_{h;yz}$	2-1	19.618	54.49			
	2-2	18.920	52.56			
	2-3	17.700	49.17	51.04	6.02	36.21
	2-4	15.000	41.67			
	2-5	20.630	57.31			

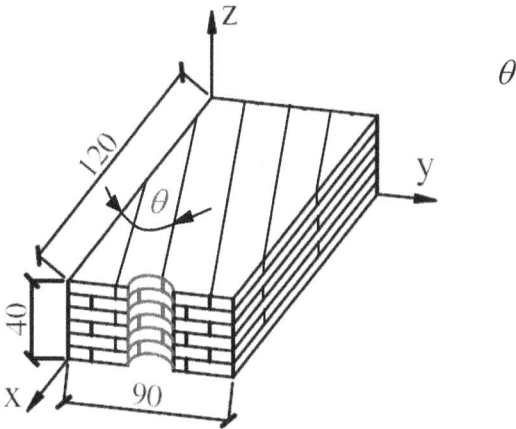

Figure 5.11 Embedment test in off-axis direction.

5.2.5 Embedment Strength of Glubam in Arbitrary Directions

Current studies are all based on directions either perpendicular or parallel to one of the main axes of the glubam. In a bolted connection, the loading direction may be off these axes; therefore, embedment behavior in arbitrary directions needs to be investigated. Experiments on embedment behavior with the loading direction at an angle with the bamboo fiber direction, as shown in Figure 5.11, are currently being planned. Before further testing data become available, the Tsai–Wu failure theory[21] may be adopted for estimating the embedment strength combined with Eq.5.2, similar to the suggestions by Yang et al. for the off-axis compressive strength of glubam.[22] Alternatively, Hankinson's equation[23] for the off-grain compressive strength of timber may be used,

$$f_{c,\alpha} = \frac{f_{h,xz} \times f_{h,yz}}{f_{h,xz} sin^2\alpha + f_{h,yz} cos^2\alpha} \tag{5.3}$$

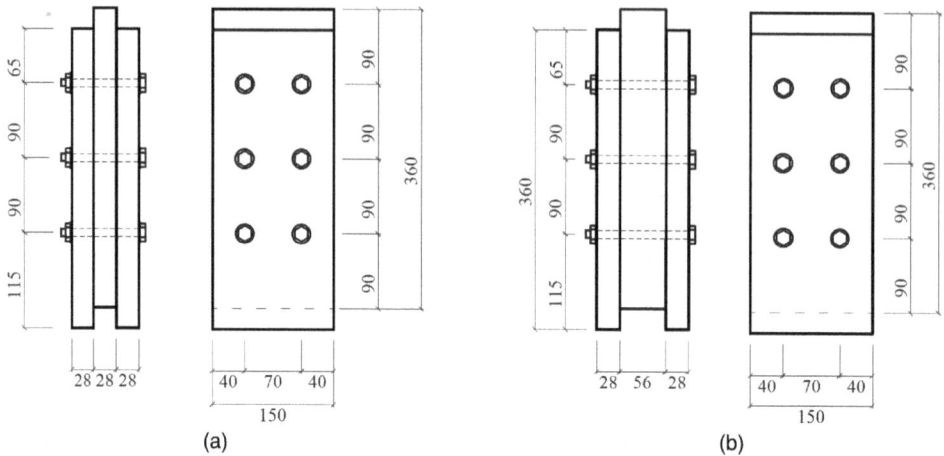

Figure 5.12 Specimen details: (a) for first and second groups of specimens; (b) for third group of specimens.

Table 5.3 Specimen list of glued bamboo bolt joints

Groups	Number of specimens	Thickness of main board	Bolting condition
BGJ Group 1	3	28 mm	Tight
BGJ Group 2	3	28 mm	Snug-tight
BGJ Group 3	3	56 mm	Snug-tight

The American Wood Council's National Design Specification[24] also adopts Hankinson's equation for calculating the compressive strength of timber loaded in the direction of α angle with grain direction.

5.3 Compressive Behavior of Bolted Glubam Joints

5.3.1 Compression Test Program

One of the simpler and more reliable ways to study the group effects of bolted connection in timber structures is to test the joint in compression. A pilot study on testing three groups of bolted glubam under compressive loading was carried out by Yang and Xiao,[25] as shown in Table 5.3.

Specimens: Details of the specimens are presented in Figure 5.12 (a) and (b), which were designed according to the design specification for wood structures. The diameter d of the bolts is 12 mm. The end distance of 90 mm along the main bamboo fiber direction (grain direction) is larger than the required minimum end distance of $7d = 84$ mm. The 90 mm pitch is larger than the minimum pitch of $7d = 84$ mm. The 40 mm edge distance exceeds the required minimum edge distance of $3d = 36$ mm. The size of the gauge is 70 mm, larger than the minimum gauge of $3.5d = 42$ mm.

As shown in Table 5.3, the research parameters of the first and second groups of specimens are whether the bolts are tightened using a wrench or only made snug-tight by hand. In the tightened case, slippage between the center and the side plates is essentially prohibited, while the snug-tight bolt allows such slippage. The testing parameters of the second and third groups of specimens are the thickness of the main board. Three specimens were tested for each testing parameter. All the specimens were prepared by professional carpenters, with particular attention given to planing the bamboo board surfaces and the loading ends. The hole diameter was drilled 1 mm larger than the diameter of the bolts. The drilling speed should not be more than 120 mm/min. The rotation speed of the electric drill should not be too slow, and it can take 300 r/min.

Loading methods: A 1000 kN universal testing machine is used for testing the bolted glubam joints. As shown in Figure 5.13, a dial indicator is used to measure the relative slip between the main board and the side boards. The dial indicator is fixed on the specimen with special devices and arranged symmetrically on both sides. Based on the Chinese standard GB/T 50329 for testing timber structures,[26] the loading procedure, as shown in Figure 5.14, involves first loading the specimen to 0.3 *F* for 30 seconds (s), then it is unloaded to 0.1 *F* for another 30 s. After that, the load is increased every 30 s, and the increment load per stage is 0.1 *F*. After exceeding 0.7 *F*, the loading speed is slowed down to 0.1 *F* per min., until the specimen is broken, as shown in Figure 5.14. Here, *F* is the pre-estimated external load on the specimen when the bolt reaches its yield, and is taken as 67.0 kN, based on the following equation from the Chinese code for design of wood structures,[19]

$$N_v = k_v d^2 \sqrt{f_c}$$

(5.4)

where N_v is the design value of the bearing capacity of each shear surface of a bolt; k_v is the calculation coefficient of the bearing capacity of the bolt connection, taken as 5.5;

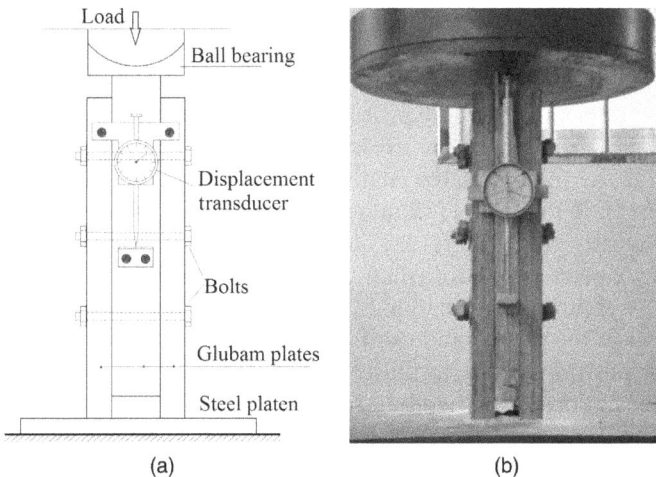

(a)   (b)

Figure 5.13 Test setup.

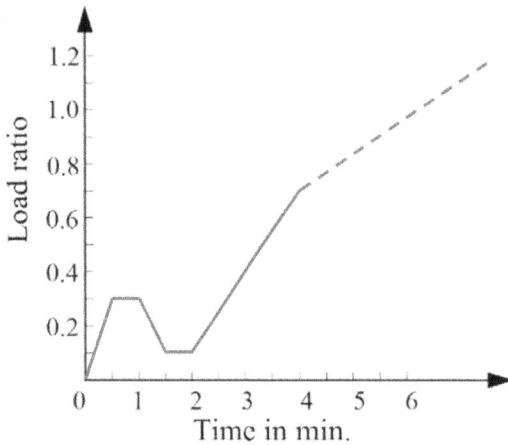

Figure 5.14 Loading process.

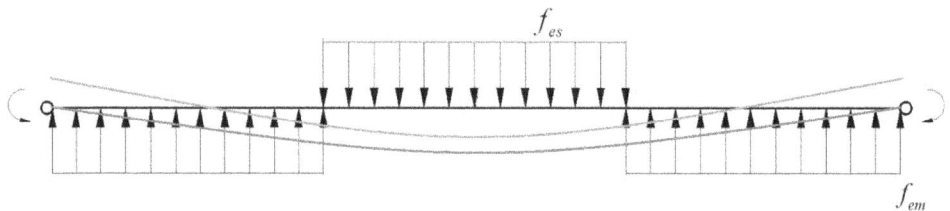

Figure 5.15 Schematic loading condition of bolt.

d is the bolt diameter (in mm); f_c is the designed value of the compressive strength of the wood.

5.3.2 Experimental Results

Failure patterns: During testing, the bolts are pushed down in the middle and supported by the two side plates, as schematically shown in Figure 5.15. Bending of the bolts results in the slippage of the middle plates relative to the side plates (shown by the lower curve in Figure 5.15). If the relative slippage exceeds 15 mm, the joint is judged as a failure. The bending may cause an upturn deformation of the bolt ends and the holes of the side plates are pressed upward (shown by the upper curve in Figure 5.15). In this case, if the relative slippage deformation is more than 10 mm, it is considered a failure. All the specimens in the second and third groups had the deformation pattern with the bolt ends being pushed upward, as illustrated in Figure 5.16. During loading, sharp sounds indicating the rupture of bamboo fibers are audible and become inten-sive corresponding to the increase of the load. For specimens in Group 1, due to the restriction of slippage by tightening the bolts, the target slippage of 10 mm could not be reached and the specimens essentially failed through local crushing in compression.

Figure 5.16 Push up of bolt ends due to bending.

Load deformation relationships: The load and slippage deformation relationships for all the specimens are shown in Figure 5.17. As shown in Figure 5.17 (a), two of the three specimens had high initial stiffness during preloading to 0.3 F, indicating the effect of tightening bolts. However, one of them had lower initial stiffness. Such a discrepancy reveals the somewhat random and undependable control of the tightness of the bolts. The specimens failed essentially due to local crushing of one of the side boards in compression, so the load was terminated without reaching the target failure deformation of 10 mm. One of the specimens in Group 2 failed prematurely at a load level of 100 kN, due to the local crushing of one of the side glubam plates. The other two specimens behaved well and the loading was terminated when the deformation exceeded the target failure deformation at 10 mm. Specimens in Group 3 behaved reasonably consistently, reaching the target failure deformation 10 mm, as shown in Figure 5.17 (c). It should be clarified that the failure of some specimens due to compression crushing of the side plates implies the unsuccessful testing of the joint, rather than damage to the bolted joints. Therefore, in the discussions of the load carrying capacities and yielding mechanism of the bolted connection, the specimens with premature compression failure should not be used.

5.3.3 Failure Analysis of Bolted Glubam Joints

Based on the 5% offset yield strength definition shown in Figure 5.9, the yield strength of specimens is calculated. The average yield capacity of the two specimens with slippage deformation reaching 10 mm in Group 2 is calculated as 134.4 kN; whereas, the average yield capacity of the three specimens in Group 3 is 125.0 kN. Apparently, according to Feng, the results are higher than the capacity F, estimated based on shear failure according to the Chinese code.[13] Therefore, other failure mechanisms need to be considered.

The yield capacities of dowel-type bolted joints can be analyzed based on the mechanisms originally proposed by Johansen.[7] As shown in Figure 5.18, for the double shear joint, four failure mechanisms are considered.

Mode-I: Failure is caused by the damage to the center main board, while the side boards and the bolt remain intact.

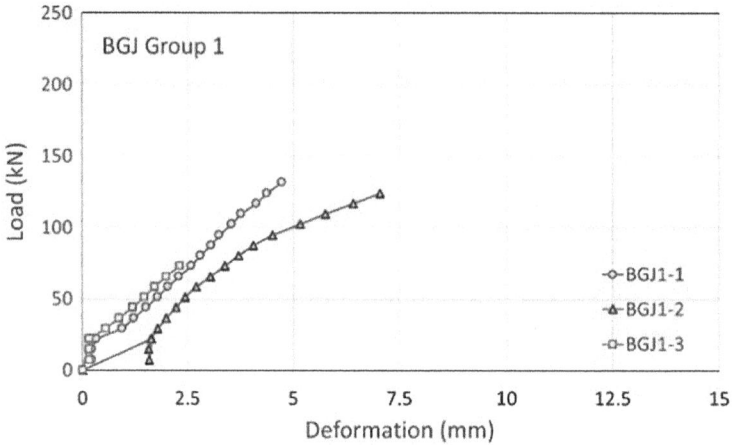

(a)

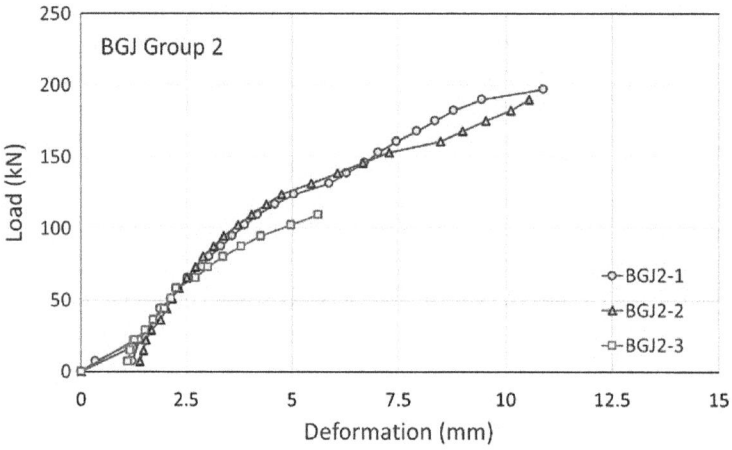

(b)

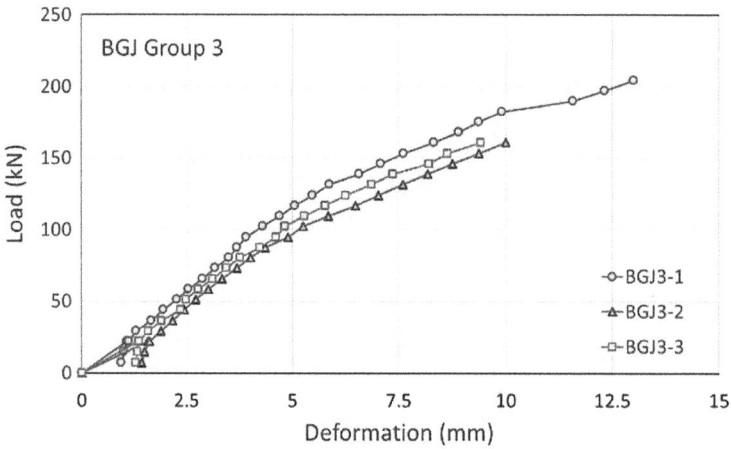

(c)

Figure 5.17 Load and slippage deformation relationships of specimens in: (a) Group 1; (b) Group 2; and (c) Group 3.

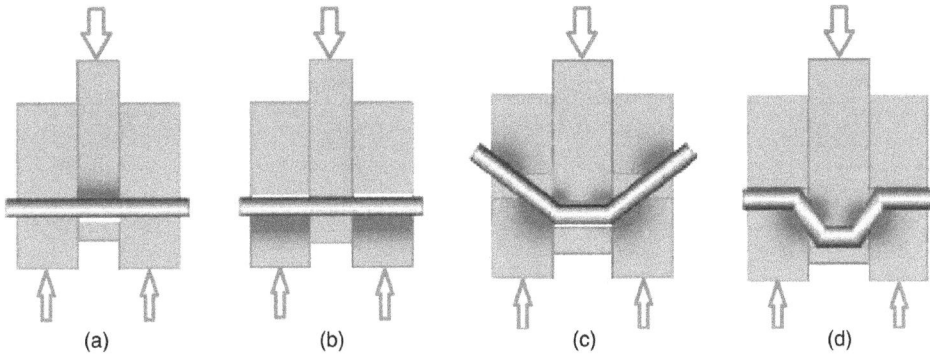

Figure 5.18 Johansen's failure mechanisms: (a) mode-I; (b) mode-II; (c) mode-III; (d) mode-IV.

Mode-II: Failure is caused by the damage to the side boards, while the center main board and bolt remain intact.

Mode-III: Failure occurs in the wood boards with the yielding in the middle portion of the bolt.

Mode-IV: Bearing failure in the wood boards with the yielding of the bolt in three locations.

The yield capacities of the connection based on the four failure modes have been established in design codes, based essentially on Johansen's theory,[7] and its refinements,[27,28] however, with some modifications. In this book, the equations given in the U.S. NDS, Table 12.3.1A and B[24] are reproduced and listed as Table 5.4 (a) and Table 5.4 (b).

Based on the overreaction of the specimens in Group-2 and Group-3, the mode-IV failure patterns are assumed for these specimens, therefore the capacities can be calculated as follows, neglecting the reduction factor R_d,

$$Z_4 = 2D^2 \sqrt{\frac{2F_{em}F_{yb}}{3(1+R_e)}} = 2 \times 12^2 \times \sqrt{\frac{2 \times 59 \times 235}{3\left(1+\dfrac{50}{235}\right)}} = 24.76 \text{kN}$$

Since the bolt group satisfies the requirements for all the spacing, the capacity can be calculated as 6 × 24.76 = 148.6 kN. The result is reasonably close to the 5% offset yield capacities of the specimens in Group-2 and Group-3.

Zhang et al.[29] carried out compressive loading tests on similar engineered bamboo joints, but with 8 bolts configured in a staggered fashion. They also observed the mode-IV failure patterns for the specimens and concluded that the NDS specifications could obtain the design-bearing capacity of bolted laminated bamboo joints more accurately.

5.4 Tensile Behavior of Bolted Glubam Joints

Bolted joints subjected to tension may behave differently from those in compression. Different failure modes, such as the rupture of the base material, need to be addressed. Yang et al. experimentally investigated the tensile behavior of single bolt pin connections for glubam.[22]

Table 5.4a Yield limit equations in NDS Table 12.3.1A

Yield mode	Single shear		Double shear	
I_m	$$Z = \dfrac{D\,\ell_m F_{em}}{R_d}$$	(12.3-1)	$$Z = \dfrac{D\,\ell_m F_{em}}{R_d}$$	(12.3-7)
I_s	$$Z = \dfrac{D\,\ell_s F_{es}}{R_d}$$	(12.3-2)	$$Z = \dfrac{2D\,\ell_s F_{es}}{R_d}$$	(12.3-8)
II	$$Z = \dfrac{k_1 D\,\ell_s F_{es}}{R_d}$$	(12.3-3)		
III_m	$$Z = \dfrac{k_2 D\,\ell_m F_{em}}{(1+2R_e)R_d}$$	(12.3-4)		
III_s	$$Z = \dfrac{k_3 D\,\ell_s F_{em}}{(2+R_e)R_d}$$	(12.3-5)	$$Z = \dfrac{2k_3 D\,\ell_s F_{em}}{(2+R_e)R_d}$$	(12.3-9)
IV	$$Z = \dfrac{D_2}{R_d}\sqrt{\dfrac{2F_{em}\,F_{yb}}{3(1+R_e)}}$$	(12.3-6)	$$Z = \dfrac{2D_2}{R_d}\sqrt{\dfrac{2F_{em}\,F_{yb}}{3(1+R_e)}}$$	(12.3-10)

Notes:

$$k_1 = \frac{\sqrt{R_e + 2R_e^2(1+R_t+R_t^2)+R_t^2 R_e^3 - R_e(1+R_t)}}{(1+R_e)}$$

$$k_2 = -1+\sqrt{2(1+R_e)+\frac{2F_{yb}(1+2R_e)D^2}{3F_{em}\,\ell_m^{\,2}}}$$

$$k_3 = -1+\sqrt{\frac{2(1+R_e)}{R_e}+\frac{2F_{yb}(2+R_e)D^2}{3F_{em}\,\ell_s^{\,2}}}$$

D = diameter, inches (in.). (see NDS Table 12.3.7)
F_{yb} = dowel bending yield strength, psi
R_d = reduction term (see NDS Table 12.3.1B)
$R_e = F_{em}/F_{es}$
$R_t = \ell_m/\ell_s$
ℓ_m = main member dowel bearing length, in.
ℓ_s = side member dowel bearing length, in.
F_{em} = main member dowel bearing strength, psi (see NDS Table 12.3.3)

Source: American Wood Council (2018).[24]

5.4.1 Tensile Test Programs

In order to simplify the research problems, the research focused on the single bolt connections of glubam, and selected steel as the side plate whereas glubam as the main board to ensure that the damage occurred in the main board. As shown in Figure 5.19, the experimental parameters included the edge and end distances (*b* and *e*) measured from the center of the bolt hole.

The thin-strip bidirectional glubam board measured with a nominal thickness of 28 mm was adopted as the main board in the single pin joint. The ratio of the bamboo fibers in the directions of longitudinal grain orientation and transverse orientation is 4:1. As shown in Figure 5.20, the testing side of the glubam element is the single pin joint with a bolt of 12 mm diameter, whereas the other end is connected to three 14 mm diameter bolts for fixing the specimen to the testing machine. The holes in the glubam elements are about 1 mm larger than the bolt diameter.

Table 5.4b Reduction term R_d in NDS Table 12.3.1B

Fastener size	Yield mode	Reduction term, R_d
$0.25 \leq D \leq 1$"	I_m, I_s	$4 K_\theta$
	II	$3.6 K_\theta$
	III_m, III_s, IV	3.2" K_θ
$D < 0.25$"	$I_m, I_s, II, III_m, III_s, IV$	K_D[1]

Notes:
$K_\theta = 1+0.25(\theta/90)$
θ = maximum angle between the direction of load and the direction of grain $(0° \leq \theta \leq 90°)$ for any member in a connection
D = diameter, in. (see NDS Table 12.3.7)
$K_D = 2.2$ for $D \leq 0.17$"
$K_D = 10D + 0.5$ for 0.17" $< D < 0.25$"

1 For threaded fasteners where nominal diameter (see Appendix L) is greater than or equal to 0.25" and root diameter is "less than 0.25", $R_d = K_D K_\theta$.

Source: American Wood Council (2018).[24]

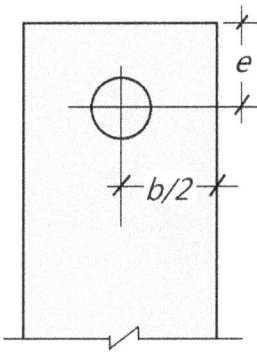

Figure 5.19 Single bolt pinned connection.

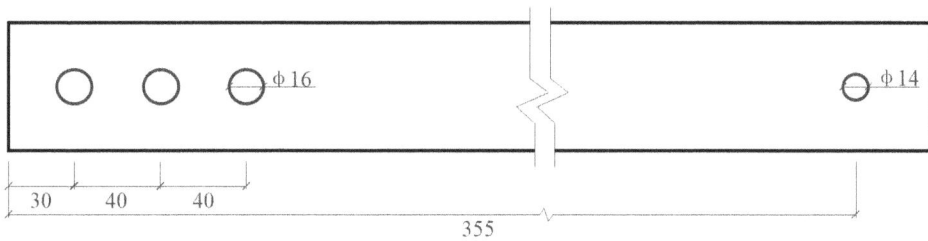

Figure 5.20 Specimen size.

Table 5.5 Tensile testing matrix for pin joints

Longitudinal loading				Transverse loading			
Specimen name	b/2 (mm)	e (mm)	Number	Specimen name	b/2 (mm)	e (mm)	Number
LAa-1–5	12	24	5	TAa-1–5	12	24	5
LAb-1–5	12	36	5	TAb-1–5	12	36	5
LAc-1–5	12	48	5	TAc-1–5	12	48	5
LBa-1–5	18	24	5	TBa-1–5	18	24	5
LBb-1–5	18	36	5	TBb-1–5	18	36	5
LBc-1–5	18	48	5	TBc-1–5	18	48	5
LCa-1–5	24	24	5	TCa-1–5	24	24	5
LCb-1–5	24	36	5	TCb-1–5	24	36	5
LCc-1–5*	24	48	5	TCc-1–5*	24	48	5

Note: *Specimens in LCc and TCc groups are also used for embedment strength evaluation, described in section 5.3.4.

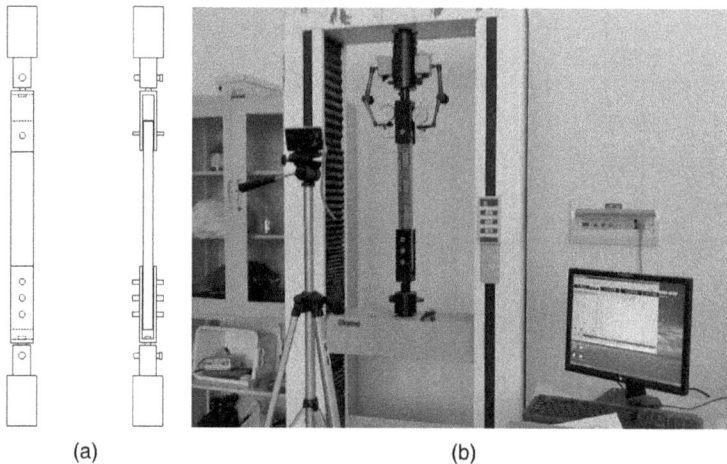

(a)   (b)

Figure 5.21 Test setup: (a) specimen fixtures; (b) test setup.

In Yang et al.'s experiment, the specimens were designed for longitudinal tensile loading and transverse tensile loading. For each loading direction, the specimens were further divided into 9 groups according to different end and edge spacing. The number of specimens in each group is 5, with a total of 90 specimens. Table 5.5 shows the size and grouping of all specimens.

The tests were conducted in accordance with ASTM D5652.[30] The loading equipment was a universal test machine with a maximum loading capacity of 200 kN. In the test, displacement control was adopted to apply the load, and the loading speed was 3mm/min., so as to ensure that the time from loading to failure was no less than 5 minutes and no more than 20 minutes. Figure 5.21 (a) shows the specially designed specimen fixture. The upper end is the tensile testing end and the lower end is the fixed end. Figure 5.21 (b) shows the basic situation during the test.

5.4.2 Failure Modes and Load-displacement Relationships

In the tests, the failure of the specimens is strongly influenced by the loading directions. During the tests with longitudinal loading in the direction of the 80% bamboo strips (fibers), the specimens typically failed in the plug shear mode from the pin to the end of the specimen, as shown in Figure 5.22 (a). When the edge distance increases, the plug shear failure tends to occur after the significant enlargement of the hole along the loading direction due to embedment deformation. Similar trends can also be seen in the specimens with an edge of 24 mm and an end distance of 48 mm under transverse loading, in which failure occurred after some enlargement deformation. However, most specimens subjected to transverse loading in the direction along the 20% bamboo fiber arrangement appear to have a complex failure mode, with rupture initiated typically from the net section on one side of the bolt hole, as shown in Figure 5.22 (b). In both cases, the 12 mm diameter bolts did not develop sufficient yield deformation; therefore, the specimens are deemed to have failed in terms of the base material.

The applied load and deformation (pin movement) relationships for the specimens under longitudinal and transverse loading are summarized in Figure 5.23 and Figure 5.24, respectively. For each testing group, only the curve of the specimen that approximately represents the average curve of the group is shown.

As shown in Figure 5.23, for the same edge distance $b/2$, increasing the end distance e results in a significant improvement of the load and deformation behavior with increased load carrying capacity and deformability. By comparing Figure 5.23 (a), (b), and (c), one can notice that the curves with the same end distance e but different edge distance $b/2$ are essentially similar. This provides the evidence that the end distance controls the failure pattern for specimens subjected to longitudinal loading. Both the U.S. NDS specification[24] and the Chinese GB code[19] require an end distance ($7d$ for soft wood and $5d$ for hardwood) for dowel type bolted joints. Based on the test results for glubam, a $4d$ (48 mm) end distance is enough to effectively delay the plug shear failure. For the edge distance, a $1.0d$ ($b/2 = 12$ mm) is sufficient to prevent rupture of the net area across the main bamboo fiber direction.

| (a) | (b) |

Figure 5.22 Typical failure modes: (a) longitudinal loading; (b) transverse loading.

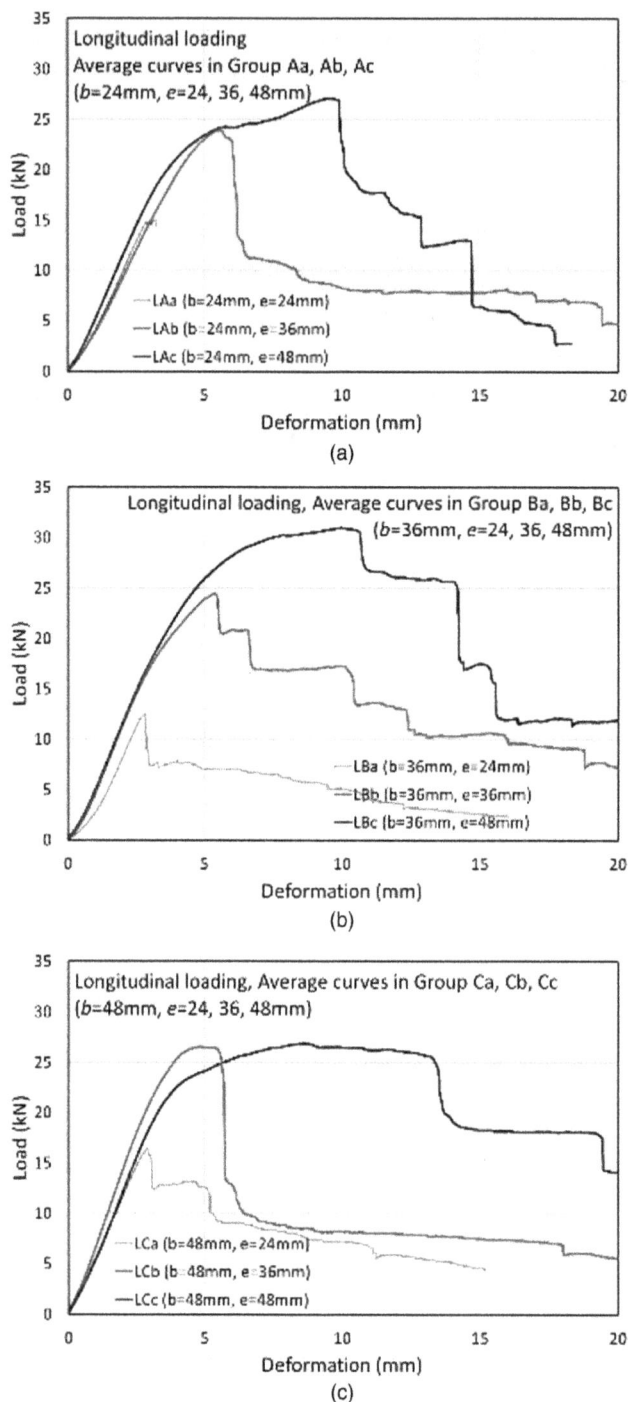

Figure 5.23 Tensile load versus deformation relationships for longitudinal loading tests: (a) specimens with $b/2 = 12$ mm and different end distances; (b) specimens with $b/2 = 18$ mm and different end distances; (c) $b/2 = 24$ mm and different end distances.

As shown in Figure 5.24 (a), for an edge distance of $b/2 = d$ (12 mm), increase of end distance e does not result in any change in the behavior of the bolted joint subjected to loading in the transverse direction, indicating that the failure is controlled by the rupture of the net area perpendicular to the tensile force. By comparing the broken curves in Figure 5.24 (a), (b), and (c), it can be seen that the increase in the edge distance does not result in distinct strength enhancement for specimens with the smallest end distance $e = 2d$. This is considered to be due to the small end distance: The area between the hole and the end edge is more flexible, therefore the rupture is localized near the hole. When the end distance is larger, the area between the hole and the end edge is stiffer; as a result, more materials can be engaged in resisting the rupture in the net area. This is evidenced in Figure 5.24 (b) and (c), in which an increase in the end distance is shown to be effective in enhancing the tensile capacity. Echavarría et al.[31] developed an analytical approach to explain the failure modes in bolted timber joints. Their research findings indicate that the shorter end distance may result in more intensive stress concentration around the bolt hole, increasing the tendency to failure, particularly when e is less than $4d$.

5.4.3 Discussions on Strength of Bolted Connections

The 5% yield capacity and the maximum load are obtained for the specimens tested by Yang and Xiao[25] and shown in Table 5.6. The following calculated capacities are also shown in Table 5.6.

Tensile capacity of net section, T_{nt}:

$$T_{nt} = (b-d)tf_t \tag{5.5}$$

Plug-shear capacity, T_{ps}:

$$T_{ps} = 2etf_v \tag{5.6}$$

where f_t, and f_v are the tensile strength along the direction of loading, and shear strength in the section of shear parallel to the loading direction, respectively.

As discussed previously for the longitudinal loading, test results indicate that the edge distance $b/2 = d = 12$ mm is sufficient to provide enough tensile strength for the net section to prevent rupture. This is reflected by the much larger calculated net tensile capacities compared with the experimentally obtained yield force and maximum load-carrying capacities of the specimens in the longitudinal loading in the direction along the main bamboo fiber direction. On the other hand, the calculated plug shear capacities yield conservative predictions in relation to the yield capacities of the specimens subjected to longitudinal loading.

The net section rupture capacities calculated based on Eq.5.5 can predict conservatively the capacities for most of the specimens subjected to the transverse tension in the

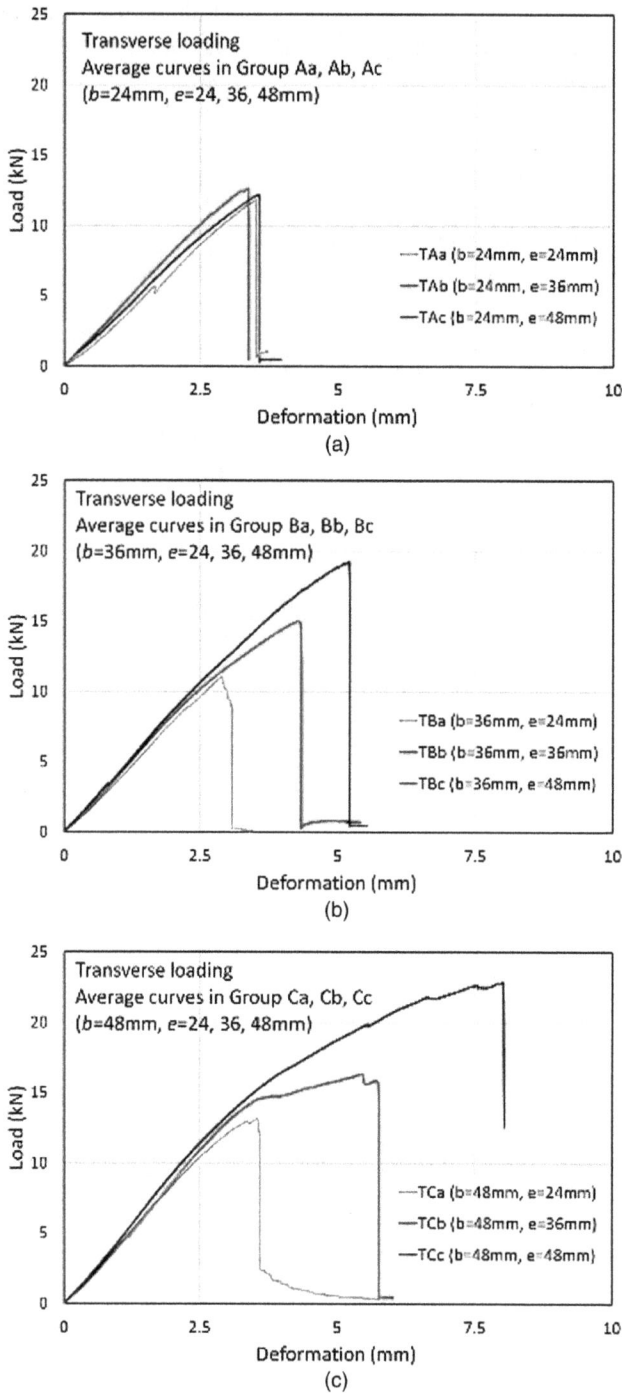

Figure 5.24 Tensile load versus deformation relationships for transverse loading tests: (a) specimens with $b/2 = 12$ mm and different end distances; (b) specimens with $b/2 = 18$ mm and different end distances; (c) $b/2 = 24$ mm and different end distances.

Table 5.6 Tensile test capacities of bolted joints

	b/2 (mm)	e (mm)	Yield load (kN)	Maximum load (kN)	Net tension capacity (kN)	Plug shear capacity (kN)
LAa	12	24	13.5	15	27.6	9.9
LAb	12	36	21	20.2	27.6	14.9
LAc	12	48	23.8	26.3	27.6	19.9
LBa	18	24	12	13.9	55.1	9.9
LBb	18	36	22.8	24.1	55.1	14.9
LBc	18	48	26.5	28.9	55.1	19.9
LCa	24	24	16.3	17.5	82.7	9.9
LCb	24	36	25.1	25.9	82.7	14.9
LCc	24	48	23.2	26.9	82.7	19.9
TAa	12	24	10.6	12.3	5.7	9.9
TAb	12	36	11.6	13.1	5.7	14.9
TAc	12	48	10.8	12.3	5.7	19.9
TBa	18	24	11.1	12.1	11.4	9.9
TBb	18	36	14.3	15.1	11.4	14.9
TBc	18	48	17.8	18.6	11.4	19.9
TCa	24	24	11.8	12.6	17.0	9.9
TCb	24	36	14.5	16.3	17.0	14.9
TCc	24	48	17.5	20.6	17.0	19.9

less bamboo fiber direction, except for specimens with $b/2 = 2d = 24$ mm with an end distance of $2d$ or $3d$. For these two cases, the plug shear equation Eq.5.6 provides a closer but slightly lower prediction of the capacities. Therefore, as a result, the following minimum equation can provide the conservative predictions as to the test results, though not necessarily very accurately.

$$T = \min[(b-d)tf_t, 2etf_v]$$ (5.7)

It should be pointed out that cross-grain loading (transverse direction to the wood fibers) is not permitted for timber in structural elements, as shown in Figure 5.19. For glubam used in structural elements, the typical bamboo fiber ratio is 4:1 in the longitudinal and transverse directions. Therefore, a pinned connection with loading in the transverse direction can resist a certain level of tension force. However, except for special circumstances, pinned glubam connection in transverse loading is not recommended.

5.5 Design Considerations of Bolted Glubam Connections

Structural wood and bamboo joint design is complex, as many possible failure modes are involved; however, the preference is to rely on the resisting mechanism that can provide a certain degree of deformability, such as bolt yielding. Typically, the geometry of the connection is selected based on the design of connecting members, available bolt choices, and various design constraints. Then various capacities based on different failure modes, such as bolt yielding, embedment failure, net section rupture, and plug

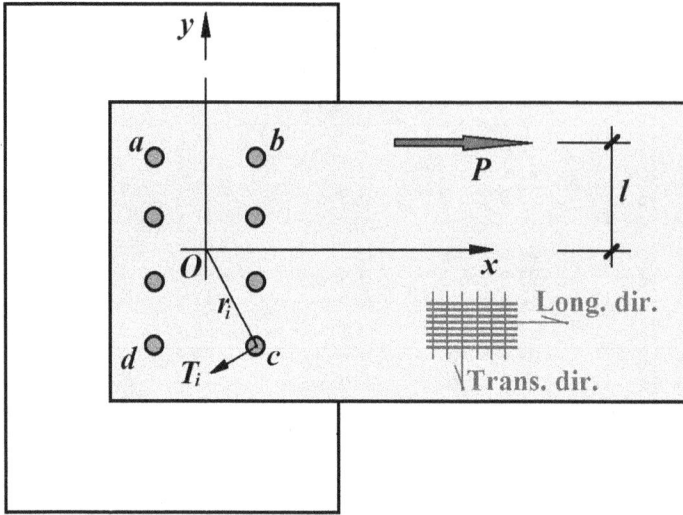

Figure 5.25 Example of an eccentric connection.

shear, etc., can be calculated and compared to find the smallest value for the design capacity.

Taking the eccentric connection shown in Figure 5.25 as an example, the design considerations for bolted glubam connection are shown below.

First the applied force P is transferred to the centroid of the bolts group, and replaced by a torsional moment of Pl and a concentric force P_i. Then the concentric force P_i and the torsional force T_i applied on each bolt is calculated as follows:

$$P_i = P/n \tag{5.8a}$$

$$Ti = A_b \frac{Plr_i}{\sum A_b r_i^2} = \frac{Plr_i}{\sum r_i^2} \tag{5.8b}$$

where n is the number of bolts in the bolted connection; A_b is the cross-sectional area of one bolt; r_i is the centroidal distance of bolt i.

The critical bolt is identified as one of the corner bolts that has the maximum vector resulting from the concentric force and the torsional force; the bolt is at the corner close to the edge and end of the element. For the case shown in Figure 5.25, bolt a is the most critical one. Then, various strengths, such as bolt shear, bolt yielding, embedment, plug shear, and shear rupture should be checked. For the strength of the bolt, the calculation should be carried out using the resultant force. However, for the base material glubam, the strength analyses may be conducted using the projected components of the resultant force in each coordinate axis. Some assumptions for the rupture areas may be needed and should be made conservatively.

5.6 Toothed Metal Mending Plate Connected Glubam Joints

Lightweight metal mending plate connected wood trusses or joists are popular in timber structure construction.[32-37] The author's research group has carried out a program intended to combine the advantages of metal plate connections with bamboo-based glubam structures.[38-41] This section summarizes the main findings from the extensive testing studies on the behaviors of toothed metal mending plate glubam connections with various loading conditions in tension and in shear.

5.6.1 Materials and Tensile Strength of the Toothed Metal Mending Plate

The toothed metal mending plate used for glubam joints is shown in Figure 5.26 (a) for its front and back (teeth-side) faces. Definitions of the coordinate systems and loading directions for the metal plate and the glubam base material are shown in Figure 5.26 (b) and (c), respectively. Thin-strip glubam with a bamboo fiber ratio of 4:1 in longitudinal to transverse directions is used. The metal plate thickness is 1.04 mm with the depth and width of the teeth being 8.5 mm and 3.2 mm, respectively. The density of the teeth is 1.3/cm^2. Based on the testing method for metal plates with a thickness of 0.1–3 mm,[42] tensile stress-strain curves for three samples are obtained and shown in Figure 5.27. The average yield strength and the ultimate tensile strength are 560 MPa and 608 MPa, respectively.

Following JGJT 265 – "Technical code for light wood trusses",[43] the tensile strength of the toothed metal plate was first obtained, using the specimens for the loading in parallel ($\beta = 0°$) and perpendicular ($\beta = 90°$) to the metal tooth direction, as shown in

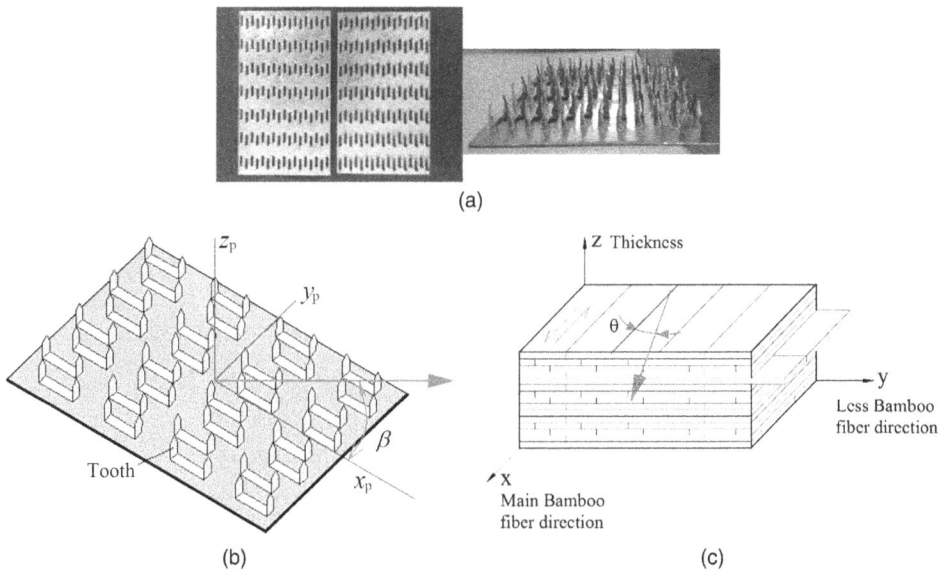

Figure 5.26 Toothed metal mending plate: (a) front and back view; (b) coordinate system for metal plate; (c) coordinate system for glubam.

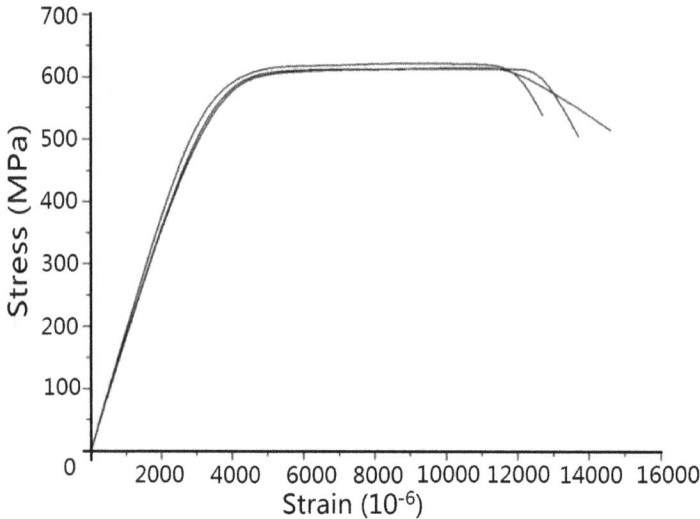

Figure 5.27 Stress-strain curves for metal plate samples.

Figure 5.28 (a) and (b), respectively. The tensile tests were conducted on a universal testing machine. The upper and lower glubam elements were attached to the loading jigs, as shown in Figure 5.29. The loading was controlled by the loading head displacement control with 1 mm/min., so the specimen can be tested to failure in about 10 min. For each direction, three identical specimens were prepared and tested.

Table 5.7 summarizes the testing parameters and results. Based on the codes,[19,43] the experimental tensile strength needs to be modified by the ratio of the nominal tensile strength of the steel (in this case, 490 MPa for the Q345 steel) and the material test strength (608 MPa of the average tensile strength in Figure 5.27). The design value is the average of the two smaller modified tensile strength divided by 1.75,[19,43] and is also computed in Table 5.7.

5.6.2 Tensile Behaviors of Toothed Metal Mending Plated Glubam Joints

In order to study the anchorage strength of the metal teeth for glubam connection, a large number of tensile tests were carried out by Wu,[38,39] using the test setup shown in Figure 5.29. The loading direction related to the metal plate (angle β) and the glubam (angle θ) is taken as the main testing parameter. As shown in Table 5.8, altogether, 10 groups of specimens with combinations of the angles β and θ are tested. For each testing group, 10 identical specimens were prepared and tested.

The specimens are prepared following Chinese codes GB 50005 – "Code for design of timber structures"[19] and JGJ/T 265 – "Technical code for light wood trusses".[43] The design details of the specimen are shown in Figure 5.30 (a). The teeth adjacent to the connection seam need to be removed, and the remaining area is considered as the effective metal plate area, as illustrated by the shaded section in Figure 5.30 (b). The edge distance b is taken as 6 mm, and the end distance e is set as 12 mm. The teeth of the metal plate were compressed evenly into the glubam elements using a 100

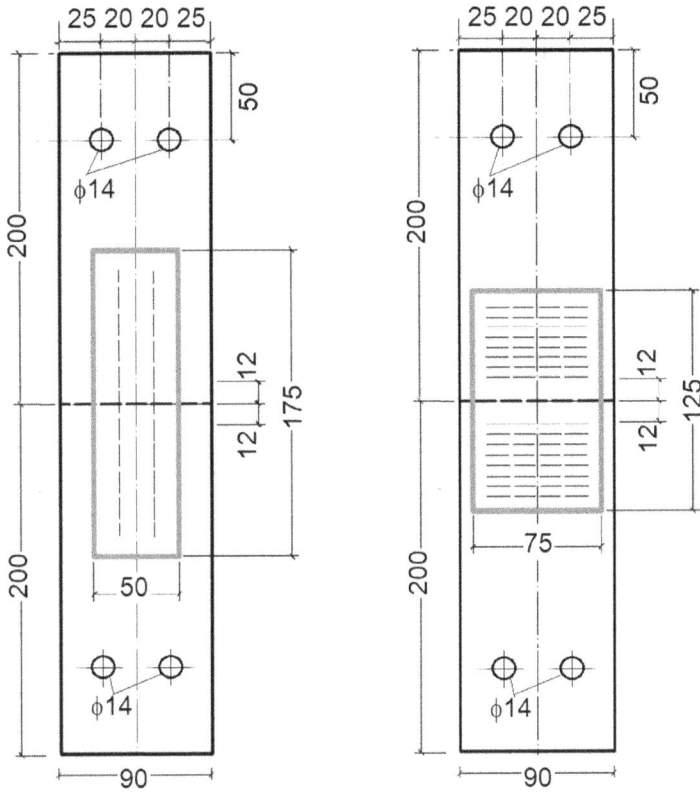

Figure 5.28 Specimens for tensile strength of metal plates: (a) $\beta = 90°$; (b) $\beta = 90°$.

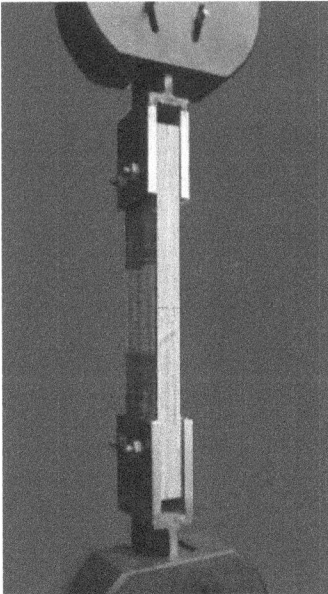

Figure 5.29 Test setup.

Table 5.7 Tensile strength tests of metal plates

Specimen	Width x length (mm²)	Ultimate tensile load (kN)	Width perpendicular to load (mm)	Tensile strength (N/mm)	Modified tensile strength (N/mm)	Design tensile strength (N/mm)
MPT-0-0-1	50×175	10.64	50	221.8	178.5	
MPT-0-0-2	50×175	14.52	50	290.4	233.8	117.8
MPT-0-03	50×175	18.63	50	372.6	299.9	
MPT-0-90-1	125×75	28.25	75	376.7	303.2	
MPT-0-90-2	125×75	24.44	75	325.9	262.3	154.4
MPT-0-90-3	125×75	25.92	75	345.6	278.2	

Note: specimen name designation example: MPT-0-90-1, Metal Plate Tension in angular combination of $\theta = 0°$ and $\beta = 90°$, specimen Number 1.

Source: Echavarría et al. (2007).[31]

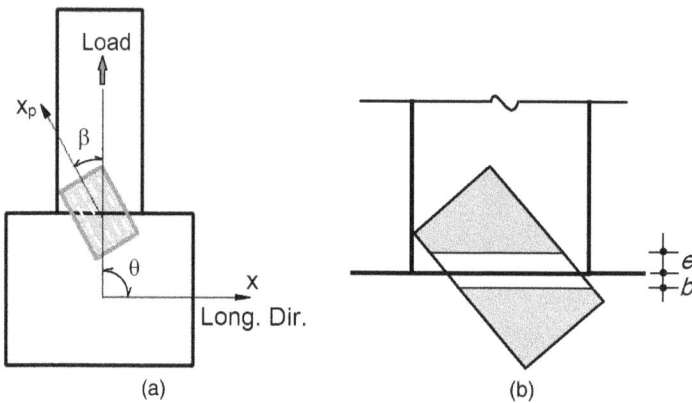

Figure 5.30 Specimen details: (a) configuration; (b) effective area.

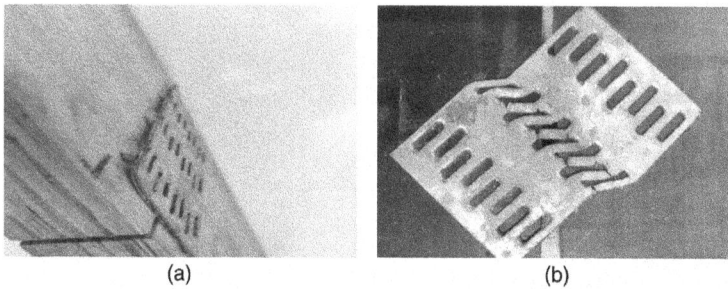

Figure 5.31 Failure patterns: (a) pull out of metal teeth from glubam; (b) rupture of plate.

Table 5.8 Testing matrix for toothed metal plate glubam connections

Specimen group	θ	β	Number of specimens	Teeth plate width x length (mm x mm)[1]	Glubam width x length (mm x mm)[2]
GBJ-0-0	0°	0°	10	75×100	90×200
GBJ-0-30	0°	30°	10	50×75	90×200
GBJ-0-45	0°	45°	10	50×75	90×200
GBJ-0-60	0°	60°	10	50×75	90×200
GBJ-0-90	0°	90°	10	50×75	90×200
GBJ-90-0[3]	90°	0°	10	75×100	90×200, 150×200
GBJ-90-30[3]	90°	30°	10	50×75	90×200, 150×200
GBJ-90-45[3]	90°	45°	10	50×75	90×200, 150×200
GBJ-90-60[3]	90°	60°	10	50×75	90×200, 150×200
GBJ-90-90[3]	90°	90°	10	50×75	90×200, 150×200

1 Thickness of metal plate is 1.04 mm.
2 Nominal thickness of glubam is 30 mm.
3 For specimens with $\theta = 90°$, the upper and lower glubam elements have different dimensions.

kN pneumatic compression machine. The water absorption ratio of the glubam was about 16.5%.

Most specimens failed due to the pull out of the toothed plate, as shown in Figure 5.31 (a); however, some specimens with loading in a twisted direction ($0°<\beta<90°$) had metal plate rupture, as shown in Figure 5.31 (b). The examples of load and displacement curves for the orthogonal loading conditions are shown in Figure 5.32. The load carrying capacity is seen to be most influenced by the orientation of the metal plate in relation to the loading direction. The direction with the largest capacity is loading in alignment with the tooth slot direction (x_p direction), as shown in Figure 5.32 (a) and (c); whereas the lowest capacity is in loading perpendicular to the tooth slot direction, as shown in Figure 5.32 (b) and (d).

The ultimate strength is defined as the tensile loading capacity divided by the effective area of the toothed plate, which is equal to the area of one of the shaded portions shown in Figure 5.30 (b). The tensile strength thus obtained for each group is provided in Table 5.7, along with the characteristic values using Eq.4.1. It is possible to derive the design strength based on the approaches described in Chapter 4. However, the current code approaches are introduced here. Based on the design codes,[14,26] the design strength for the metal teeth anchorage is determined using the average of the lowest 3 test results of the 10 specimens divided by an ultimate strength adjustment factor, k. In Appendix A of JGJ/T 265,[43] k is taken as 1.89, while in Appendix M of GB 50005,[19] k is determined based on flame retardant treatment and water content. Following GB 50005,[19] considering that the glubam is not flame retardant and has a water content of 16.5%, k should be 3.61. The discrepancy between these two specifications is apparently too big; therefore the value $k = 2.37$, recommended by Pan and Zhu[44] is followed for determining the design strength shown in Table 5.9.

The experimental average strength, characteristic strength, and design strength are depicted in Figure 5.33 (a) and (b), for longitudinal and transverse loading, respectively. As shown in Figure 5.33, there is a general trend of strength increase following the increase of the angle between the load and the tooth slot direction, which is opposite with the load as discussed previously. This reflects the difference in the metal plate

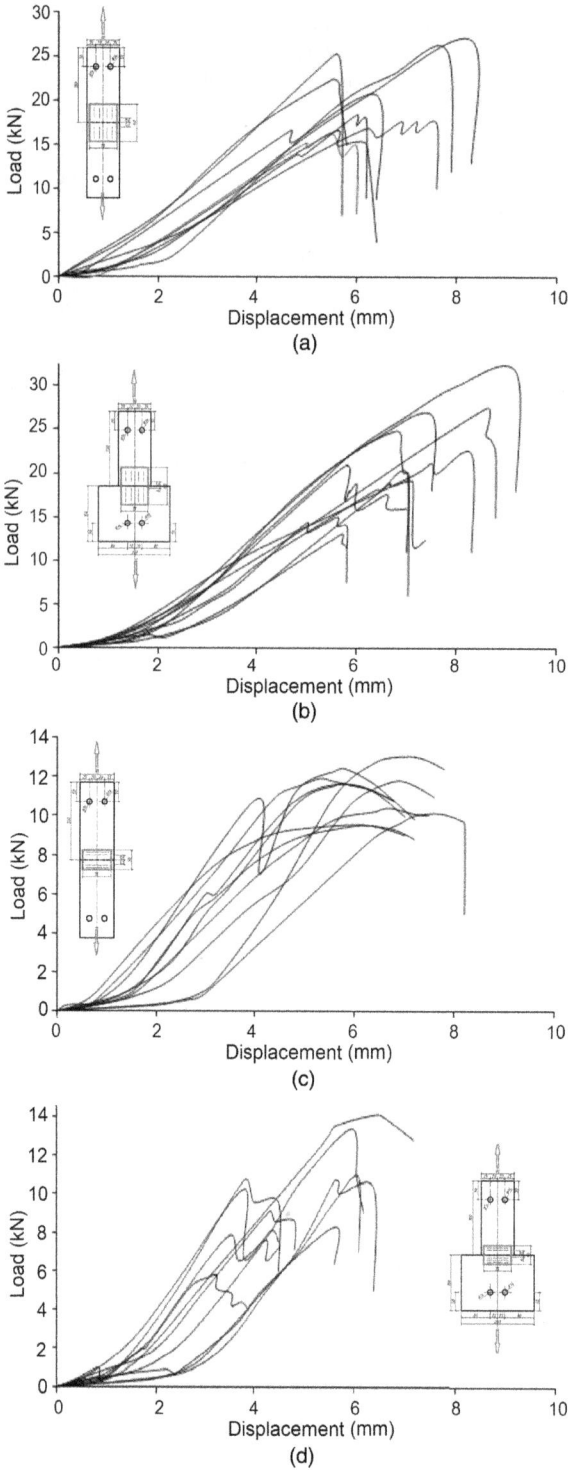

Figure 5.32 Load – displacement curves: (a) $\theta = 0°$, $\beta = 0°$; (b) $\theta = 0°$, $\beta = 90°$; (c) $\theta = 90°$, $\beta = 0°$; (d) $\beta = 90°$, $\theta = 90°$.

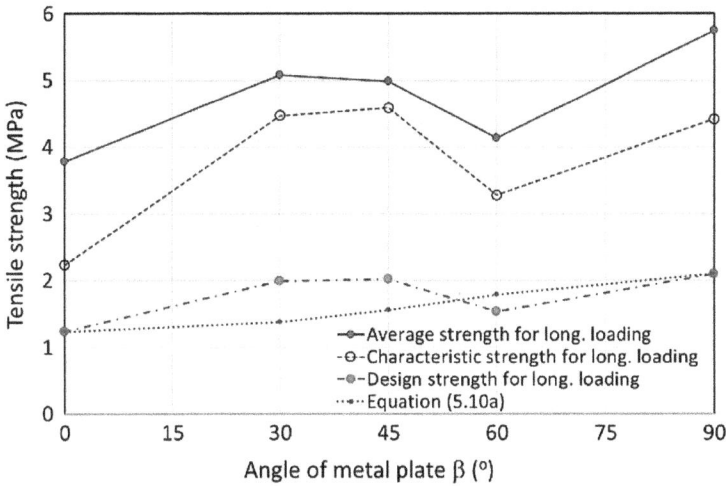

(a)

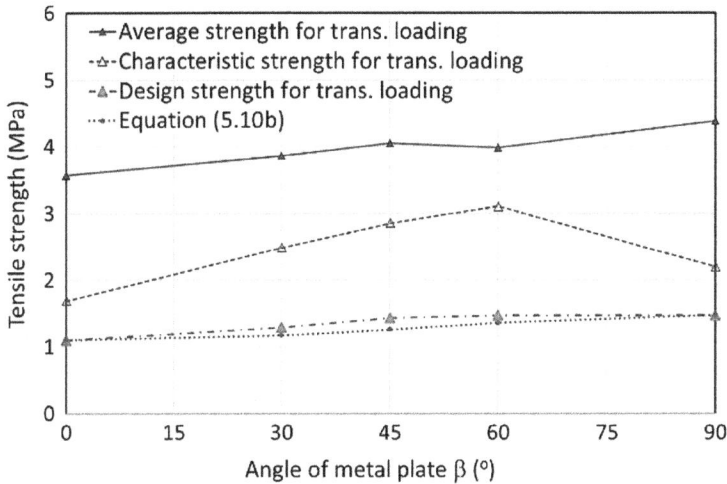

(b)

Figure 5.33 Tensile strength of toothed metal plate connected glubam joints: (a) in longitudinal loading ($\theta = 0°$); (b) in transverse loading ($\theta = 90°$).

effective areas. Since the study is still limited and the results vary quite significantly, it is suggested that the approximate average design values of 1.78 MPa be used for the loading along the longitudinal direction of the glubam; and 1.20 MPa in the transverse direction, respectively, regardless of the angle variation related to the metal plate.

For loading along a direction of an angle θ with the grain direction (longitudinal direction for glubam), the GB 50005 code[19] provides the following equations for calculating the pull-out tensile strength of the joint.

Table 5.9 Experimental results and design values for metal teeth anchorage strength

θ, β (°)	Average strength (MPa)	Standard deviation (MPa)	Characteristic strength (MPa)[1]	Design strength (MPa)	Failure mode and percentage among specimens
0, 0	3.780	0.736	2.23	1.243	Pull out of teeth 100%
0, 30	5.083	0.292	4.47	1.994	Pull out of teeth 70%, rupture 30%
0, 45	4.984	0.189	4.59	2.020	Pull out of teeth 70%, rupture 30%
0, 60	4.134	0.407	3.28	1.540	Pull out of teeth 90%, rupture 10%
0, 90	5.748	0.632	4.42	2.101	Pull out of teeth 100%
90, 0	3.576	0.896	1.69	1.103	Pull out of teeth 100%
90, 30	3.863	0.654	2.49	1.291	Pull out of teeth 70%, rupture 30%
90, 45	4.046	0.565	2.86	1.435	Pull out of teeth 70%, rupture 30%
90, 60	3.980	0.415	3.11	1.470	Pull out of teeth 90%, rupture 10%
90, 90	4.379	1.037	2.20	1.473	Pull out of teeth 100%

1 Characteristic value f_k is determined using Eq.4.1, $f_k = m - kS$, where m is average strength, S is standard deviation, and k is taken as 2.104, for a specimen number of 10.

For loading parallel to main axis of metal plate ($\beta = 0°$):

$$n_r = \frac{n_{r1}\, n_{r2}}{n_{r1} \sin^2 \theta + n_{r2} \cos^2 \theta} \tag{5.9a}$$

For loading perpendicular to main axis of metal plate ($\beta = 90°$):

$$n_r' = \frac{n_{r1}'\, n_{r2}'}{n_{r1}' \sin^2 \theta + n_{r2}' \cos^2 \theta} \tag{5.9b}$$

where, n_{r1}, n_{r2} are the strengths at ($\theta = 0°, \beta = 0°$) and at ($\theta = 90°, \beta = 0°$), respectively; n_{r1}', n_{r2}' are the strengths at ($\theta = 0°, \beta = 90°$) and at ($\theta = 90°, \beta = 90°$), respectively. For loading in direction β ($0°<\beta<90°$), the code[19] requires the linear interpolation using strength n_{r1} ($\theta = 0°, \beta = 0°$), and n_{r1}' ($\theta = 0°, \beta = 90°$), or n_{r2}' ($\theta = 90°, \beta = 0°$), and n_{r2}' ($\theta = 90°, \beta = 90°$). However, Wu[39] showed that the linear interpolation provides significant underestimation to the test results in Table 5.7 for the specimens in loading directions with $0°<\beta<90°$. Here, if the same angular equation approach of Eq.5.9 is attempted, it can be rewritten for the tensile strength in loading direction β:

For loading parallel to longitudinal direction of glubam ($\theta = 0°$):

$$n_r = \frac{n_{r1}\, n_{r1}'}{n_{r1} \sin^2 \beta + n_{r1}' \cos^2 \beta} \tag{5.10a}$$

For loading perpendicular to longitudinal direction of glubam ($\theta = 90°$):

$$n_r' = \frac{n_{r2}\, n_{r2}'}{n_{r2} \sin^2 \beta + n_{r2}' \cos^2 \beta} \tag{5.10b}$$

Figure 5.34 Shear test setup.

Table 5.10 Shear testing matrix for metal plated glubam joints

Angle β	Side plate dimension (mm³)	Center plate dimension (mm³)	Metal plate width x length (mm²)	Shear plane length (mm)	Average capacity (kN)	Average strength (kN/mm)	Design strength (MPa)
0°	30×90×200	30×200×200	150×50	100	26.73	267.3	121.1
90°	30×90×200	30×150×200	25×125	50	13.67	273.4	121.0
30°SC	30×90×200	30×200×200	150×25	57.7	9.33	161.6	73.5
60°SC	30×90×200	30×200×200	150×25	100	12.47	124.7	56.0
120°SC	30×90×200	30×200×200	25×125	57.7	9.91	171.6	78.1
150°SC	30×90×200	30×200×200	25×125	100	19.28	192.8	87.7
30°ST	30×90×200	30×200×200	25×125	100	40.25	402.5	181.1
60°ST	30×90×200	30×200×200	25×125	57.7	29.48	510.6	233.3
120°ST	30×90×200	30×200×200	150×25	100	19.88	198.8	89.6
150°ST	30×90×200	30×200×200	150×25	57.7	17.77	307.8	136.3

Note: SC and ST refer to loading condition of shear and compression, shear and tension, respectively.

where n_{r1}, n_{r2} are the strengths at ($\theta = 0°$, $\beta = 0°$) and at ($\theta = 90°$, $\beta = 0°$), respectively; n_{r1}', n_{r2}' are the strengths at ($\theta = 0°$, $\beta = 90°$) and at ($\theta = 90°$, $\beta = 90°$), respectively. The design strength in different loading directions of β calculated using Eq.5.10 are also presented in Figure 5.33, showing reasonably close similarities with the test results.

5.6.3 Shear Behaviors of Toothed Metal Mending Plated Glubam Joints

Using the same materials, Wu[39] also carried out shear tests on toothed metal plated glubam joints. Figure 5.34 shows the test setup on a universal testing machine with 200 kN capacity. As shown in Table 5.10, altogether, ten groups of specimens with different loading angle configurations related to the main axis of the metal plate (along the direction of tooth slots) were tested, and each testing group had three identical specimens. Figure 5.35 (a) and (b) exhibit the specimens in orthogonal loading directions related to the metal plate main axis. Figure 5.35 (c) and (d) show the shear and compression loading specimens at angle β and $\beta + 90°$, respectively, and β is taken as 30° and 60°;

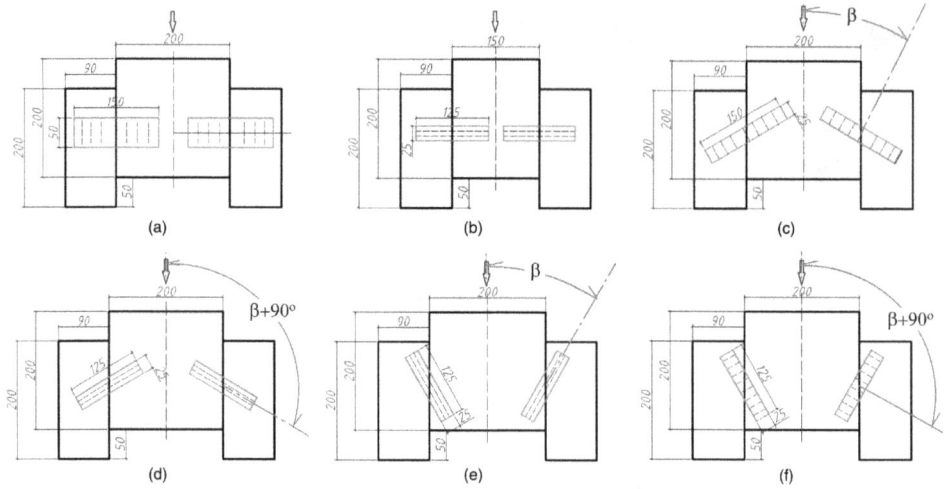

Figure 5.35 Shear test specimens for glubam joints with toothed metal plates: (a) β = 0°; (b) β = 90°; (c) shear and compression at β; (d) shear and compression at β + 90°; (e) shear and tension at β; (f) shear and tension at β + 90°.

Figure 5.36 Typical failure pattern.

thus, four groups were tested ($\beta = 30°$, $\beta = 120°$, $\beta = 60°$, $\beta = 150°$). Figure 5.35 (e) and (f) show the shear and tensile loading specimens at angle β and $\beta + 90°$, respectively. Since β is taken as 30° and 60°, thus, four groups were tested for the shear and tension combination loading ($\beta = 30°$, $\beta = 120°$, $\beta = 60°$, $\beta = 150°$). The specimens with the loading angle β and $\beta + 90°$ are the pairs with the same metal plate size and plate orientation with the loading; however, their tooth slot directions (main axis of metal plate) are 90° apart.

As shown in Figure 5.36, the typical failure of the joints is the shear rupture of the toothed metal plates along one of the shear planes. The main results of the shear tests are shown in Table 5.10, including the average load capacity, the adjusted average strength, and the calculated design shear strength. Based on the Chinese design codes,[19,43] the adjusted shear strength is obtained by multiplying the average experimental shear strength by the ratio of the nominal tensile strength of the steel (in this case, 490 MPa for the Q345 steel) and the material test strength (608 MPa of the average tensile strength in Figure 5.27). The design value is the average of the two smaller modified shear strength values divided by 1.75 among the three specimens within the same testing group.[19,43]

Comparing the counterpart shear and compression specimens with loading angle of β and $\beta + 90°$, their design strengths are generally in the same order. However, for the shear and tension loading, the design strength of the specimen with loading angle β is 1.7 to 2.0 times higher than its counterpart specimen with loading angle $\beta + 90°$. The angle between the plate longitudinal direction and the shear direction is introduced and defined as α, to further explain the plate orientation effects. For specimens with shear and compression loading, the plate main axis (tooth slot direction) is aligned with the plate longitudinal direction when $\alpha = \beta - 180°$; however, it is perpendicular with the plate longitudinal direction when $\alpha = \beta - 90°$. For specimens with a shear and tension loading combination, the plate main axis (tooth slot direction) is aligned with the plate longitudinal direction when $\alpha = \beta$; however, it is perpendicular with the plate longitudinal direction when $\alpha = \beta - 90°$. Figure 5.37 (a) and (b) compare the effects of metal plate orientation on the design strengths, for specimens with shear-compression loading and shear-tension loading, respectively. If the loading is along the orthogonal direction related to the metal plate, the strength is basically the same for aligning the metal plate longitudinal direction either parallel or perpendicular to the slot direction. For loading in all other angles, aligning the two axis results in a larger design shear strength.

The design codes[19,43] specify the linear interpolation for the loading angle β as between 0° and 90°; however, based on the data shown in Table 5.10, this apparently is not a very accurate estimation. As shown in Figure 5.37, it might be simple and reasonable to use the linear interpolation between the strengths corresponding to longitudinal angle $\alpha = 30°$ and $\alpha = 90°$. Alternatively, a more conservative approach can be adopted by using the lower strength values for design.

5.7 Glued-in Rebar Glubam Joints

The bonded-in rod joint is a new type of connection method used in timber structures. Compared with ordinary timber joints, bonded-in rod joints have some advantage such as better performance in strength, stiffness, and fire resistance; meanwhile, the weight is smaller.

Bonding of rebars with wooden members has been studied by a relatively large number of researchers. For example, Davis and Claisse investigated the behavior of glulam using an epoxy-injected doweled connection.[45] Gattesco and Gubana studied the performance of glued-in threaded rod wood joints subjected to axial force and bending moments.[46] All these studies indicate that the stiffness and strength of glued-in rod joints outperform the bolted joints. Using steel as reinforcement materials to repair cracked spruce beams,[47] Alam et al. proved that the stiffness and strength of the fractured spruce beam can be increased by 114% and 255% respectively. A large number of research studies show that factors such as the type of rods, joint geometry, bond length, diameter of rods, adhesive type, adhesive layer thickness, edge distance,

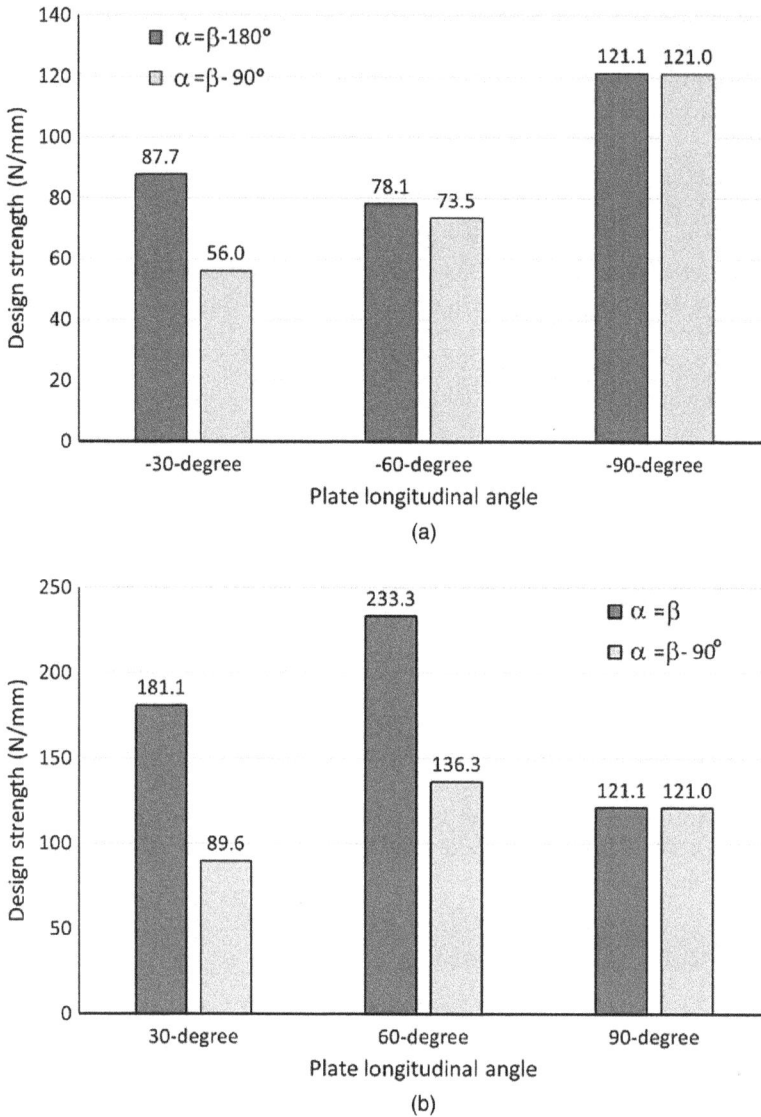

Figure 5.37 Effects of plate orientation on design shear strength: (a) shear and compression loading; (b) shear and tension loading.

and the angle between bar and grains have some influencing effects on the bonded-in rod joints.[48–53]

Due to the similarity in mechanical properties with engineered wood, glubam may be also suitable for the application of connections with glued-in rods. Two experimental programs on glued-in rebar (or rods) glubam joints were carried out by the author's research group.[54,55] The motivation of the studies is for the design of glued-rod joints in

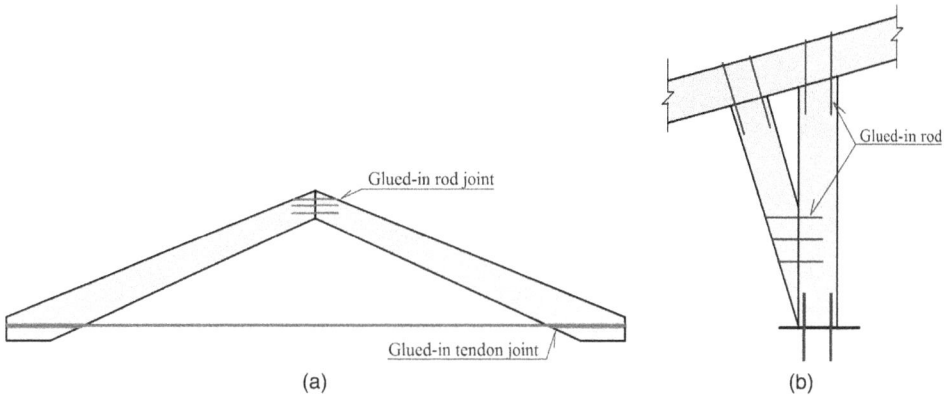

Figure 5.38 Glued-in rod joints: (a) tie rod truss; (b) beam column frame joint.

potential glubam structures, as illustrated in Figure 5.38. Besides the author's research group, another study has been carried out on glued-in steel rods in bamboo scrimber joints.[56] So far, these appear to be the only three research investigations on engineered bamboo with glued-in rebar or rod joints.

5.7.1 Materials and Testing Methods

In Li et al.'s studies, the thin-strip glubam and threaded rods were used for the joints,[55] whereas He and Xiao tested thick-strip glubam with glued-in deformed rebars.[54] Similar to reinforced concrete studies, two types of testing method can be adopted to study the bond and slip behavior of the joints.

As shown in Figure 5.39 (a), He and Xiao's research[54] adopts the pull-out method for testing prismatic thick-strip glubam block specimens with the same cross-sectional size (130x130 mm) and glued-in 10 mm diameter deformed bars (yield strength: 583 MPa; tensile strength: 726 MPa). The length of the specimen was the anchorage length of the bar plus 10 mm in all cases. At the center of each specimen, holes of different diameters were drilled, and one end was filled with epoxy resin to about 3/4 of the volume of the hole. Then, the deformed rebar with 10mm nominal diameter was slowly inserted into the hole to reach its bottom. During the process, the bar was gently rotated to exclude possible air bubbles that may affect the pull-out performance. Each hole was covered with a rubber gasket to ensure that the rods were in the center of the holes. All specimens were placed in a room at 20°C and 65% relative humidity (RH) for at least 7 days after the rod was inserted into the hole.

During testing, as shown in Figure 5.39 (a), the glubam block was fixed to a steel plate by four steel threaded bars at the corners, and the steel plate was welded with a steel bar so as to be clamped by the machine grip. The steel plate on the block was designed to eliminate the effect of compressive forces on the pull-out strength.

As detailed in Table 5.11, the experimental variables in He and Xiao's pull-out tests of rebar glued-in glubam include: the bond length (l_a) which varied between 4 and 16 times the diameter of the rebar; the glue-line thickness (t) of 3 mm to 5 mm; and the

Table 5.11 Testing matrix for glued-in rebar thick-strip glubam joints

Specimen	Glue-line thickness (mm)	Anchorage length (mm)	Angle between bar and bamboo fibers (°)	No. of specimens
S3-40-0	3	40	0	3
S4-40-0	4	40	0	3
S5-40-0	5	40	0	3
S5-40-30	5	40	30	3
S5-40-60	5	40	60	3
S5-40-90	5	40	90	3
S5-40-90⊥	5	40	90	3
S3-80-0	3	80	0	3
S4-80-0	4	80	0	3
S5-80-0	5	80	0	3
S5-80-30	5	80	30	3
S5-80-60	5	80	60	3
S5-80-90	5	80	90	3
S5-120-0	5	120	0	3
S5-120-90	5	120	90	3
S5-160-0	5	160	0	3
S5-160-90	5	160	90	3

Note: Specimen name designation example: S3-40-0, steel bar with 3 mm thick glue-line, 40 mm anchorage length, and an angle of 0° between the bar and bamboo fibers; S5-40-90⊥, steel bar with 5 mm thick glue-line, 40 mm anchorage length, an angle of 90° between the bar and bamboo fibers, and rebar perpendicular to the width direction of bamboo strips.

Source: He & Xiao (2020).[54]

angle between bar and bamboo fibers (θ), with a range of 0°, 30°, 60°, and 90°. The angles of 0° and 90° imply that the bar is bonded parallel to the glubam fibers and perpendicular to the glubam fibers, respectively, as exhibited in Figure 5.40.

As shown in Figure 5.39 (b), the direct tension tests were conducted by Li et al. on 17 groups of glubam joint specimens with glued-in threaded rods in parallel to the longitudinal bamboo strip (fiber) direction.[55] For each group, 6 identical specimens were prepared and tested. As shown in Table 5.12, the experimental parameters included anchorage length, drill-hole diameter, bonding length–diameter ratio, edge distance, and glue-line thickness. Epoxy resin was used to bond the threaded rods in the predrilled holes, which had different lengths for the supporting end and the testing end as shown in Figure 5.41 (a). Special attention was paid to align the threaded rod with the center line of the predrilled holes by using a plastic ring disk, as illustrated in Figure 5.41 (b). During testing, the rods extending out from the glubam specimen were gripped by the upper and lower mechanical gripping jigs of the universal testing machine, as shown in Figure 5.39 (b).

5.7.2 Experimental Results

In the two testing programs, the identified failure modes are shown in Figure 5.42. Three failure modes are common for the thick-strip and thin-strip glubams. These are the pull out of the rebar or rod along the hole interface, yielding of the bar or rod, and

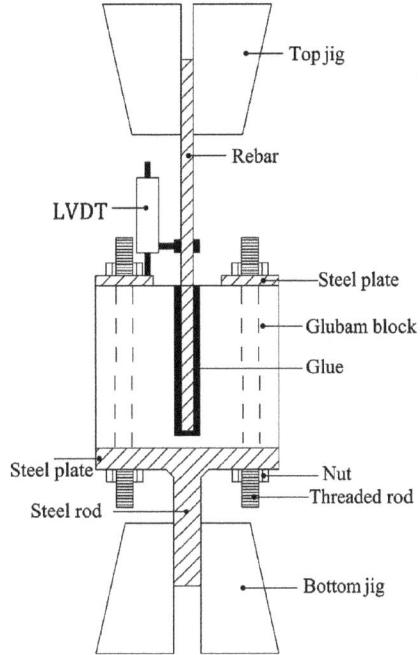

(a)

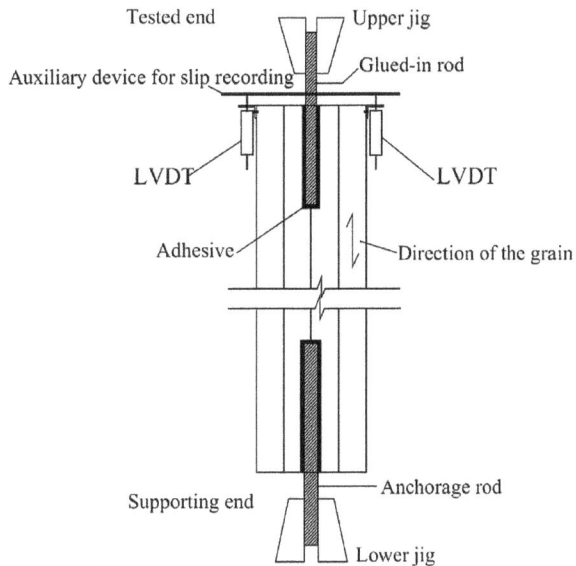

(b)

Figure 5.39 Bond testing methods: (a) pull-out test in He and Xiao; (b) direct tension test adopted in Li et al.

Sources: (a) He & Xiao (2020);[54] Li et al. (2020).[55]

Table 5.12 Testing matrix and specimen details in Li et al.'s tests of thin-strip glubam with glued-in threaded rods

Specimen	e (mm)	d_a (mm)	e/d_a	l_a (mm)	d_h (mm)	λ	t (mm)	No. of replicates
S12-7-1E	28	12	2.3	105	14	7.5	1	6
S12-10-1E	28	12	2.3	140	14	10	1	6
S12-12-1E	28	12	2.3	175	14	12.5	1	6
S12-15-1E	28	12	2.3	210	14	15	1	6
S16-7-1E	42	16	2.6	135	18	7.5	1	6
S16-10-1E	42	16	2.6	180	18	10	1	6
S16-10-1L	28	16	1.8	180	18	10	1	6
S16-10-1H	56	16	3.5	180	18	10	1	6
S16-10-1MH	70	16	4.4	180	18	10	1	6
S16-10-2E	42	16	2.6	180	20	10	2	6
S16-10-3E	42	16	2.6	180	22	10	3	6
S16-12-1E	42	16	2.6	225	18	12.5	1	6
S16-15-1E	42	16	2.6	270	18	15	1	6
S20-7-1E	56	20	2.8	165	22	7.5	1	6
S20-10-1E	56	20	2.8	220	22	10	1	6
S20-12-1E	56	20	2.8	275	22	12.5	1	6
S20-15-1E	56	20	2.8	330	22	15	1	6

Note: single parameters include: e = edge distance; d_a = rod diameter; l_a = anchorage length; d_h = drill-hole diameter; slenderness ratio = $\lambda = l_a/d_h$; t = glue-line thickness.

Source: Li et al. (2020).[55]

Figure 5.40 Specimens with different loading angles in He and Xiao.
Source: He & Yiao (2020).[54]

the splitting of the base glubam material. It is observed that the pull-out failure can also be coupled with the splitting of glubam. Another failure pattern – plug pull out of glubam with the rod – was observed in some of the specimens of Li et al.,[55] but was not seen in He and Xiao.[54] One reason might be the confinement to the glubam block by the clamping bolts used in the pull-out test configuration.

As shown in Figure 5.39 (a), during testing, the applied tensile load and the slip displacement of the bar or rod relevant to the end of the glubam specimen is measured

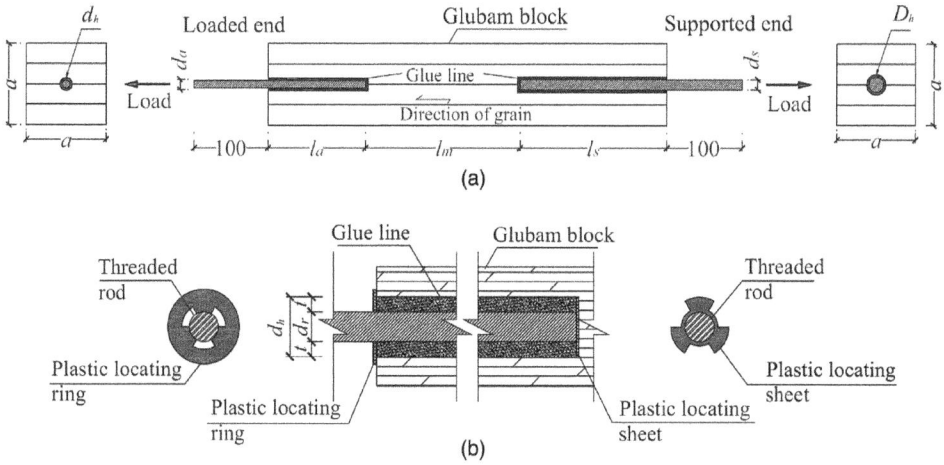

Figure 5.41 Thin-strip glubam specimens with glued-in rods: (a) specimen details; (b) testing end details.

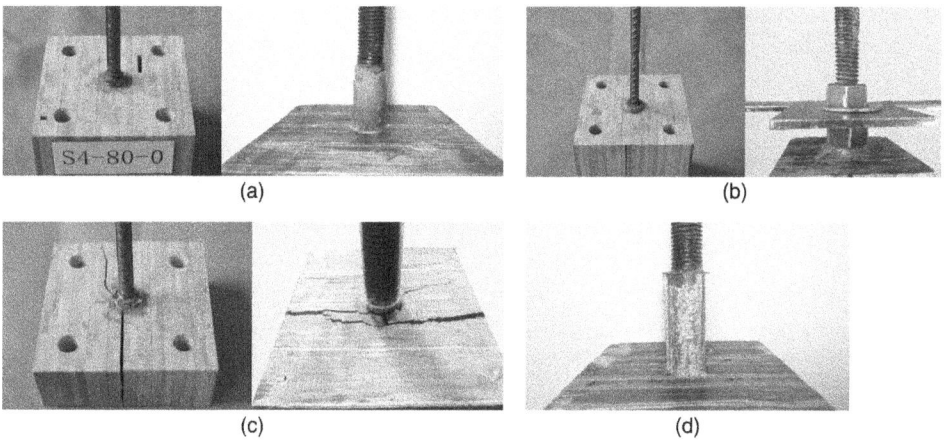

(a)

(b)

(c)

(d)

Figure 5.42 Failure modes: (a) pull out of rebar or rod; (b) yielding of rebar or rod; (c) split of glubam; and (d) plug pull out of glubam with rebar or rod.

Source: He and Xiao;[54] and Li et al. (2020).[55]

using a linear variable displacement transducer (LVDT). The load and slip relationships of the specimens are examined. Figure 5.43 compares the load and slip curves for the glued-in rebar thick-strip glubam specimens with different anchorage length. Apparently, the length of anchorage has a significant influence on the behavior of the glued-in rebar glubam joints. The results show that with an anchorage length equal to 120 mm, or $12d$ (diameter of rebar or rod), the specimens can develop the rebar yield strength, but not all can sustain the full yield plateau. Brittle behavior as shown in Figure 5.43 (a) is typical when the pull-out or split failure occurs, whereas a ductile

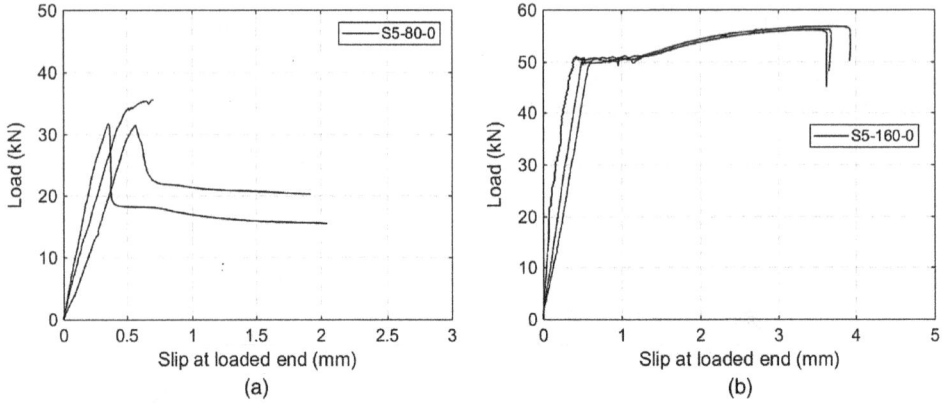

Figure 5.43 Glued-in rebar thick-strip glubam with different anchorage lengths: (a) 8*d* = 80 mm; (b) 16*d* =160 mm.

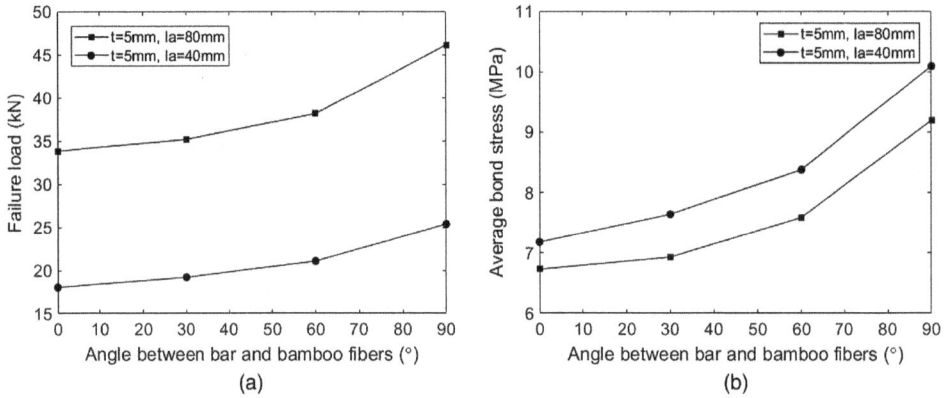

Figure 5.44 Effects of angle between bar and bamboo fibers: (a) failure load; and (b) average bond stress.

behavior dominated by the rebar behavior can be expected if the anchorage length is 160 mm or 16*d*, as shown in Figure 5.43 (b).

Figure 5.44 (a) shows the effect of the angle between bar and bamboo fibers on the loading capacity, indicating the increase of the failure load capacity corresponding to the increase of angle between bar and bamboo fibers. Similar trends are also seen in Figure 5.44 (b), with the average bond stress defined as follows,

$$\tau_u = T_u / (\pi d_h l_a) \tag{5.11}$$

where, T_u is the ultimate load, l_a is the anchorage length, and d_h represents the hole diameter. Such results may only be specific to the pull-out loading method adopted

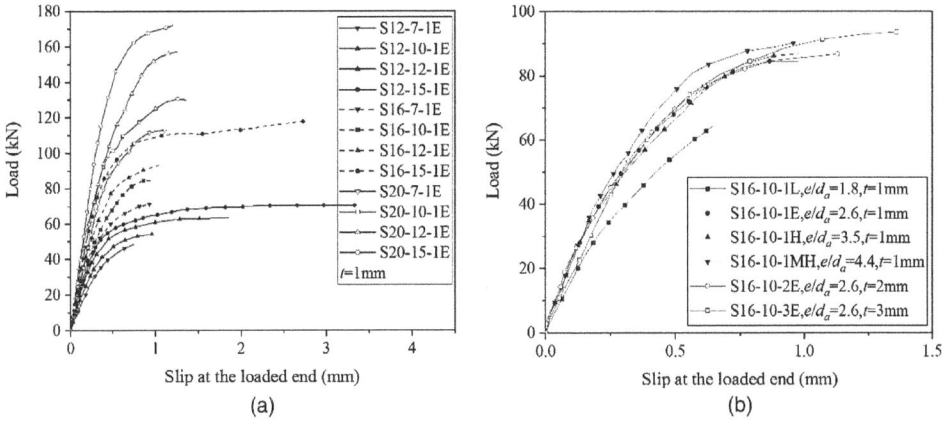

Figure 5.45 Load and slip relationships from Li et al.: (a) average curves for specimens with different rod diameter; (b) curves with different edge distances.

Source: Li et al. (2020).[55]

in He and Xiao,[54] in which the glubam block is essentially subjected to out-of-plane bending. The increase of the angle between the bar and the bamboo fiber implies the greater participation of bamboo fiber in bending resistance.

Li et al.'s results[55] on thin-strip glubam joints with glued-in threaded rods are summarized in Figure 5.45 (a). Specimens with a longer anchorage length or larger anchorage to bar diameter ratio can behave in a ductile manner with yielding of the rod. Similar to the results of thin-strip glubam, an anchorage length of $12d$ seems to be critical for developing yielding in the glued-in rod or rebar. As shown in Figure 5.45 (b), the comparison of three groups in the S16 series (rod diameter = 16 mm) with different edge distance to diameter ratio e/d indicates that an increase of the edge distance can enhance the behavior and result in yielding of the glued-in threaded bar. Li et al. suggest an edge distance to rod diameter ratio of $2.3d$ as the lower limit.[55]

In both Li et al.'s and He and Xiao's research,[55,54] the influence of the glue-line thickness on load capacity and bond strength is investigated. In general, the increase of glue-line thickness tends to reduce the bond strength, and Li et al. suggestlimiting the glue-line for the glued-in rod to a thickness of 2 mm. However, a 3 mm thickness is adequate for the glue-line of a glued-in rebar in thick-strip glubam as exhibited in He and Xiao's testing program. This may reflect the difference between the relatively smooth rod and the deformed contour of the rebar.

5.7.3 Pull-out Strength

In the previously described testing programs of the author's research group, existing pull-out strength equations originally developed for glued-in rod or rebar timber joints are examined, and are shown to be inadequate to predict the test results of glubam joints. The following empirical equations for bond strength are then proposed.

For thick-strip glubam:[54]

$$f_v = (8 - 0.018 l_a)(0.15\theta^2 + 1) \le \frac{f_y d}{4 l_a} \quad \text{(in MPa)} \tag{5.12}$$

where l_a is the anchorage length; θ represents the angle between bar and bamboo fibers with the unit of radian; d is the diameter of the rebar and is assumed to be equal to the hole diameter d_h; f_y is the yield strength of the rebar. Eq.5.12 is suggested based on the modification of the EC-5 design equation by considering the loading angles.[57]

For thin-strip glubam:[55]

$$f_{v,mean} = 7.9(\lambda/10)^{-0.45}(e/d)^{0.1} t^{-0.08} \le \frac{f_y d}{4 l_a} \quad \text{(in Mpa)} \tag{5.13}$$

where $\lambda = l_a / d$; d is rod diameter; e is edge distance; and t is glue-line thickness (mm).

Until further research is available to establish a more robust and unified approach, the above two equations are recommended independently for the two types of glubam joints with glued-in rod or rebar.

5.7.4 Glued-in Carbon Fiber Reinforced Polymer (CFRP) Rebar Glubam Joints

As a follow up study, the author's group also tested glued-in CFRP rebar glubam joints, using the pull-out testing methods shown in Figure 5.39 (a), similar to the conditions for testing glued-in steel rebar joints.[58,59] Table 5.13 shows the testing matrix. Preparation of specimens, testing methods, and parameters are similar to those for glued-in steel rebar glubam joints. The CFRP rebar had a diameter of 9.98 mm measured using the

Table 5.13 Testing matrix

Specimen	Glue-line thickness (mm)	Anchorage length (mm)	Angle between bar and bamboo fibers (°)	No. of specimens
C3-100-0	3	100	0	3
C4-100-0	4	100	0	3
C5-100-0	5	100	0	3
C5-100-30	5	100	30	3
C5-100-60	5	100	60	3
C5-100-90	5	100	90	3
C5-100-90Per	5	100	90	3
C5-200-0	5	200	0	3
C5-200-90	5	200	90	3
C5-300-0	5	300	0	3
C5-300-90	5	300	90	3

Note: Specimen name designation examples: C3-100-0, steel bar with 3 mm thick glue-line, 100 mm anchorage length, and an angle of 0° between the bar and bamboo fibers; C5-100-90Per, CFRP bar with 5 mm thick glue-line, 100 mm anchorage length, an angle of 90° between the bar and bamboo fibers, and a CFRP bar perpendicular to the width direction of the bamboo strips.

Figure 5.46 Failure mode of glued-in CFRP rebar joints: (a) pull out of rebar; (b) split of glubam; and (c) rupture of CFRP bar.

submerge method, and its tensile strength, modulus, and ultimate tensile strain are 1510 MPa, 118 GPa, and 1.28%, respectively.

At the initial stage of testing, the specimens were basically in an elastic stage with no obvious signs of damage. With the load gradually increased, the specimens emitted continuous crisp sounds. When loaded to the ultimate load, the specimen was damaged and made a loud noise. The failure modes of the specimens were affected by the anchorage length and the angle between the CFRP bar and the bamboo fiber. Based on observations of 33 specimens, three typical failure modes were identified as shown in Figure 5.46, including pull out of the CFRP bar, splitting of the glubam block, and rupture of the CFRP bar.

For the specimens with a shorter anchorage length, the failure modes of the specimens were pull out of the CFRP bar and splitting of the glubam block. The bearing capacity of the specimens with the CFRP bar bonded perpendicular to the bamboo fiber was larger than that of other specimens; however, splitting of the glubam block occurred more frequently. As discussed previously, for steel rebars, a critical anchorage length of $16d$ can afford the development of steel yielding. The failure modes of specimens with an anchorage length of $30d$ (300 mm) all comprised a CFRP bar rupture, and it is suspected that the critical anchorage length to prevent the CFRP rebar rupture is between $20d$ and $30d$. Needless to say, rupture of the CFRP bar is a brittle failure mode, which is different from a yielding of the rebar within glued-in rebar glubam joints, as discussed previously.

The applied load and slip displacement curves of most specimens show only an ascending section followed by brittle failure. During the loading process, specimens were basically in the elastic stage, and the load was increased linearly with the slip at the loading end. When loaded to the ultimate load, the load instantly dropped to zero. The damage of the specimens occurred in an instant, reflecting brittle characteristics.

Figure 5.47 shows the ultimate loads of specimens with different anchorage length, glue-line thickness, and angle between bar and bamboo fibers. It can be seen from Figure 5.47 (a) that the ultimate load increased linearly with the increase of anchorage length. It can be observed from Figure 5.47 (b) that as the glue-line thickness increased, the ultimate load of the specimen also increased gradually, mainly due to the increase in the bond area between glubam hole and adhesive with the increase of glue-line thickness, but the increase was small. As shown in Figure 5.47 (c), with the increase of angle between bar and bamboo fiber, the ultimate load increased, and the increase rate also increased, similar to the results for glued-in steel rebar joints.

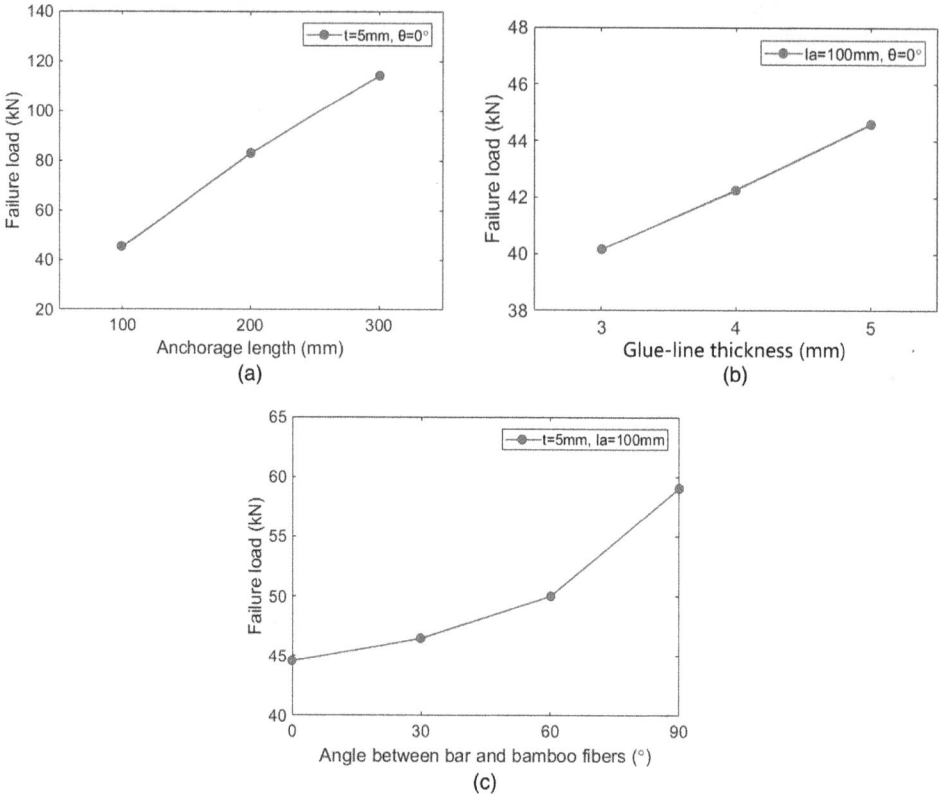

Figure 5.47 Failure load versus anchorage length (a), glue-line thickness (b), and angle between bar and bamboo fibers (c).

The EC-5 design equation[57] can be used to describe the pull-out strength for glued-in single rod timber joints. The EC-5 design equation was modified by taking into account the density of glubam and the angle between the bar and bamboo fibers, the pull-out strength of glued-in CFRP bars glubam joints can be estimated by Eq.5.14, which can also be used to predict the pull-out strength of glued-in rebar glubam joints.

$$F_{ax,k} = f_{v,k} \pi d_h l_a (0.15\theta^2 + 1) \quad \text{for} \quad 0 \le F_{ax,k} \le A_r f_y \tag{5.14}$$

where, d_h is the hole diameter, l_a is the anchorage length, θ represents the angle between the bar and bamboo fibers, the unit of θ is radian, A_r is the cross-sectional area of the CFRP bar, f_y is the tensile strength of the CFRP bar, and $f_{v,k}$ is the bond strength at glubam/adhesive interface, given by:

$$f_{v,k} = 0.44 \times 10^{-3} \times d_h^{-0.2} \rho^{1.5} \tag{5.15}$$

where ρ represents the density of glubam elements.

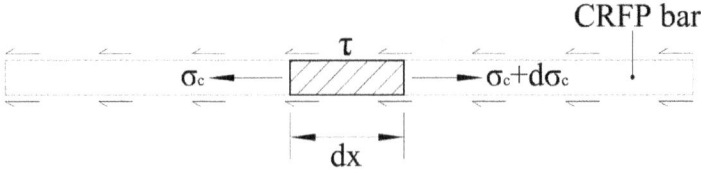

Figure 5.48 Rebar interface model.

5.7.5 Elastic Analysis of Pull-out Mechanisms

Theoretical approach: Due to the nature of the elasto-brittle behavior of most timber and bamboo structures, in most cases, their design is based on the service load condition; thus, only the first branch of the bond stress-slip curve is important. Therefore, attention is given to the elastic analysis of the bonding mechanism. As shown in Figure 5.48, a micro-segment dx at any position along the bar or rod is analyzed, based on the assumption that the interface bonding is in the state of pure shear, regardless of the normal stress effect.

The equilibrium equation of force should be satisfied in the interface analysis model, as shown in Eq.5.16.

$$\frac{d\sigma_c}{dx} - \frac{4}{d_c}\tau = 0 \tag{5.16}$$

Based on the deformation compatibility relationship and the elastic stress-strain relationship of the bar, the following equation is established.

$$\frac{\sigma_c}{E_c} = \frac{S_x}{dx} = \varepsilon \tag{5.17}$$

Substituting Eq.5.17 into Eq.5.16, we have,

$$\frac{E_c d_c}{4} S_x'' - \tau_x = 0 \tag{5.18}$$

In the elastic stage, bond stress is assumed to be in a linear relationship with slip S_x, then,

$$\tau_x = k S_x \tag{5.19}$$

then,

$$S_x'' - \beta S_x = 0 \tag{5.20}$$

where $\beta = \dfrac{4k}{E_c d_c}$.

The non-trivial solution of the second-order homogeneous differential equation, Eq.5.20, is solved with the boundary conditions as the following Eq.5.21.

$$S_x = S_{la}e^{\sqrt{\beta}(x-l_a)} \tag{5.21}$$

where, S_{la} = slip at the loaded end. The axial stress of the CFRP bar at the loaded end can then be given by,

$$\sigma_{la} = \frac{4k}{d_c}\int_0^{l_a} S_{la}e^{\sqrt{\beta}(x-l_a)}dx \tag{5.22}$$

In the above analysis, the parameter k is the bonding stiffness between the bond stress and slip, which is contributed by both the adhesive layer and the base material. In He's study,[58] the value of k is estimated at 32.5 MPa/mm, based on the tests with the tested smallest anchorage length of l_a = 40mm (Specimen S4-40-0 group in Table 5.11). As an example, the analyzed bar stress and anchorage end slip compares reasonably well with the test results for glued-in CFRP rebar glubam joints in Figure 5.49.

Using the analytical model, the bond slip distributions along the bonding length are analyzed and shown for different anchorage lengths l_a, in Figure 5.50, in which both the vertical axis and the horizontal axis are normalized by the end slip S_{la} and the anchorage length l_a, respectively. As exhibited in Figure 5.50, the distribution of CFRP bar slippage tends to be flattened when the anchorage length becomes shorter.

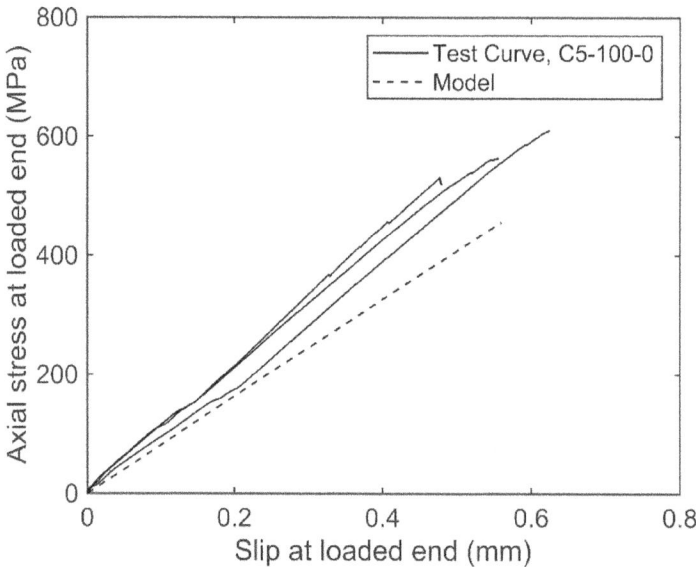

Figure 5.49 Comparison between analyzed and test results of CFRP rebar stress-slip relationships.

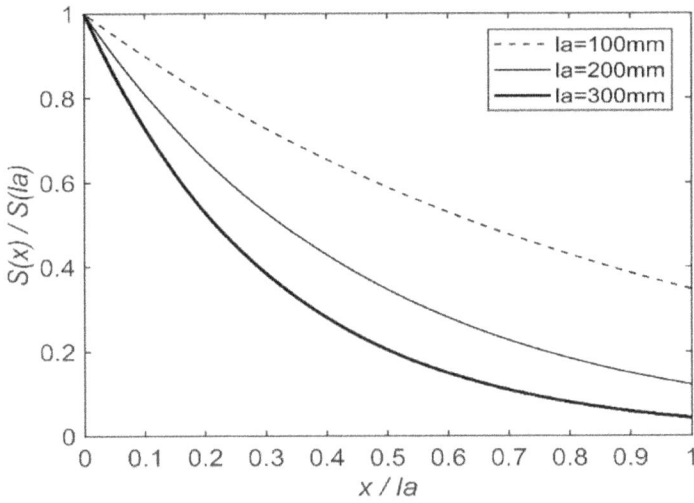

Figure 5.50 Analyzed slip distributions along anchorage length.

5.7.5.1 Numerical Approach

Another numeric analysis approach has also been established to simulate the bond stress-slip relationship of glued-in rebar glubam joints (Figure 5.51). In the elastic analysis model, the rebar and glue were divided into a series of segments, and each segment was replaced by an equivalent spring.[54] The equilibrium equation of force and displacement compatibility on the interface should be satisfied in the elastic analysis model, as shown in Eq.5.23 to Eq.5.26. According to the equilibrium equation of force and displacement compatibility on the interface, the bond force-slip relationship and the bond stress-slip relationship can be expressed as Eq5.27 to Eq.5.29, respectively and can be calculated using MATLAB software.

$$F_1 = Q_1 = k_a S_1, k_a = k\frac{l_a}{n} \tag{5.23}$$

$$Q_i = k_a S_i \tag{5.24}$$

$$F_i = F_{i-1} + Q_i \tag{5.25}$$

$$S_i = S_{i-1} + \delta_{i-1}, \delta_{i-1} = \frac{F_{i-1} \cdot \frac{l_a}{n}}{EA} \tag{5.26}$$

where, F_1, Q_1, and S_1 represent the tension, shear, and slip of the first element, respectively; F_i, Q_i, and S_i represent the tension, shear, and slip of ith element, respectively; δ_i is

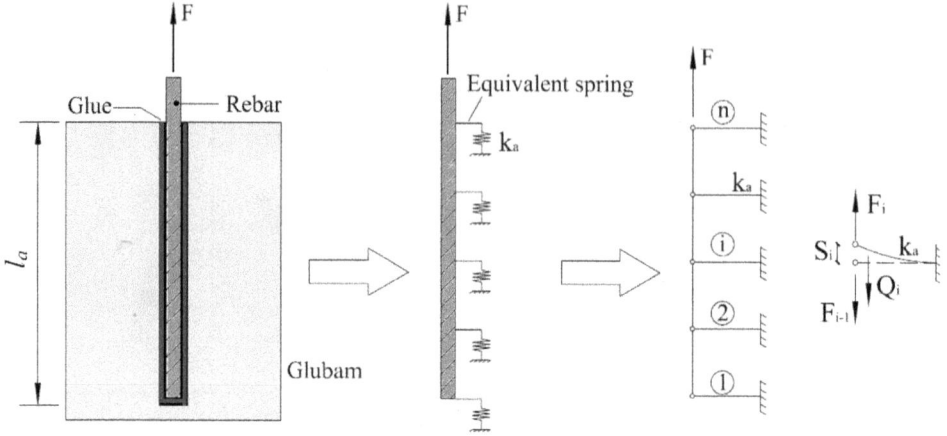

Figure 5.51 Elastic analysis model of glued-in rebar glubam joints.

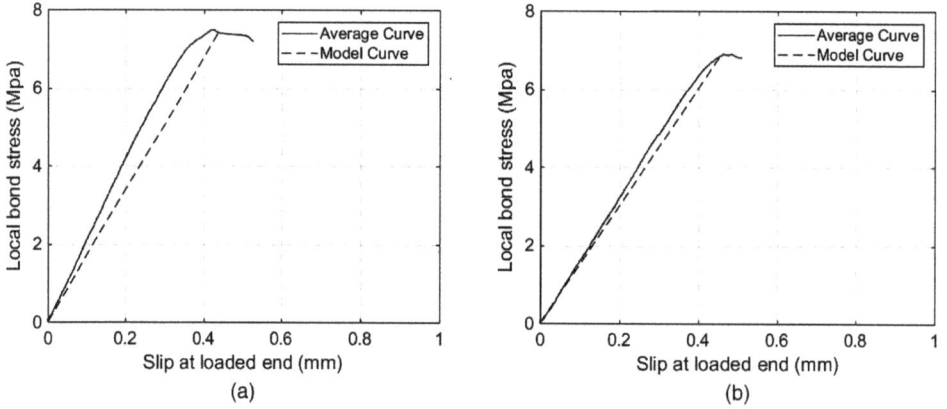

Figure 5.52 Comparison between model curves and average experimental curves of specimens with different glue-line thickness: (a) S3-80-0; (b) S4-80-0.

the elastic extension of rebar in ith element, k is the equivalent stiffness of element per unit length; k_a is the equivalent stiffness of an element; l_a is the anchorage length; n is the number of segments of anchorage length; E is the elastic modulus of the rebar; and A is the cross-sectional area of the rebar.

$$F = \frac{\alpha^{n-1} + \cdots + \dfrac{n(n-1)(n+1)}{6}\alpha + n}{\alpha^{n-1} + \cdots + \dfrac{n(n-1)}{2}\alpha + 1} \cdot k\frac{l_a}{n}S_n, \alpha = \frac{k\left(\dfrac{l_a}{n}\right)^2}{EA} \tag{5.27}$$

$$\tau = \frac{F}{\pi d_h l_a} \tag{5.28}$$

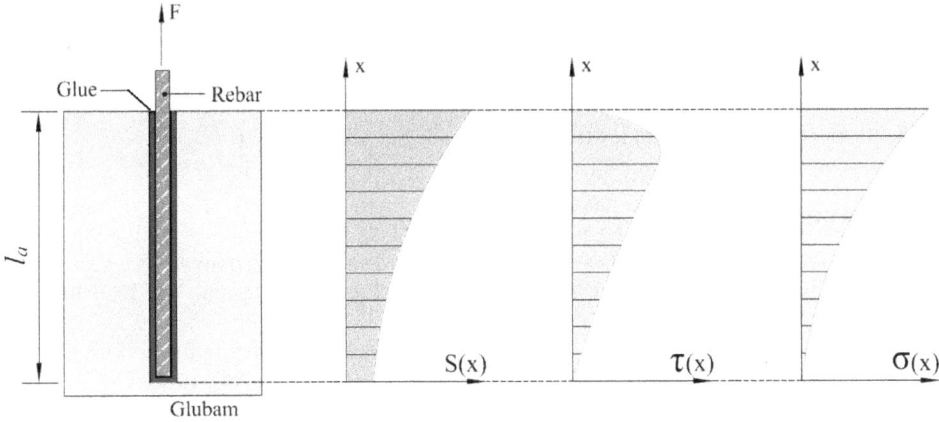

Figure 5.53 Distribution of interface parameters along anchorage length.

$$\tau = \frac{\alpha^{n-1} + \cdots + \dfrac{n(n-1)(n+1)}{6}\alpha + n}{\alpha^{n-1} + \cdots + \dfrac{n(n-1)}{2}\alpha + 1} \cdot \frac{k}{n\pi d_h} S_n \text{ for } \tau \le \tau_m \tag{5.29}$$

where d_h is the hole diameter; τ_m is the bond stress corresponding to the ultimate load and can be calculated by Eq.5.30, which is a regression value of test results.

$$\tau_m = (8 - 0.018 l_a)(0.15\theta^2 + 1) \tag{5.30}$$

where θ represents the angle between the bar and bamboo fibers, the unit of θ is radian. By averaging the bond stress-slip curves of specimens with $\theta = 0°$ and different anchorage lengths, the value of spring factor k can be calculated to be about 780 N/mm. Substituting this value into Eq.5.29, the bond stress-slip curves of the glued-in rebar glubam joints can be numerically simulated, and are plotted in Figure 5.52. It is obvious that the average curves can be simulated well by the model curves, demonstrating that the initial ascending branch of the bond stress-slip relationship can be approximately analyzed using the numerical model. According to the elastic analysis model of glued-in rebar glubam joints, the distribution of interface parameters along the anchorage length can be obtained, as shown in Figure 5.53. In order to satisfy the boundary conditions, the distribution of interlaminar shear stress was calculated by the fitting method proposed by Hong and Zhang based on the elastic analysis model.[60]

5.8 Summary of Glubam Connections

Research on connections in engineered bamboo structures is still rare, and the experimental studies carried out by the author's research group are summarized in this chapter. It is promising that many types of joints and connections used in timber structures can be extended into use with engineered bamboo structures. In the case of bidirectionally designed glubam, the behavior may even outperform timber structures due to the existence of bamboo fibers in the transverse direction as well as the longitudinal main direction. In order to fully establish the design method and specifications for glubam connections, further research is still needed.

References

[1] ASCE Task Committee on Fasteners. (1997). *Mechanical connections in wood structures* (Edited by L.A. Soltis). American Society of Civil Engineers, Reston, VA, DOI:10.1061/9780784401101.

[2] Salenikovich, A.J., Loferski, J.R., & Zink, A.G. (1996). Understanding the performance of timber connections made with multiple bolts. *Wood Design Focus*, 7(4), 19–26.

[3] Aicher, S., Reinhardt, H.W., & Garrecht, H. (Eds.). (2014). *Materials and joints in timber structures: Recent developments of technology*. RILEM Bookseries Vol. 9, Springer, Dordrecht, The Netherlands.

[4] Disén, K., & Clouston, P. (2013). Building with bamboo: A review of culm connection technology. *Journal of Green Building*, 8, 83–93, DOI:10.3992/jgb.8.4.83.

[5] Hong, C., Xiong, Z., Lorenzo, R., Corbi, I., Corbi, O., Wei, D., ... Zhang, H. (2020). Review of connections for engineered bamboo structures. *Journal of Building Engineering*, 30, 101324, DOI:10.1016/j.jobe.2020.101324.

[6] Xiao, Y., & Xie, L. (2003). *Seismic behavior of base-level diaphragm anchorage of hillside woodframe buildings*. CUREE Publication No. W-24. Consortium of Universities for Research in Earthquake Engineering, Richmond, CA.

[7] Johansen, K.W. (1949). Theory of timber connections. Publication No. 9. International Association of Bridge and Structural Engineering, pp. 249–262.

[8] De Jong, T. (1977). Stresses around pin-loaded holes in elastically orthotropic or isotropic plates. *Journal of Composite Materials*, 11(3), 313–331, https://doi.org/10.1177/002199837701100306

[9] Zhang, K.-D., & Ueng, C.E.S. (1984). Stresses around a pin-loaded hole in orthotropic plates. *Journal of Composite Materials*, 18(5), 432–446, https://doi.org/10.1177/002199838401800503

[10] Zhou, T., & Guan, Z. (2006). Review of existing and newly developed approaches to obtain timber embedding strength. *Progress in Structural Engineering and Materials*, 8(2), 49–67, https://doi.org/10.1002/pse.213

[11] Li, Z., Zhang, J.Y., Wang, R., Monti, G., & Xiao, Y. (2020). Design embedment strength of ply-bamboo panels used for GluBam. *ASCE Journal of Materials in Civil Engineering*, 32(5), 04020082.

[12] Wang, R. (2020). Research on lightweight glubam frame structures (in Chinese). Doctoral Thesis, Hunan University.

[13] Feng, L. (2015). Bolt connection of modern bamboo-wood structures: Theoretical analysis and experimental study (in Chinese). Master of Engineering Thesis, Hunan University.

[14] ASTM International. (2018). ASTM D5764-97a: Standard test method for evaluating dowel-bearing strength of wood and wood-based products. ASTM International, West Conshohocken, PA.

[15] European Committee for Standardization (CEN). (2007). EN 383: Timber structures – Test methods – Determination of embedment strength and foundation values for dowel type fasteners. European Committee for Standardization, Brussels.

[16] Leijten, A.J.M., & Kohler, J. (2004). *Evaluation of embedment strength data for reliability analyses of connections with dowel type fasteners*. Technical Report No. COST E24. Delft University of Technology, Delft, The Netherlands.

[17] European Committee for Standardization (CEN). (2002). EN 1990: Basis of structural design. European Committee for Standardization, Brussels.

[18] ASTM International. (2007). ASTM D143-07: Standard test methods for small clear specimens of timber. ASTM International, West Conshohocken, PA, www.astm.org

[19] Ministry of Housing and Urban–Rural Development of the People's Republic of China. (2017). GB 50005-2017: Code for design of timber structures (in Chinese). China Architecture & Building Press, Beijing.

[20] ASTM International. (2010). ASTM D2915-10: Standard practice for sampling and data-analysis for structural wood and wood-based products. ASTM International, West Conshohocken, PA.

[21] Tsai, S.W., & Wu, E.M. (1971). A general theory of strength for anisotropic materials. *Journal of Composite Materials*, 5(1), 58.

[22] Yang, R.Z., Xiao, Y., & Lam, F. (2014). Failure analysis of typical glubam with bidirectional fibers by off-axis tension tests. *Construction and Building Materials*, 58, 9–15.

[23] Hankinson, R.L. (1921). Investigation of crushing strength of spruce at varying angles of grain. Air Force Information Circular No. 259. U.S. Air Service.

[24] American Wood Council. (2018). *National design specification for wood construction*. American Wood Council, Leesburg, VA.

[25] Yang, R.Z., & Xiao, Y. (2021). Experimental studies on bolted glubam connections. *Advances in Structural Engineering*, 24(13), 3010–3020..

[26] Ministry of Housing and Urban–Rural Development and General Administration of Quality Supervision, Inspection and Quarantine of the People's Republic of China. (2012). GB/T 50329-2012.Standard for test methods of timber structures (in Chinese). China Building Industry Press, Beijing.

[27] McLain, T.E., & Thangjitham, S. (1983). Bolted wood-joint yield model. *Journal of Structural Engineering*, 109(8), 1820–1835.

[28] Soltis, L.A., Hubbard, F.K., & Wilkinson, T.L. (1986). Bearing strength of bolted timber joints. *Journal of Structural Engineering*, 112(9), 2141–2154.

[29] Zhang, D.S., Fei, B.H., Ren, H.Q., & Wang, Z. (2008). The research of joint composed by laminated bamboo lumber. In Y. Xiao, M. Inoue & S.K. Paudel (Eds.), *Modern bamboo structures: Proceedings of First International Conference on Modern Bamboo Structures (ICBS-2007), Changsha, China, 28–30 October 2007*. CRC Press/Balkema, Leiden, The Netherlands.

[30] ASTM International. (2007). ASTM D5652-95: Standard test methods for bolted connections in wood and wood-base products. ASTM International, West Conshohocken, PA.

[31] Echavarría, C., Haller, P., & Salenikovich, A. (2007). Analytical study of a pin-loaded hole in elastic orthotropic plates. *Composite Structures*, 79(1), 107–112.

[32] Gupta, R. (1994). Metal-plate connected tension joints under different loading conditions. *Wood and Fiber Science*, 26(2), 212–222.

[33] Gupta, R., & Gebremedhin, K.G. (1990). Destructive testing of metal-plate-connected wood truss joints. *ASCE Journal of Structural Engineering*, 116(7), 1971–1982.

[34] Vatovec, M., Gupta, R., & Miller, T. (1996) Testing and evaluation of metal-plate-connected wood truss joints. *Journal of Testing and Evaluation*, 24, 63–72.

[35] Guo, W., Song, S., Zhao, R. Ren, H., Jiang, Z., Wang, G., ... Fei, B. (2013). Tension performance of metal-plate connected joints of Chinese larch dimension lumber. *BioResources*, 8(4), 5666–5677.

[36] Lau, P.W.C. (1987). Factors affecting the behaviour and modelling of toothed metal-plate joints. *Canadian Journal of Civil Engineering*, 14(2), 183–195.

[37] Via, B.K., Zink-Sharp, A., Woeste, F., & Dolan, J.D. (2001). Influence of specific gravity on embedment gaps in metal-plate-connected truss joints. *Forest Products Journal*, 51(10), 88–92.

[38] Wu, J.M., Xiao, Y., & Shea, E. (2016). Ultimate strength of metal-plate-connected glubam joints (in Chinese). *Industrial Construction*, 46(7), 118–123.

[39] Wu, J.M. (2013). Test research on ultimate strength of metal-plate-connected GluBam joints (in Chinese). Master of Engineering Thesis under supervision of Y. Xiao, Hunan University.

[40] Peng, Q., Xiao, Y., & Wu, J.M. (2018). Strength of metal-plate-connected glubam joints (in Chinese). *Building Structure*, 048(019), 91–96.

[41] Peng, Q. (2017). Research on mechanical behavior of metal-plate-connected GluBam floor trusses (in Chinese). Master of Engineering Thesis under supervision of Y. Xiao, Hunan University.

[42] General Administration of Quality Supervision, Inspection and Quarantine, and Standardization Administration of the People's Republic of China. (2010). GB/T 228.1: China national standard – Metallic materials – Tensile testing – Part 1: Method of test at room temperature. Standards Press of China, Beijing.

[43] Ministry of Housing and Urban–Rural Development of the People's Republic of China. (2012). JGJT 265-2012: Technical code for light wood trusses. China Architecture and Building Press, Beijing.

[44] Pan, J.L., & Zhu, E.C. (2019). *Principles of timber structure design* (in Chinese). China Architecture and Building Press, Beijing.

[45] Davis, T.J., & Claisse, P.A. (2001). Resin-injected dowel joints in glulam and structural timber composites. *Construction and Building Materials*, 15, 157–167.

[46] Gattesco, N., & Gubana, A. (2000). Studio sperimentale sulle unioni incollate di elementi in legno lamellare (in Italian). In *Proceedings of the 14th CTE Meeting, Mantua, Italy*.

[47] Alam, P., Ansell, M.P., & Smedley, D. (2010). Mechanical repair of timber beams fractured in flexure using bonded-in reinforcements. *Composites Part B: Engineering*, 40, 95–106.

[48] Broughton, J.G., & Hutchinson, A.R. (2001a). Adhesive system for structural connections in timber. *International Journal of Adhesion and Adhesives*, 21, 177–186.

[49] Broughton, J.G., & Hutchinson, A.R. (2001b). Pull-out behaviour of steel rods bonded into timber. *Material Structures*, 34(2), 100–109.

[50] Ling, Z.B., Yang, H.F., Liu, W.Q., Lu, W.D., Zhou, D., & Wang, L. (2014). Pull-out strength and bond behaviour of axially loaded rebar glued-in glulam. *Construction and Building Materials*, 65, 440–449.

[51] Yeboah, D., Taylor, S., McPolin, D., & Gilfillan, R. (2011). Behaviour of joints with bonded-in steel bars loaded parallel to the grain of timber elements. *Construction and Building Materials*, 25(5), 2312–2317.

[52] Steiger, R., Gehri, E., & Widmann, R. (2007). Pull-out strength of axially loaded steel rods bonded in glulam parallel to the grain. *Material Structures*, 40(1), 69–78.

[53] Widmann, R., Steiger, R., & Gehri, E. (2007). Pull-out strength of axially loaded steel rods bonded in glulam perpendicular to the grain. *Material Structures*, 40(8), 827–838.

[54] He, Z.W., & Xiao, Y. (2020). Experimental study on axial pull-out behavior of steel rebars glued-in glubam. *ASCE Journal of Materials in Civil Engineering*, 32(3), 04020021.

[55] Li, T.Y., Shan, B., Xiao, Y., Guo, Y.R., & Zhang, M.P. (2020). Axially loaded single threaded rod glued in glubam joint. *Construction and Building Materials*, 244, 118302.

[56] Luo, X., Ren, H.Q., & Zhong, Y. (2020). Experimental and theoretical study on bonding properties between steel bar and bamboo scrimber. *Journal of Renewable Materials*, 8(7).

[57] European Committee for Standardization (CEN). (1997). EN 1995-2: Eurocode 5: Design of timber structures – Part 2: Bridges.. European Committee for Standardization, Brussels.

[58] He, Z.W. (2020). Bond-anchorage behaviors of glubam joints with glued-in rods (in Chinese). Master of Engineering thesis supervised by Y. Xiao, Nanjing Tech University.

[59] He, Z.W., Shan, B., & Xiao, Y. (2021). Pull-out behavior of CFRP bars glued-in glubam joints. *ASCE Journal of Composites in Construction*, 25(4).

[60] Hong, X.J., & Zhang, Y. (2000). The fitting method of the smooth bond stress in the bond-slip test. *Structural Engineers*, 3, 44–48.

Chapter 6

Performance of Glubam Structural Members

In previous chapters, the material properties, design strengths of engineered bamboo, and connecting methods have been introduced. This chapter focuses on introducing the behavior and design of glubam structural components, including members in axial tension, bending members, and axially loaded columns. Experimental results are discussed and design methods for various capacities are described. For the load carrying capacities of various glubam members, the discussion follows the various timber design codes with necessary modifications.

It should be clarified that in the Chinese GB 50005 code[1] or the U.S. National Design Specification (NDS),[2,3] the design of a member is actually based on checking the stress under load so as not to exceed the design strength. To reflect the design for structural members, the main focus of this chapter is on the discussion of the ultimate capacities, which are used in both the Chinese GB code, the load and resistance factor design (LRFD) standard in the United States, and elsewhere. In other word, the emphasis is on derivations of design capacity R in Eq.4.6 (or nominal resistance R_n in Eq.4.13 for LRFD). For allowable stress design (ASD), the general design framework laid out in Chapter 4 can be followed, modifying the ultimate member capacities discussed in this chapter.

6.1 Glubam Bending Members

Bending, or flexural, members subject to loads acting perpendicularly to their longitudinal axis are used in most structural systems. Beams, girders, joists, purlins, etc., are typical bending members. Bending action also exists in other members such as slabs and walls, with bending along more than one axis. However, in most cases in timber structures, slabs and walls are designed as composite systems with elements resisting bending actions such as individual beams.

This section mainly discusses the bending design of linear beams and girders, which are relatively slender with a longitudinal measurement that is much larger than the transverse dimension. In this book, the typical definitions are followed, in which girders are main bending elements in bridges, or those supporting other beams; while beams are somewhat smaller in size and are often referred to as bending members in buildings. For simplicity, beam is used as the main term for discussions of flexural and bending members in this book.

DOI: 10.1201/9781003204497-6

6.1.1 Design Concepts of Glubam Beams

As discussed in Chapter 2, glubam structural components are typically made with a two-step manufacturing approach. The first step is to manufacture glubam boards or plybamboo using either thick or thin bamboo strips. Due to the constraint of the hot-pressure process, the thickness of the boards is typically less than 40 mm. The second process is a cold-pressure process to laminate the elements cut from the bamboo boards into components based on structural design need, similar to that for making timber-based glulam. The two processes can be completed at the same facility, but it is more typical for them to be executed separately, due partly to the need for transportation and logistics efficiency as well as to suit construction flexibility. Figure 6.1 illustrates examples of glubam beams designed and manufactured by the author's research group for constructing bridges and building girders.[4,5]

For wood-based glulam beams,[6] the lamination is typically made using a horizontal lamination (flatwise layup) in which the lumbers are laminated with their wider sides facing each other and the load is applied perpendicularly to the width of the lumbers, as shown in Figure 6.2 (a). Glubam beams can also be made in a similar way to glulam beams. However, for glubam, due to the fact that the bamboo boards are typically manufactured as a larger sheet with a width dimension of up to 1.22 m and a length of more than 2.0 m, so it is more efficient to produce glubam beams using a vertical lamination method (sidewise layup), as shown in Figure 6.2 (b). In this case, the load is applied in parallel to the lamination line. In Chapter 2, the advantages and disadvantages of the two lamination layup methods are also discussed. Moreover, for thin-strip glubam beams, the sidewise layup vertical lamination can be beneficial in terms of shear resistance due to the existence of transverse bamboo fibers.

In the last 15 years, the author's research group has carried out extensive tests on glubam beams with different experimental parameters, such as size, layup methods, finger joint details, etc.[7–9] In the next section, a recent experimental program carried out by the author's research group is introduced, with comparison of the mechanical behaviors of glubam beams made with different lamination configurations and two types of plybamboo.

6.1.2 Summary of Experimental Behaviors

A comprehensive testing program was carried out by Yang[9] and Li et al.[10] In the testing program, two types of glubam beams made with different preprocessed plybamboo boards were tested. One type of glubam beam specimens was made with thick-strip glubam (GB-I series) and the others were thin-strip glubam (GB-II series). Table 6.1 shows the testing matrix. Note that the orientation configuration of the bamboo strips in plybamboo board and in the final beam specimens are actually opposite.

The cold-pressing layup lamination of the plybamboo elements for all the specimens used the vertical (sidewise) layup. However, there are at least three types of configuration for the thick-strip plybamboo, as shown in Figure 6.3. As discussed in Chapter 2, material tests show that the bending specimens with sidewise layup have a higher modulus of elasticity and bending strength than the specimens with flatwise layup.

The glubam beam specimens were made by pressure laminating elements cut from plybamboo boards at room temperature around 20°C. Because all the specimens were

(a)

(b)

Figure 6.1 Example of glubam beams for constructing bridges (a); and building girders (b).

longer than the length of the original plybamboo boards (less than 2.4 m), it was necessary to lengthen the elements cut from the plybamboo using finger joints. In order to reduce the impact on the flexural capacity of the glubam beam, finger joints were not arranged at mid-span. In this research, the shortest distance between the finger joint and mid-span was 1.5 m. The length of the fingers was 100 mm whereas the pitch was 50 mm. To obtain glubam beams with a specific dimension, the plybamboo boards were sized into required elements with finger joints and numbered according to the staking sequence requirements. Then following the bottom-up principle, 2-part epoxy adhesive

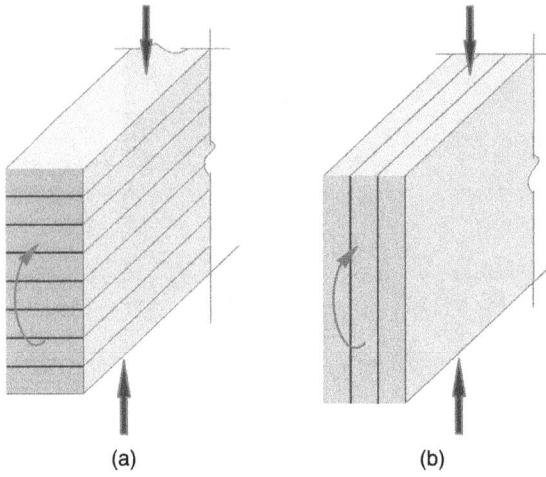

Figure 6.2 Different layup methods for lamination: (a) horizontal or flatwise layup; (b) vertical or side-wise layup.

Table 6.1 Glubam beam specimens

Type of glubam beam	Span (mm)	Dimension (mm x mm)	Bamboo strip configuration
GB-I-1-4	9000	400×100	Sidewise layup
GB-I-5-8	9000	400×100	Flatwise layup
GB-I-9	7800	400×100	Flatwise layup
GB-II-1-3	6000	450×84	Sidewise layup

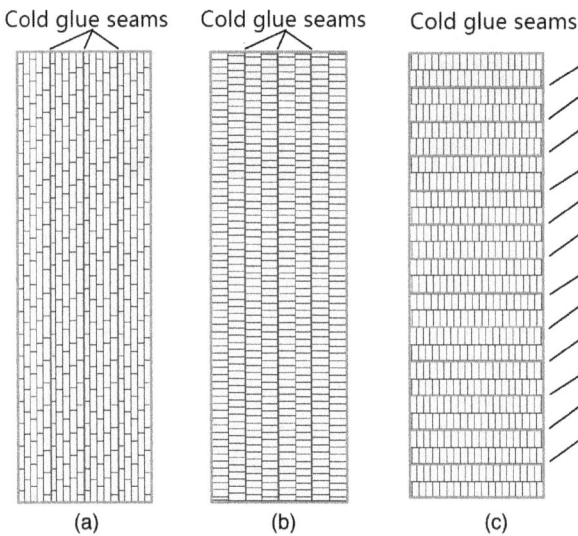

Figure 6.3 Configurations of thick-strips in plybamboos: (a) flatwise pressed formation; (b) sidewise pressed formation; (c) multi-layer layup of sidewise pressed formation.

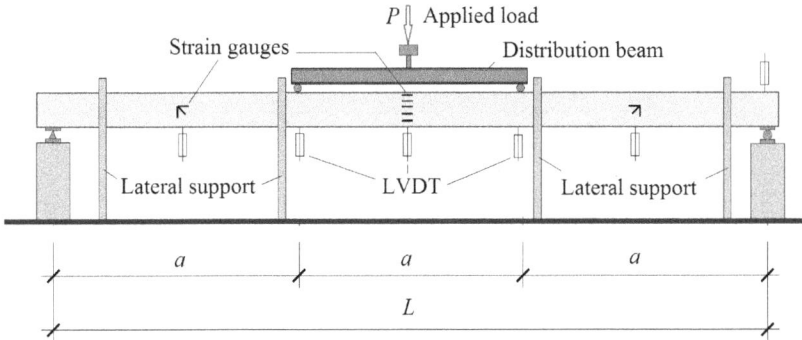

Figure 6.4 Test setup.

was applied at the finger joints and gluing surface. When all the plybamboo boards were assembled, a pressure of about 1.5 MPa was applied using hydraulic jacks and distribution beams. The pressure was kept on for 24 hours for curing of the adhesive.

As shown in Figure 6.4, four-point loading was adopted in the test setup. Displacements of the beam specimen at points every 1/6 of its length, including mid-span, were measured using linear variable displacement transducers (LVDTs). After the specimen was fixed, three preload processes were carried out to eliminate possible initial errors of equipment. At the beginning of the test, the load was kept at 7 kN per step and remained for 5 minutes in order to obtain stable deformation data. When visible cracks appeared at the bottom of the specimen, the load per step was decreased to 3.5 kN until the termination of the loading due to specimen failure.

The load mid-span deflection curves of the GB-I and GB-II specimens are shown in Figure 6.5 (a) and Figure 6.5 (b), respectively. All the specimens essentially behaved in an elasto-brittle fashion. At the beginning stage, the slopes of the load-deflection curves of the specimens are almost linear until reaching a load level of about 60% to 70% ultimate capacity. For the GB-I testing series, two failure modes were identified: Figure 6.6 (a) shows the tensile rupture at the section with a finger joint along with horizontal shear splitting; Figure 6.6 (b) shows interlamination horizontal shear splitting. Apparently, due to the unidirectional arrangement of bamboo strips (fibers) in the longitudinal direction, horizontal shear split rupture is typical for the thick-strip glubam beams. For thin-strip glubam beam specimens, two failure modes, tensile rupture in the section with a finger joint and compression crush in the compression zone are identified and shown in Figure 6.7 (a) and (b), respectively.[11] No shear failure was observed, due to the existence of bamboo fibers in the transverse direction which provide a certain resistance to shear.

6.1.3 Flexural Stiffness

The main experimental results for the glubam beams are shown in Table 6.2. The table also lists the load, $P_{L/250}$, corresponding to a mid-span displacement equal to $L/250$, which is typically the displacement limitation defined by Chinese code GB/T 51226.[12]

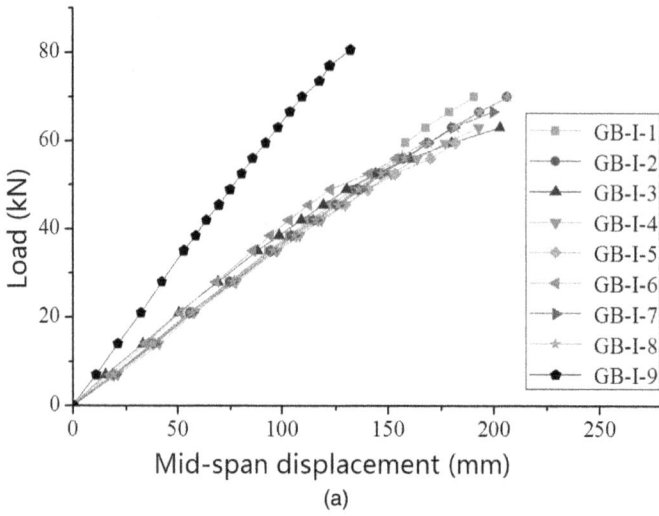

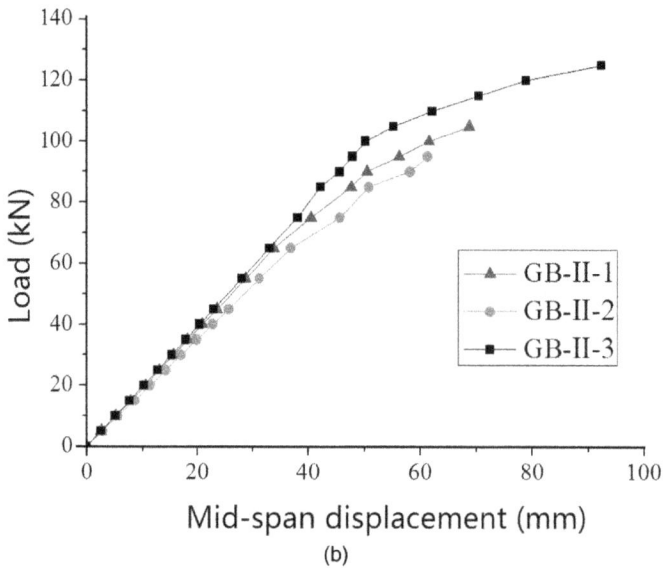

Figure 6.5 Load and displacement behaviors: (a) thick-strip glubam beams; (b) thin-strip glubam beams.

The load at $L/250$ displacement is approximately 20% of the maximum capacity, indicating that the design of glubam beams is likely controlled by deflection limitation. Using much smaller beam specimens with thick-strip glubam (also referred to as LBL), Li et al. showed that loads at a displacement of $L/250$ can be as low as less than 15% of the maximum capacity.[13]

Figure 6.6 Failure patterns of GB-I testing series: (a) tensile rupture; (b) interlamination shear split rupture in tension zone.

Figure 6.7 Failure patterns of GB-II testing series: (a) tensile rupture; (b) compression crush.

Table 6.2 Main test results

Specimen	P_{max} (kN)	δ_{max} (mm)	$P_{L/250}$(kN)	EI (10^{12}Nmm^2)	$(EI)_{cal}$ (10^{12}Nmm^2)	$(EI)_{cal}/(EI)_{ave}$	P_{cal} (kN)	$P_{cal}/(P_{max})_{ave}$
GB-I-1	72.1	190.4	13.4	5.27	5.17	1.02	85.6	1.25
GB-I-2	71.8	206.3	13.5	5.03				
GB-I-3	64.7	202.9	15.3	5.13				
GB-I-4	65.9	192.7	12.9	4.92				
GB-I-5	60.5	181.7	13.2	4.90				
GB-I-6	61.1	167.3	14.9	5.22				
GB-I-7	66.5	199.6	13	5.11				
GB-I-8	70.0	206.4	13.3	5.17				
GB-I-9	80.5	141.1	17.5	5.11	5.17	1.01	127.5	1.50
GB-II-1	105.0	68.7	49.8	7.35	5.99		96.4	
GB-II-2	95.0	61.3	45.7	6.74		0.85		0.89
GB-II-3	125.0	92.4	49.6	7.10				

Note: P_{max} is the ultimate load; δ_{max} is the deflection of mid-span; $P_{L/250}$ is the load when deflection reached L/250, L is the span of the glubam beam; EI is the initial bending stiffness obtained from test results; $(EI)_{cal}$ is the calculated initial flexural rigidity; P_{cal} is the calculation considering the weakening effect of finger joints; $(EI)_{ave}$ is average stiffness for the specimens within the same group; $(P_{max})_{ave}$ is average maximum load for the specimens within the same group.

The experimental initial stiffness shown in Table 6.2 is obtained using the following equation,

$$EI = \frac{23L^3 \Delta P}{1296 \Delta \delta} \approx \frac{L^3 \Delta P}{56 \Delta \delta} \tag{6.1}$$

On the other hand, the flexural stiffness EI can also be calculated theoretically with elastic modulus from the material bending tests in Chapter 3 (9100 MPa for GB-I, and 9400 MPa for GB-II, based on Table 3.1 (b)). The calculated initial bending stiffness is slightly larger than the experimental stiffness for GB-I series, however, underestimates the results of GB-II series specimens. The results from tests by Yang,[9] and Li et al.[10] reconfirm that the initial stiffness of glubam beams can be estimated reasonably well using the modulus of elasticity from material tests, disregarding the finger joint. Similar conclusions were also obtained from earlier studies of the author's group.[5,11]

6.1.4 Flexural Capacities

From the values shown in Table 6.2, it can be found that the average maximum load carrying capacity of specimens GB-I-1 to GB-I-4 is 68.6 kN, larger than 65.4 kN, the average capacity of specimens GB-I-5 to GB-I-8, reflecting the influence of the configurations of the bamboo strips. This is consistent with the results described in Chapter 3 for material bending tests. It is suspected that the strips with sidewise (vertical) layup may be able to resist shear force more efficiently than the strips with flatwise layup in the beam.

As an extreme case, in a previous testing program on glubam beams made of thin-strip plybamboo with a flatwise layup, Zhou observed sliding shear failure along the lamination interfaces, as shown in Figure 6.8 (a), in comparison with the bending failure of the counterpart specimen made with sidewise layup.[11]

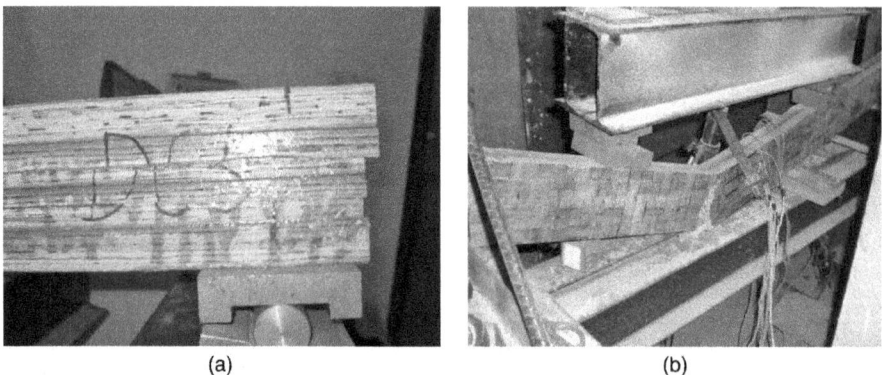

(a)   (b)

Figure 6.8 Different failure patterns for thin-strip glubam beam: (a) sliding shear failure for beam with flatwise lay up; (b) bending failure of specimen with sidewise layup.

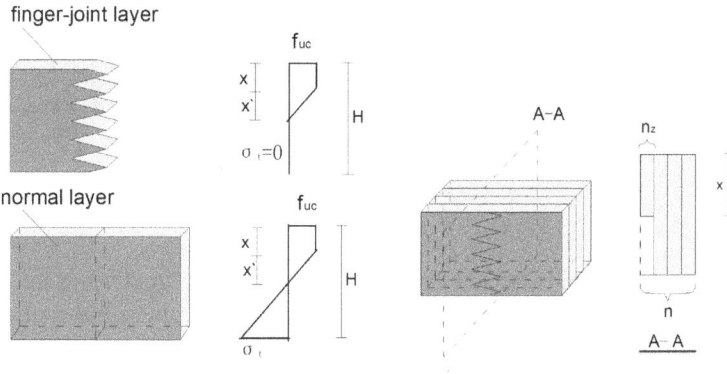

Figure 6.9 Finger joint glubam structure.

More accurate analytical methods should be developed in future based on the approach of composite mechanics. Only an engineering approach for the flexural capacity analysis is considered in this book.

Considering the section configuration of the glubam beam with sidewise layup lamination, and the experimental fact that failure began from a section with a finger joint, Zhou devised a relatively simple approach. Conventional plane assumption is adopted, along with neglecting the tension zone of the layer with the finger joint, as shown in Figure 6.9.

The calculation is mainly focused on the stress distribution of the fracture section of the test piece. When loading to the ultimate bearing capacity, it is assumed that the fracture section must be at a finger joint of the pure bend section. At the same time, it is assumed that in the limit state, the bamboo fiber at the edge of the cross-section compression zone reaches the maximum compressive stress. Since the cross section is weakened, the pressed bamboo material is peeled off layer by layer immediately after the yielding, thereby causing a sudden brittle fracture. A linear elasto-brittle stress-strain relationship is assumed for tension, whereas the elasto-plastic stress-strain model is assumed for the material in compression, and the outermost fibers both in compression and in tension reach the ultimate strength at the same time.

As shown in Figure 6.9, the compressive resultant force and the tensile resultant force are in equilibrium, therefore:

$$f_{uc} x n_0 t + \frac{1}{2} x' f_{uc} n_0 t = \frac{1}{2}(H - x - x') f_{ut}(n_0 - n_J)t \tag{6.2}$$

where, n_0 is the total number of layers of plybamboo board, and n_j is the layers with the joints in the section under consideration; t is the thickness of one board; x and x' are the depth in the compression zone; and x can be expressed as follows:

$$x = \frac{\xi h - \frac{1}{2} f_{tu}(n_0 - n_j)th}{\xi - \frac{1}{2} f_{tu}(n_0 - n_j)t - f_{cu} n_0 t} \tag{6.3}$$

where $\xi = \left(\dfrac{1}{2} f_{cu} n_0 t + \dfrac{1}{2} f_{tu} \left(n_0 - n_j \right) t \right) \dfrac{f_{cu}}{f_{cu} + f_{tu}}$. Therefore, the stress can be integrated along the height of the section, and the ultimate bending moment of the section can be obtained by counting all the internal moments about the neutral axis as follows:

$$M_u = \left(f_{uc} x \left(\frac{1}{2} x + x' \right) + \frac{1}{3} x'^2 f_{uc} \right) tn_0 + \frac{1}{3} f_{ut} (h - x - x')^2 \left(n_0 - n_j \right) t \tag{6.4}$$

Alternatively, the ultimate moment capacity can also be simply calculated using conventional beam theory by ignoring the layer with the finger joint, and taking the smaller value of tension or compression strength, as follows:

$$M_u = \min \left[\left(n_0 - n_j \right) th^2 f_{uc} / 6, \left(n_0 - n_j \right) th^2 f_{ut} / 6 \right] \tag{6.5}$$

For the loading condition, the vertical load corresponding to the moment capacity can be calculated as $P_{cal} = 2M_u/a$, where a is the shear span length. The calculation results using the simple approach are shown in Table 6.2. For the thick-strip glubam beam specimens, the calculation overestimates the test results; however, it underestimates the results for the thin-strip glubam beams.

Using smaller specimens with thick-strip glubam (referred also as LBL), Li et al.[13] investigated the behavior of glubam beams with different section width to height ratios and shear span ratios. Their specimens had an effective length of 2100 mm and had no lengthening joints. Thus, the flexural capacities of Li et al.'s specimens can be calculated using conventional beam theory without considering the reduction of finger jointed layers; i.e., $M_u = bh^2 f_{uc} / 6$. Table 6.3 shows the main testing parameters, test results, and calculated load carrying capacities.

Table 6.3 Testing parameters and results by Li et al.

Specimen	l_o (mm)	h (mm)	b (mm)	a/h	a (mm)	P_{max} (kN)	P_{cal} (kN)	P_{cal} / P_{max}
SL2.5-1	2100	100	45	2.5	250	39.7	35.2	0.89
SL2.5-2	2100	100	45	2.5	250	54.0	35.2	0.65
SL5-1	2100	100	45	5	500	21.9	17.6	0.80
SL5-2	2100	100	45	5	500	22.1	17.6	0.80
LL5-1	2100	100	55	5	500	22.4	21.5	0.96
LL5-2	2100	100	55	5	500	19.7	21.5	1.09
TL7-1	2100	100	35	7	700	9.6	9.8	1.02
TL7-2	2100	100	35	7	700	10.0	9.8	0.98
SL7-1	2100	100	45	7	700	13.7	12.6	0.92
SL7-2	2100	100	45	7	700	13.5	12.6	0.93
WL7-1	2100	100	50	7	700	13.0	14.0	1.08
WL7-2	2100	100	50	7	700	14.9	14.0	0.94
LL7-1	2100	100	60	7	700	17.2	16.8	0.97
LL7-2	2100	100	60	7	700	15.6	16.8	1.08

Source: Li et al. (2015).[13]

As shown in Table 6.3, it is clear that conventional beam theory can estimate the beams without lengthening reasonably well. From the analysis of Li et al.'s results and Yang's results,[9] it seems highly likely that the figure jointing failure may have more effects than can be accounted for by the current analysis.

6.1.5 Moment–Curvature Analysis

In Yang's tests,[9] longitudinal strains of the beam specimens were carried out for the mid-span section at different heights, as shown in Figure 6.4. The curvature can then be calculated for the section using the readings of the two strain gauges near the top and bottom edges. The moment and curvature relationships can then be presented for the specimens as shown in Figure 6.10. Alternatively, an Excel macro program based on a fiber model can be established, with plane assumption and strain and stress conditions as schematically shown in Figure 6.11, using the following stress-strain model developed in Chapter 3:

$$\text{Tension:} \quad 0 \le \varepsilon(y) \le \varepsilon_{tu}, \quad \sigma(y) = E_t \varepsilon \tag{6.6a}$$

$$\text{Compression:} \quad \varepsilon_{cu} \le \varepsilon(y) \le 0, \quad \sigma(y) = f_c \frac{xr}{r - 1 + x^r} \tag{6.6b}$$

As shown in Figure 6.11, for a given curvature, there is a compression zone with depth x, which can be determined based on the equilibrium condition $T = C$ of the section. The corresponding moment M can then be obtained by integrating the moment contributed by each fiber at a distance y to the centroidal axis..

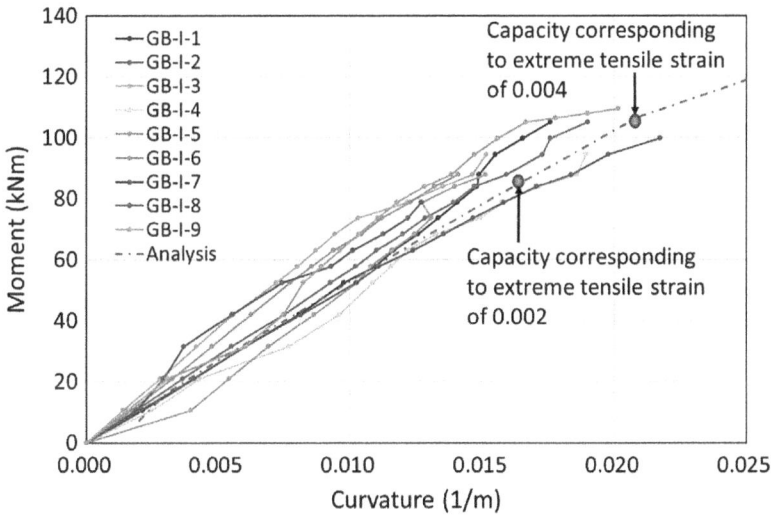

Figure 6.10 Experimental and analytical moment–curvature relationships.

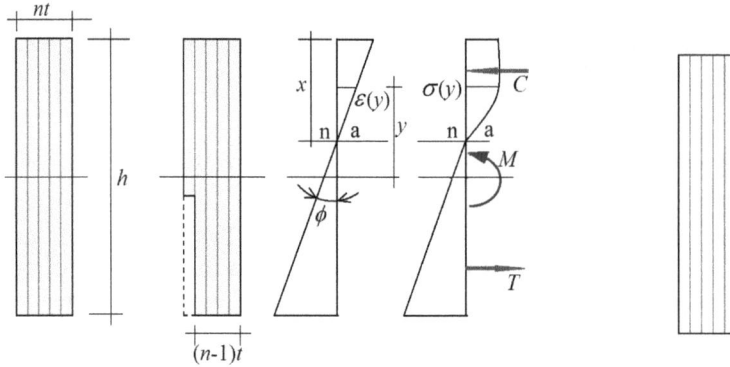

Figure 6.11 Stress-strain conditions for a section.

As can be seen in Figure 6.10, the analysis can predict the experimental moment and curvature curves reasonably well. From Yang's test results, it can be shown that the tensile strain near the soffit of the beam ranges from 0.002 to 0.0035, which is close to $0.5(f_{tu}/E_t)$. If the extreme tensile strain is set as 0.002 and 0.004, the moment capacities are estimated as $86.0\,kNm$ and $105.0\,kNm$, respectively, which are close to the testing capacities of the specimens.

6.1.6 Design Considerations for Glubam Beams

Design for glubam members subjected to bending can be summarized based on the timber codes and some modifications.

Flexural capacity based on sectional strength: For simplicity, the following equation can be used for calculating the nominal moment capacity,

$$M_n = W_n f_m \tag{6.7}$$

where M_n is the nominal flexural capacity of the section, W_n is the net sectional modulus; and f_m is the design bending strength of glubam. In calculating W_n, if the section includes a layer of glubam board with finger joint, its thickness should be excluded. More rigorous methods are given in Eq.6.5 and Eq.6.6 or based on moment curvature analysis.

Flexural capacity based on lateral torsional buckling: At the current stage, there is no specific research information available on the lateral torsional buckling of engineered bamboo members, but considering that the bamboo is essentially used within its elastic range, the conventional buckling theory can be assumed for calculating the critical moment,

$$M_{crit} = \frac{\pi}{l_{ef}} \sqrt{\frac{EI_z JG}{1 - \frac{I_z}{I_y}}} \tag{6.8}$$

where I_y, I_z are the moments of inertia of the respective axes; J is the torsional modulus; l_{ef} is the length without lateral constraint; E is the modulus of elasticity; and G is the modulus of shear. For the nominal capacity, the codified approaches can be followed,

$$M_n = \varphi_l W_n f_m \tag{6.9}$$

where, φ_l is the beam stability factor and can be calculated based on the codified equations. It should be noted that in calculating the capacity based on Eq.6.9, the full section can be counted in W_n.

Design shear capacity: Based on conventional beam theory, the design shear strength of a beam section can be calculated as,

$$V_n = f_v Ib / S \tag{6.10}$$

where f_v is the design shear strength; I is the moment of inertia of the section; b is the section width where the maximum shear stress is considered; and S is the first areal moment of the portion of the section above the location with maximum shear stress. Apparently, for rectangular section, Eq.6.10 can be simplified as,

$$V_n = 2 f_v A / 3 \tag{6.11}$$

in which A is the sectional area. It should be mentioned that the detailed design for shear should also consider the actual situation of load paths based on support details.

Deflection limit: As described previously, the design of glubam bending members may be controlled by deflection limits due to relatively low stiffness. Table 4.14 provides the deflection limits, which should not be exceeded by the deformation calculated using the design load. In calculating the deflection, the EI of the full section can be used.

This section summarizes several research topics related to fiber reinforced polymer (FRP) enhanced engineered bamboo beams, long-term loading tests, fatigue behavior of engineered bamboo beams, etc.

6.2 Research Updates on Various Glubam Beams

6.2.1 Fiber Reinforced Polymer (FRP) Enhanced Glubam Beams

As discussed in previous sections, whether it is glulam or glubam, the lower modulus of elasticity limits the performance of the material, especially in the application of large span beams and girders. Conventionally, metal materials are used to strengthen the timber beams.[14] However, metals are susceptible to corrosion and the weight is also high. Since the development of FRP in the 1960s, some scholars began to carry out research on FRP materials to enhance timber and glulam; in particular, with the continuous decline of the price of FRP and binder, the application of FRP in the field of structure has gradually increased.[15–20] To explore the possible merits of FRP to enhance engineered bamboo beams, Zhou carried out preliminary studies on carbon fiber reinforced polymer (CFRP) enhanced glubam.[11]

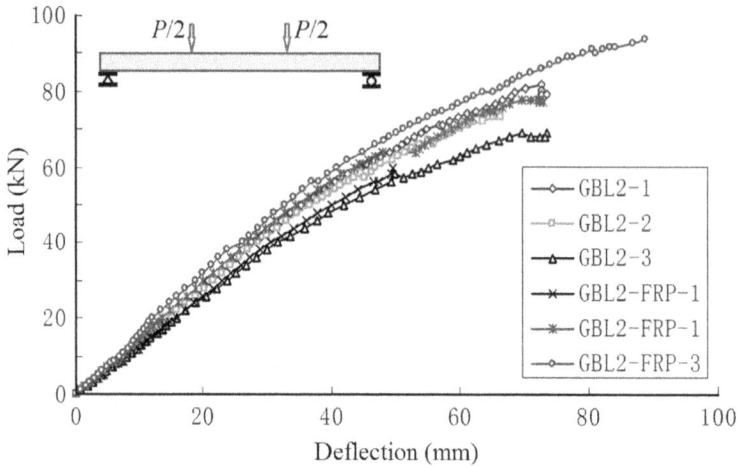

Figure 6.12 Comparison of load-deflection curves of glubam beams with and without CFRP enhancement.

Source: Zhou (2012).[11]

For full-scale glubam girders with a span of about 10 m, the CFRP strengthened specimens are shown to have apparent enhancement effects in terms of stiffness of the girders.[11] These full-scale glubam girders were designed for construction of a truck-safe bridge.[5] Smaller-size glubam beams without figure joints were also tested by Zhou[11] and the results are shown in Figure 6.12. The average results for these small-size glubam beams show slight enhancement effects when using CFRP. Based on the strain gauge readings, Zhou also confirmed that the tensile strains of the glubam beam are restrained by the CFRP, indicating an enhancement effect.

6.2.2 Long-term Creep Behavior of Glubam Girders

Similar to timber, bamboo also exhibits time dependency, and can creep or continue to deform under constant stress over time. Significant research has been carried out on the creep behaviors of engineered timber;[21,22] however, very little is known for creep behavior of engineered bamboo. Apart from the material creep testing described in Chapter 3, the author's group also conducted creep testing on a 9.4 m long bridge model, as shown in Figure 6.13.[23] The soffit of the glubam girders (depth 600 mm, width 120 mm) was further enhanced with carbon fiber reinforced polymer (CFRP) sheets. The mid-span deflections of the bamboo girders were obtained over the course of 3.7 years from creep tests under the gravity load of the bridge. The results indicate that the average creep deflection of the girders after 3.7 years was within the acceptable range.

As shown in Figure 6.13, following the construction of a real truck-safe bridge in Leiyang, China, a model bridge for long-term loading was built at Hunan University, in December 2007. The test bamboo bridge model was composed of two 10 m long glubam

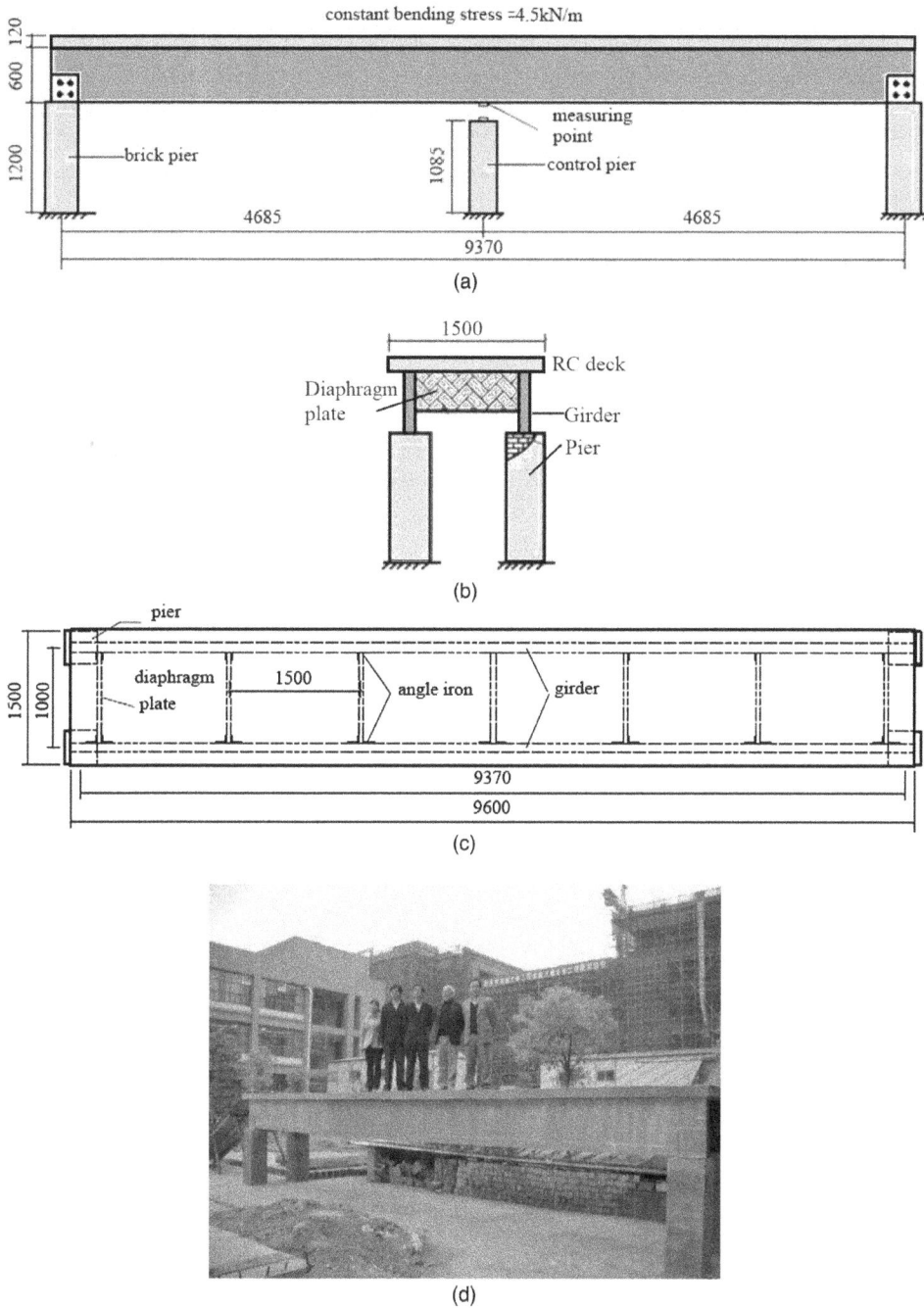

constant bending stress =4.5kN/m

600   120

1200

brick pier

4685

1085

measuring point

control pier

4685

9370

(a)

1500

RC deck

Diaphragm plate

Girder

Pier

(b)

pier

diaphragm plate

1500

angle iron

girder

1500   1000

9370

9600

(c)

(d)

Figure 6.13 Long-term performance test model bridge (unit: mm): (a) side view; (b) cross section; (c) top view; (d) actual scene.

girders with sections of 100 mm×600 mm, which support concrete slab planks (120 mm thick, 1000 mm wide, and 1500 mm long). The concrete planks were simply placed on top of the girders without any special connection; thus no composite effect was expected. The girders were supported on brick pillars, so the clear span was 9370 mm between the supports. The soffit surfaces of the glubam girders were enhanced with CFRP sheets. Seven transverse diaphragms with a section thickness of 30 mm were evenly arranged between the main girders, and 63 mm × 40 mm × 8 mm steel angles were connected with the girders.

The long-term load acting on the bamboo girders came mainly from the weight of the upper concrete deck and the weight of the bamboo beam. The self-weight of the bridge deck and glued bamboo is 2500 kg/m^3 and 890 kg/m^3, respectively. Therefore, the total creep load is about 2.3 kN/m per girder, about 0.1 times the bearing capacity of the girder based on short-term loading tests.[5] The long-term test began in October 2007 and went on until July 2011, lasting 1350 days, about 3.7 years. Since the creep rate of the glubam material gradually slows down with time, the mid-span deflection value was measured every day for the first 7 months of the test, and the deformation value was measured once a week for the next 14 months. Finally, recording was reduced to once a month until the end of the test. The results of the deflection over the time period are shown in Figure 6.14, in which the creep deformation of the supporting pillars was excluded.

As shown in Figure 6.14, the primary stage of the creep curve, or the transient creep stage, describes the first 800-day period during which the rate of deflection constantly decreases. This decrease continues until the secondary stage is attained after 800 days.

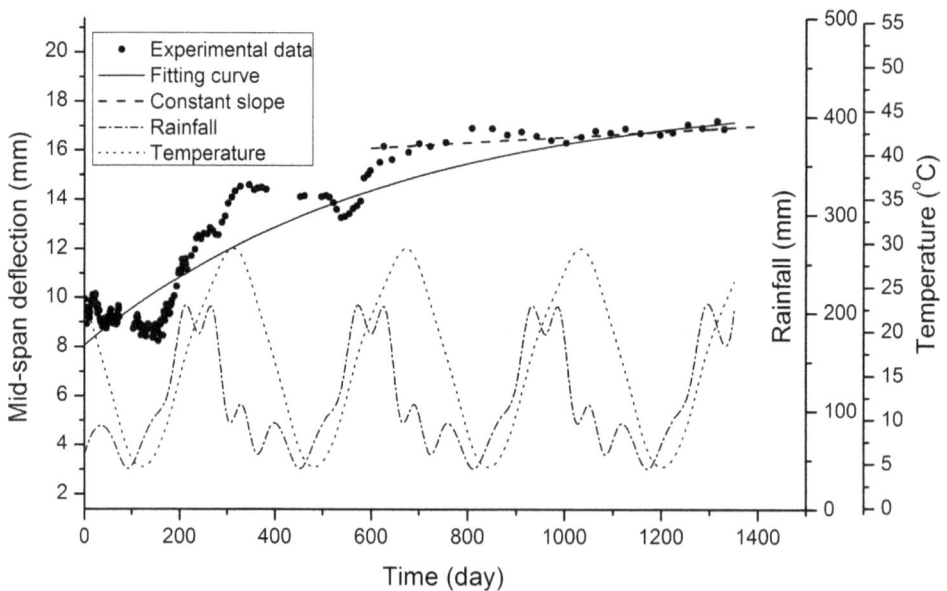

Figure 6.14 Long-term vertical deflection at mid-span of model glubam bridge.
Source: Xiao et al. (2014).[23]

The secondary stage, or steady-state stage of creep, is usually considered the period of constant rate of deflection. Thus, this region appears to be a straight line with a certain slope. The average creep deflection after 1350 days is 7.89 mm, which is below the limiting value of 1/600 of the span length, as per the Chinese bridge design code JTG 262.[24] The constant rate (the slope of the creep) of the secondary stage between 800 days and 1350 days in Figure 6.14 is approximately 1.159×10^{-3} mm per day, so the total creep of the bridge after 25 years is predicted as about 17 mm, which is slightly higher than the Chinese code's limit.

Li and the author[25] also simulated the long-term deformation of the model beam according to the creep test results for glubam material and the carbon fiber creep model, which is consistent with the long-term test results.

After 3.7 years of long-term loading, a short-term collapse test was carried out on the model bridge. Loading was executed using cast iron weight blocks and sandbags. The final failure mode was identified as being the flexural failure of a section where a finger joint of one layer of the glubam girder was located. The short-term failure loading test results show that the glubam girders had adequate ultimate load carrying capacity and stiffness, comparable with the results from short-term testing of virgin specimens, even after 3.7 years of creep loading.

6.2.3 Fatigue Performance of Glubam Beams

For structures such as bridges and beams supporting cranes subjected to dynamic loads, fatigue may cause the structural failure at a load level lower than the static capacity.[26,27] However, even for the well-established timber structures, study on their mechanical behaviors and failure under fatigue loading is not necessarily sufficient.[28] Several studies have recently been carried out on timber at the material level;[29–32] however, the study of structural performance under fatigue loading is still rare.[33–35] There are a few recent research studies on fatigue in bamboo and engineered bamboo.[36–40] This section briefly describes a preliminary study by the author's research group on full-scale engineering bamboo beams tested under fatigue loading.[11]

In this test, a total of 6 glulam beams with the same dimensions and finger joints were designed. The cross-section dimensions of the test pieces were 84×450 mm ($b \times h$), the length was 6200 mm, and the clear span between supports was 6000 mm. The test specimens were divided into two groups, and the tests were carried out with four-point loading, using the same setup shown in Figure 6.4, except that a fatigue actuator was used. Specimens L1, L2, and L3 were subjected to a 2 million-cycle fatigue loading, whereas specimens L4, L5, and L6 were tested as a counterpart by quasi-static loading.

The fatigue test was carried out using a Jinan-Shijin PMS-500 fatigue testing machine. The test was carried out in equal amplitude loading mode with a loading frequency of 4.2 Hz. Based on the Chinese code JTG D62 requirement,[24] the load corresponding to the 1/600L displacement was estimated at 15 kN. The upper load of the fatigue testing was set below the estimated design load of 15 kN. First, a small fatigue load was used for specimen L1, with a load range 1 kN–6 kN. The fatigue load amplitude of test specimen L2 was 5 kN–10 kN, and the fatigue load amplitude of the L3 specimen was 6 kN–12 kN.

Before the fatigue loading, the specimen was loaded statically to a load equivalent to the design load level and the stiffness was recorded. Similar loading was also executed when the number of load cycles reached 100,000, 300,000, 500,000, 1 million, 1.5 million, and 2 million, respectively, to measure the deflection, frequencies, and strains, and observe the crack development. At each stop, an LC1302 impulse hammer device and 13 accelerometers affixed at the mid-height of the beam were used to collect the pulse vibration mode data, and the LMS PolyMax method was used to compute the frequency.

Figure 6.15 shows the static test results at each stop for specimens L1, L2, and L3. As shown in Figure 6.15, the curves of specimens L1 and L2 are almost linear and almost duplicate the initial loading result after repetitive loading at different cycles, until 2 million cycles. The stiffness change is small for L1 and L2. However, specimen L3 is seen to have continued reduction in stiffness following the increase of the loading cycles, due probably to defects in the beam.

After 2 million repeated loadings, specimens L1 and L2 appeared to be intact; however, L3 was seen to have weakened. Therefore, the static failure tests were carried out on the three beams and results compared with those for their counterpart static virgin test specimens L4, L5, and L6. As exhibited in Figure 6.16, after 2 million cycles of loading, the stiffness of the L1 and L2 specimens is almost the same as the virgin specimens; however, the failure capacities area is about 15% smaller than the average loading capacity of L4, L5, and L6.

6.3 Glubam I-joists

I-joists are widely used in timber buildings for supporting floor and roof systems because of their cost-effectiveness, light handling, and ability to carry large loads over relatively long spans.[41,42] It is reported that approximately 50% of wood light- framed floors are built with I-joists.[43] In order to explore the advantages of glubam and the I-joint system, the author's research group has carried out a study on glubam I-joists, with spans ranging from 2.4 m to 7.5 m, as shown in Figure 6.17.[44] Experimental parameters include the types of finger joints, interface details between flanges and web, and specimen lengths with different locations relevant to the loading.

As shown in Figure 6.18, two types of web-to-flange interface connections were designed. The first type was designed and utilized with series A and B specimens, in which the upper and lower edges of the web were inserted into the pre-cut grooves in the flanges, as shown in Figure 6.18 (a). The depth of the grooves is 30 mm, equal to half of the flange thickness. Racher et al. reported that the continuous web-to-flange connection might fail due to the combination of shear and tension stresses in the glued surface induced by the plate behavior of the web.[45] Therefore, a new type of web-to-flange connection was designed for improving the behavior in the shear span and was used with series C and D specimens. As shown in Figure 6.18 (b), rectangular cuts were made in the upper and lower edges of the web within the shear span to form several rectangular shear keys, and then the toothed edges were inserted into the pre-cut grooves in the flanges.

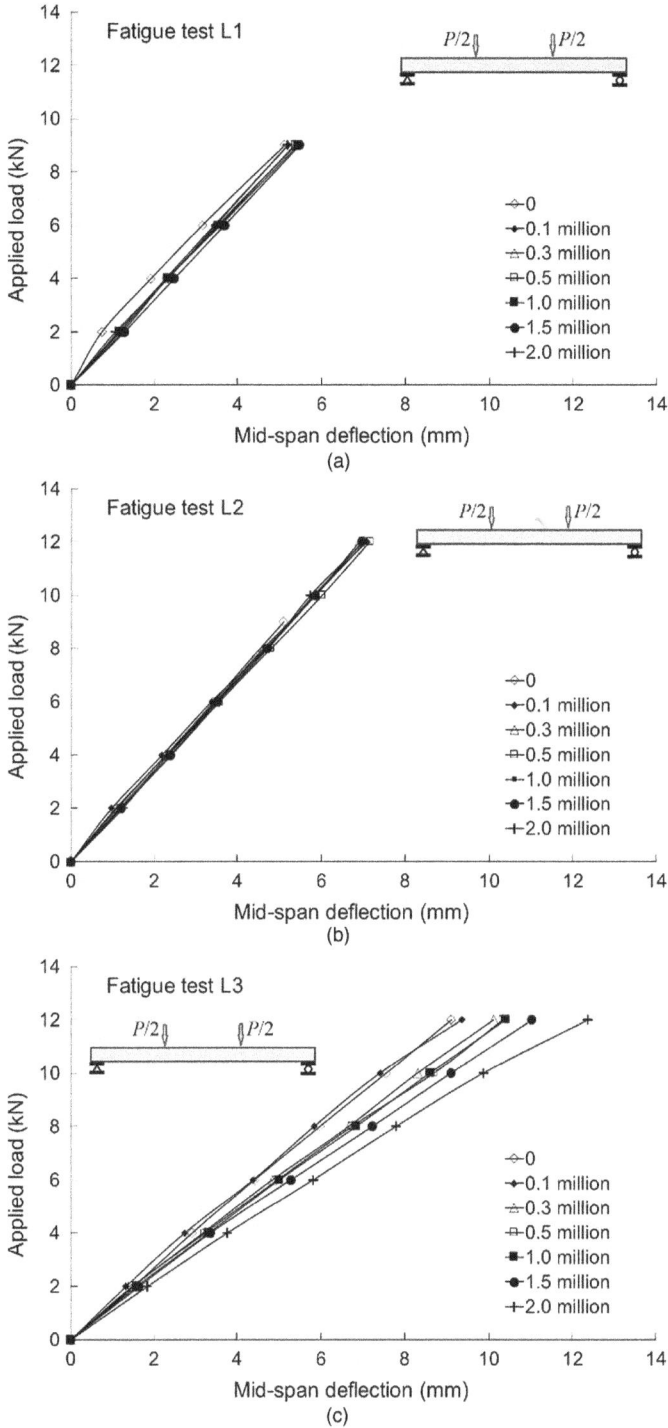

Figure 6.15 Static load–deflection curves at stops after different fatigue loading cycles: (a) specimen L1; (b) specimen L2; (c) specimen L3.

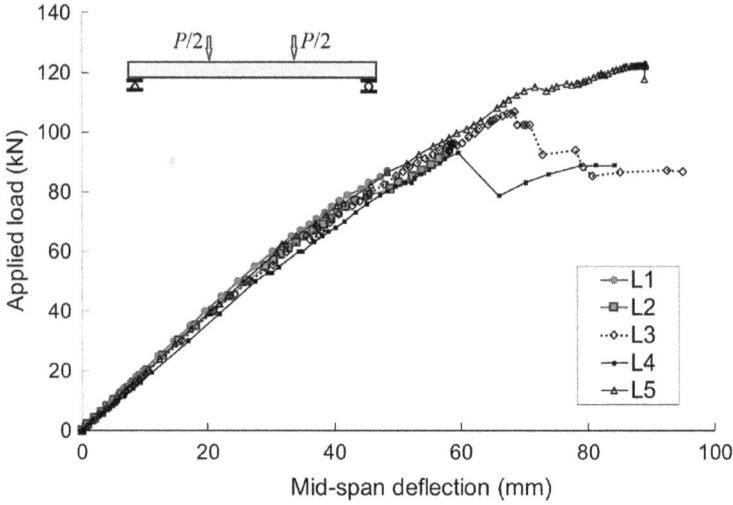

Figure 6.16 Static failure tests of glubam beams after 2 million cycles of fatigue loading (L1, L2) and virgin specimens (L4, L5, and L6).

As shown in Figure 6.17, four-point bending tests were conducted to examine the failure modes, load-deflection relationships, and load carrying capacity of glubam I-joists. Experimental results indicated that the dominant failure modes of glubam I-joists included shear failure at the finger joint in the web, bending failure at the finger joint in the bottom flange, and lateral buckling. Correspondingly, the load carrying capacity of glubam I-joists was governed by the bending strength, shear capacity, and critical bending moment. Glubam I-joists have a relatively higher mechanical performance compared with other engineered bamboo or timber I-joists with similar dimensions. Both types of web-to-flange interface connections are reliable without apparent premature failure.

The stiffness of the glubam I-joists can be estimated based on the following equation following the suggestion of APA – The Engineered Wood Association,[46]

$$\frac{1}{EI} = \left[\frac{1}{E_w I_w + E_f I_f} + \frac{24}{(3L^2 - 4a^2) G_w h_w t} \right] \tag{6.12}$$

where I_f (I_w) = moment of inertia of flange (web); E_f (E_w) = elastic modulus of the flange (web); G_w = shear modulus of the web and equal to 0.75 E_w; h_w = depth of the web. The bending capacity of the glubam I-joists with both types of web-to-flange connections meet the requirement specified in Chinese code GB/T 28985.[47] However, for the specimens with toothed interface connections, the shear capacity specified by the Chinese code was not satisfied, probably due to the disconnection of shear flow due to the teeth. For specimens that failed in bending, the moment capacity can be

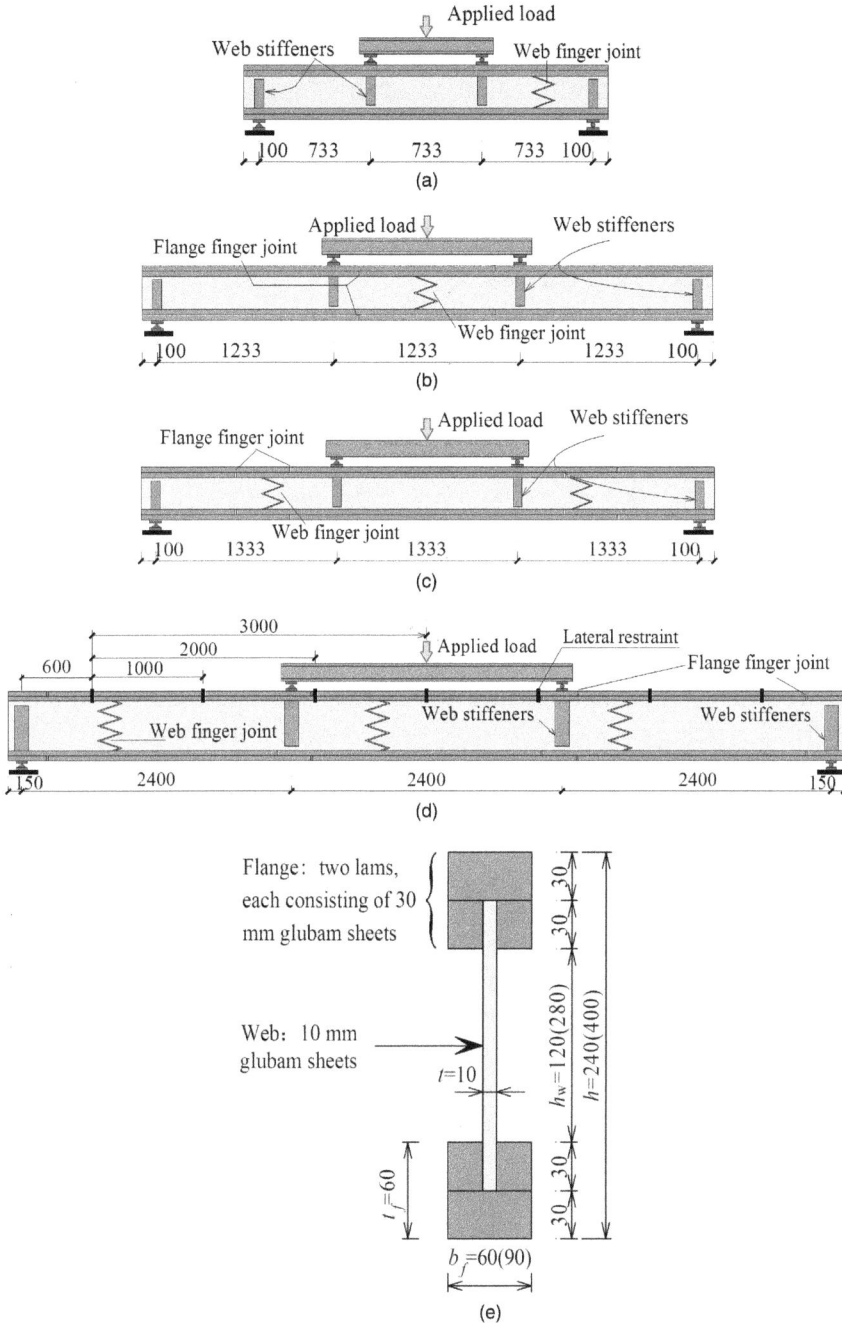

Figure 6.17 Glubam I-joist specimens: (a) testing series A; (b) testing series B; (c) testing series C; (d) testing series D; (e) cross section.

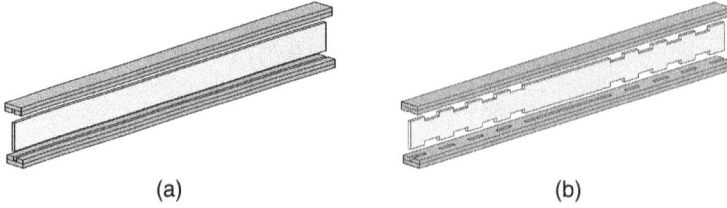

Figure 6.18 Web-to-flange interface connections: (a) for testing series A and B specimens; (b) for testing series C and D specimens.

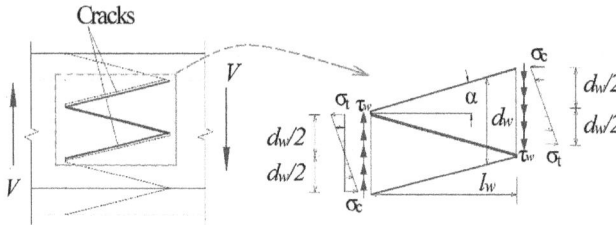

Figure 6.19 Assumption for stress condition in finger joint.

estimated based on the method specified in the Canadian Standards Association (CSA) standard:[48]

$$M_{th} = f_a \frac{E_f I_f + E_w I_w}{E_f c_f} \tag{6.13}$$

where c_f = the greatest distance from the neutral axis to the outer edge of the flange; f_a = the minimum value between the compression and tensile strength of flange material and the tensile strength of the flange finger joint. For shear strength, Tang et al.[44] proposed the following theoretical equation for web shear strength based on modification of the CSA standard:[48]

$$V_{th} = \left(\frac{f_{tw} d_w}{3 l_w}\right) h_w = \left(\frac{f_{tw} d_w}{3 f_{\tau w} l_w}\right) V_P h_w \tag{6.14}$$

where f_{tw} is the in-plane material tensile strength; $V_P = f_{\tau w} t$, the sum of strengths of all panel webs in shear-through-thickness; $f_{\tau w}$ = strength of web material in shear at the plane, t = thickness of the web; l_w = length of a fingertip; d_w = depth of one fingertip, as defined in Figure 6.19.

It is also confirmed that the following critical lateral torsional buckling strength can be used to predict the specimens that failed in buckling mode.

$$M_{cr} = \beta_b \frac{\pi^2 h E I_y}{2 l_y^2} \tag{6.15}$$

$$\beta_b = \frac{12.5 M_{max}}{2.5 M_{max} + 3 M_A + 4 M_B + 3 M_C} \tag{6.16}$$

where β_b = equivalent bending moment coefficient; EI_y = lateral bending stiffness; l_y = length of the span between points of supports or lateral restraints; h = depth of I-joist; M_{max} = absolute value of the maximum bending moment in the span without lateral resistance; M_A, M_B, and M_C = the absolute value of the bending moments at the 1/4 span, at the mid-span, and at the 3/4 span without lateral resistance, respectively.

Other types of engineered bamboo I-joists were also investigated by Chen et al.,[49,50] including the box type of I-joists in which the double webs are connected with flanges on both sides, and OSB (oriented strand bamboo) webbed I-joints. The combination of OSB and engineered bamboo is probably a good way to achieve further cost-effectiveness. In Chen et al.'s testing, the interface connections between webs and flanges are nailed instead of being glued. In particular, they also studied the I-joints with openings with different shapes in the web.

6.4 Bamboo Concrete Composite (BCC) Beams

Timber and concrete composite (TCC) beams have been shown as an efficient form of structure and found wide applications.[51–53] In a TCC composite beam, glulam beams or girders are connected with reinforced concrete slabs. A similar composite system using glubam beams and reinforced concrete slabs was investigated by Shan et al.[54,55] As shown in Figure 6.20, Shan et al. tested 9 full-scale BCC beams with a T-shape section with a length of 8000 mm. The same cross-sectional dimensions consisted of a 900×100 mm (width × depth) reinforced concrete slab and a 112×380 mm (width × depth) glubam beam.

Four types of glubam beam and concrete slab interface connection details were designed and tested: continuous steel mesh (SM), screw connector (SC), notch connector (NC), and post-tightened notched connector (PNC), as shown in Figure 6.21 (a) to (d), respectively. The former three types are for construction of cast-in-place composite beams and the PNC detail is designed for installing precast concrete slabs. Prior to the studies of the composite beams, Shan et al. also conducted shear tests to investigate different connection details.[54] Experimental results indicated that all BCC beams with different connections exhibited satisfactory performance under short-term loading conditions. The 200 mm long notch connection is recommended for composite beams for higher load carrying capacity and composite action as well as the merit of fewer connectors. The semi-prefabricated composite beam using the PNC shows a similar bending capacity compared with the corresponding cast-in-place BCC, indicating a good prospect for prefabricated and assembled structures.

(a)

(b)

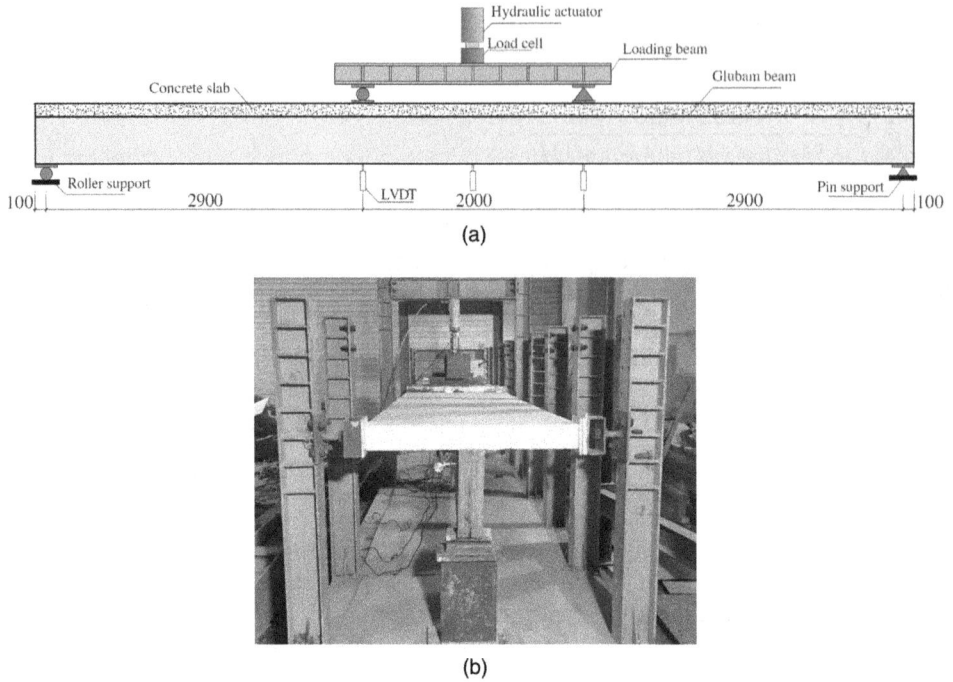

Figure 6.20 Shan et al.'s tests of BCC beams: (a) test setup; (b) loading frame.

Source: Shan et al. (2020).[55]

6.5 Glubam Members Subjected to Tension

Tension members are the most fundamental units in many structural systems, particularly in trusses. Natural materials such as wood and bamboo have very good tensile properties; however, due to difficulties in joining the tensile members with other supporting systems, their usage is somewhat limited. Figure 6.22 shows a suspended garage storage shelf with tension members connecting the shelf to the roof truss.

For timber tension members with force parallel to the grain direction, the design is based on the smallest net section, A_n. The design tensile capacity T_n is thus,

$$T_n = A_n f_t \tag{6.17}$$

where f_t is the design tensile strength along the grain direction. For timber, loading to induce tension stresses perpendicular to the grain should be avoided whenever possible; otherwise, mechanical reinforcement should be designed. For engineered bamboo, the tension member design can basically follow the same approach as for timber design, using the strength values given in Chapter 4 or based on experiment. For glubam with only longitudinal bamboo fibers provided (typical for thick-strip glubam), tension in

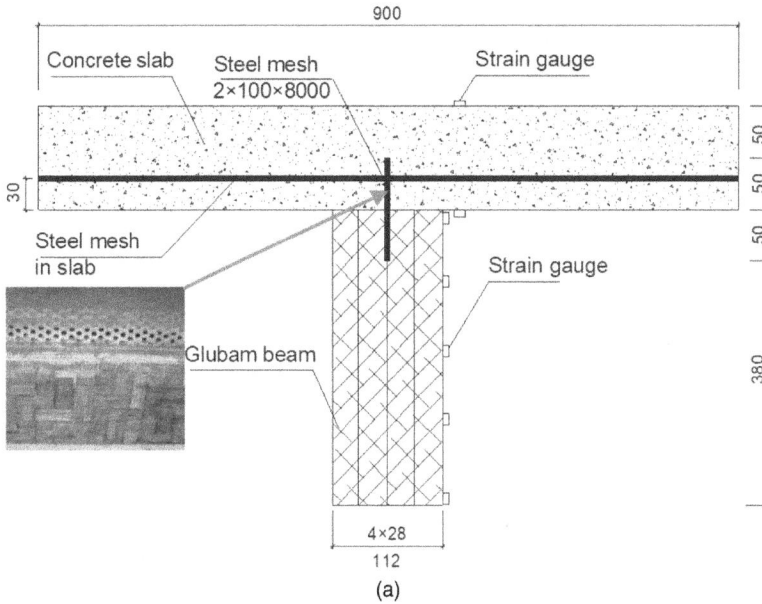

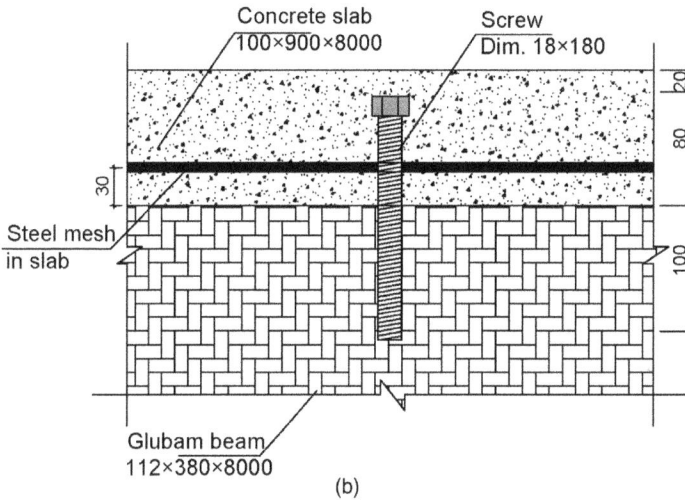

Figure 6.21 Details of connections (dimensions in mm): (a) steel mesh (SM); (b) screw connector (SC); (c) notched connection (NC); and (d) post-tightened notched connection (PNC).

the direction perpendicular to longitudinal should be avoided. For bidirectional glubam (such as the typical 4:1 thin-strip glubam), some tensile capacity does exist in the transverse direction; however, caution is needed, as the capacity is generally small.

Needless to say, for tension members, the design for connections is most important, as discussed in Chapter 5.

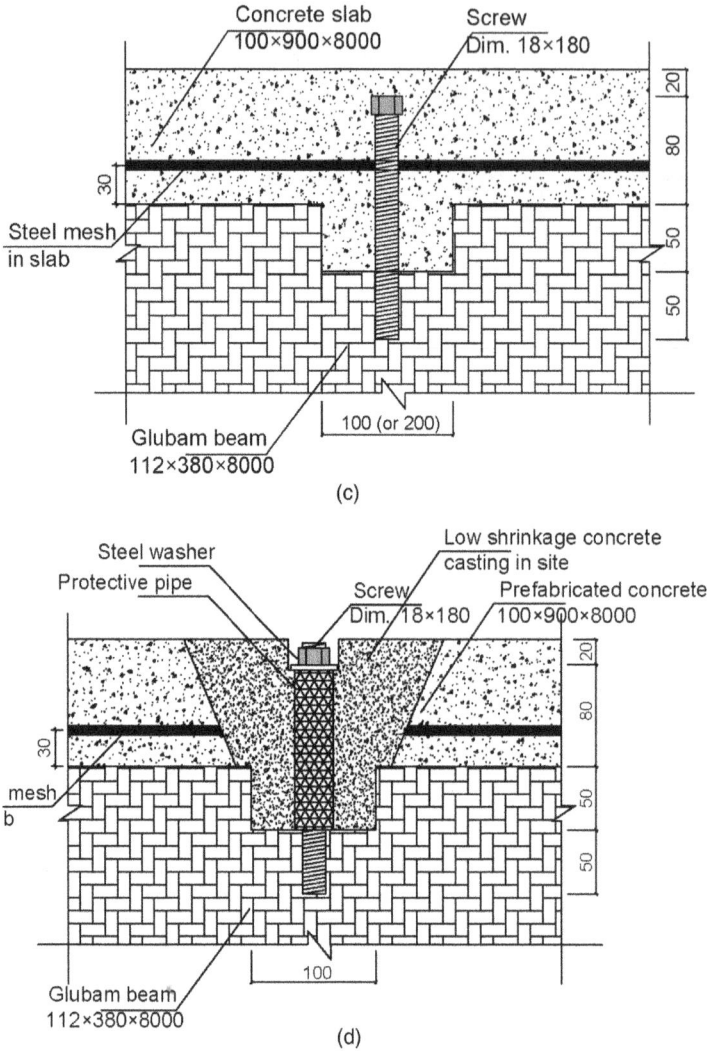

Figure 6.21 Continued

6.6 Glubam Members Subjected to Compression

In most structural systems, the column is the most typical structural member subjected to compression, and is the most important structural component, as it holds the fundamental key function to transfer the weight of the upper structure to the foundation. Columns meanwhile also need to resist lateral loadings from wind, earthquake, etc. Therefore, research on structural columns is one of the major research areas in structural engineering.

Figure 6.22 An example of tension members in a timber structure.

6.6.1 Columns and Members under Axial Compression

Unless special end support conditions are designed and provided, columns under pure axial compression are rare: Figure 6.23 (a) and (b) show examples of timber columns photographed by the author. However, compressive members in trusses can usually be treated as subjected to axial compression, as shown by one of the diagonal members in a spatial bamboo–steel hybrid truss, exhibited in Figure 6.23 (c).[56]

Design of members under compression for engineered bamboo can follow similar procedures to those for timber structures, based on the smaller of the following two design criteria:

Design for strength: When a member subjected to compression is squat enough or laterally supported throughout its length, the capacity may be determined by its material compressive strength, and the compressive capacity N_n is given as,

$$N_n = A_n f_c \tag{6.18}$$

where A_n is the net sectional area; and f_c is the design compressive strength along the grain direction.

Design for buckling: When a member is slender, the compressive capacity may be determined by critical buckling strength N_{cr},

$$N_n = N_{cr} = \varphi A f_c \tag{6.19}$$

(a)

(b)

(c)

Figure 6.23 Examples of compression members: (a) pinned column; (b) columns with fixed base and pinned top; (c) compression members in a spatial truss.

where A is the cross-sectional area if there is no weakening – otherwise, the code-specified calculation specifications should be followed; φ is the stability factor, and will be discussed in the following sections.

6.6.2 Buckling Theory and Research Background on Timber Columns

6.6.2.1 Euler Elastic Buckling Theory

When a member is subjected to pure axial load, the stress in its section can be taken as $f = P/A$. If a member is sufficiently long (slender), it may not develop its material strength-based capacity, AF_y. The member may become unstable in maintaining its original shape and geometry under a smaller load in compressions ($P < AFy = Pu$). This phenomenon is called buckling, or instability. When a member buckles, its shape

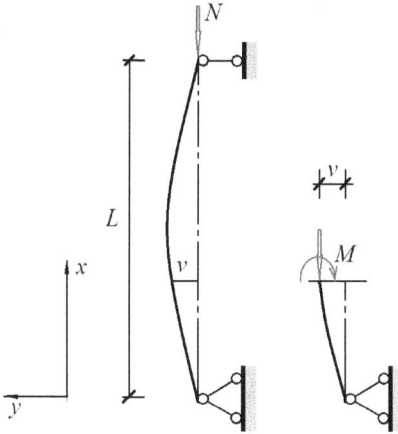

Figure 6.24 Pinned column.

changes to a new geometric form, while the equilibrium is maintained under a specific load, which is the critical buckling load.

As shown in Figure 6.24, let us consider the column pinned at its ends (pin support at one end and roller support at the other). For elastic buckling, a new equilibrium can be achieved by the member in the buckled shape. Thus,

$$M - Pv = 0 \tag{6.20}$$

On the other hand, the moment–curvature relationship of the elastic beam can be found from Bernoulli's equation,

$$\frac{M}{EI} = -\frac{dv^2}{dx^2} \tag{6.21}$$

(Note: For the positive deflection shown in Figure 6.24, the current sign convention assigns the moment as negative. Because actual direction is marked in Figure 6.24 and used in establishing the equilibrium, and because the curvature shown in Figure 6.24 is also negative, a minus sign "-" has to be used in the above Bernoulli's equation.) Further letting $\xi = \sqrt{\dfrac{P}{EI}}$, the following second-order homogeneous differential equation can be derived,

$$\frac{dv^2}{dx^2} + \xi^2 v = 0 \tag{6.22}$$

This fundamental elastic buckling differential equation has a general solution,

$$v(x) = A \sin \xi x + B \cos \xi x \tag{6.23}$$

Using boundary conditions, the constants of integration can be determined as follows,

$v(0) = 0 \rightarrow B = 0$

$v(L) = 0 \rightarrow \sin \xi L = 0$

The non-trivial solution is $\xi = \dfrac{n\pi}{L} = \sqrt{\dfrac{N}{EI}}$, therefore,

$$N = \frac{n^2 \pi^2 EI}{L^2} \tag{6.24}$$

For engineering problems, only the least value of P is important; thus, taking $n = 1$, we have the following critical load or Euler load,

$$N_{cr} = \frac{\pi^2 EI}{L^2} \tag{6.25}$$

Though the critical buckling load is the most important in dealing with engineering stability problems, other cases of integer n taking values larger than 1 are also meaningful. As shown in Figure 6.25, when n takes 2, or 3, or other values, the buckled shape of the column takes more than one half sine wave, or, in other words, the column buckles in higher modes with a larger buckling load. The higher buckling modes cannot be realized unless intermediate supports are provided. Considering practical design merits, this can be restated as a higher buckling load capacity of a member can be achieved, if proper intermediate supports are provided.

In general, with a member with any end supports, the buckling load can be expressed using its effective length, kL, and k is called the effective length factor, thus, Eq.6.25 can be rewritten as,

$$N_{cr} = \frac{\pi^2 EI}{(kL)^2} \tag{6.26}$$

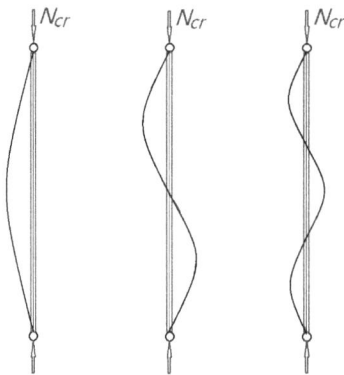

Figure 6.25 Different buckling modes for pinned column.

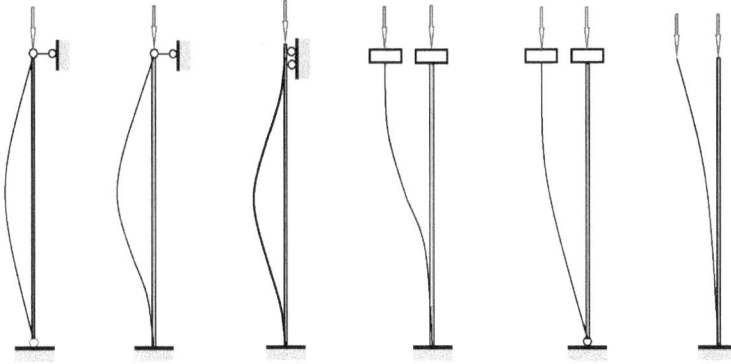

| Theory | $k = 1.0$ | $k = 0.7$ | $k = 0.5$ | $k = 1.0$ | $k = 2.0$ | $k = 2.0$ |
| Design | $k = 1.0$ | $k = 0.8$ | $k = 0.65$ | $k = 1.2$ | $k = 2.24$ | $k = 2.1$ |

Figure 6.26 Theoretical and design values for effective length factor.

Note that two values are provided for the effective length factor k for each case, with the first one as the theoretical value and the second as the recommended design value.

The effective length kL is measured as the distance between the points of inflection or end, as shown in Figure 6.26. It should be mentioned that some of the k factors for design shown in Figure 6.26 are proposed herein as the average of the typical values conventionally used for steel and timber structures.

6.6.2.2 Inelastic Buckling

In order to discuss the subject further, let us reorganize some of the parameters for the buckling load and represent it using buckling stress, $f_{cr} = \dfrac{N_{cr}}{A}$. Since the radius of gyration is $r = \sqrt{\dfrac{I}{A}}$, the buckling stress can be rewritten as,

$$f_{cr} = \frac{\pi^2 E}{\left(\dfrac{kL}{r}\right)^2} = \frac{\pi^2 E}{\lambda^2} \tag{6.27}$$

where, $\lambda = \dfrac{kL}{r}$ is called the slenderness ratio.

The Euler buckling strength is only applicable for a constant modulus of elasticity E, and is no longer true beyond the proportional limit stress, f_p. By substituting f_{cr} for f_p in Eq.6.27, a slenderness ratio limit can be defined as follows,

$$\lambda_p = \sqrt{\frac{E}{f_p}} \tag{6.28}$$

When the slenderness ratio of a compressive member is less than λ_p, non-elastic buckling may occur. Two theories, the tangent modulus theory and the double modulus theory credited to Engesser can be used to evaluate non-elastic buckling of axially loaded compressive members.

Using the tangent modulus theory, the critical buckling load can be calculated by replacing the modulus of elasticity E by the tangent modulus E_t, therefore,

$$N_{cr} = \frac{\pi^2 E_t A}{\lambda^2} \qquad (6.29)$$

On the other hand, based on the fact that the two sides of a buckled section should each follow a different modulus, the so-called double modulus (or reduced modulus) theory (with the same expression as the Euler buckling load except for replacing E with E_r) is developed and conceptually appears to be more rational. For rectangular sections, the reduced modulus E_r can be expressed as,

$$E_r = \frac{4EE_t}{\left(\sqrt{E} + \sqrt{E_t}\right)^2} \qquad (6.30)$$

The tangent modulus has been proved by Shanley[57] to provide lower boundary predictions for the experimental results, and thus is more practically useful.

The earliest comprehensive documentation on testing a large number of full-size timber columns (12 in. × 12 in., or 305 mm × 305 mm) may be the report published by Newlin and Gahagan in 1930.[58] They particularly focused on the axial loading capacities of columns with an intermediate slenderness ratio whose strength depends on both material strength and stiffness. Based on the testing results, a power function of the slenderness ratio with coordinates of $(0, f_u)$ and (λ_p, f_p) was formulated for the buckling strength, as follows,

$$f_{cr} = f_u \left\{ 1 - \left(1 - \frac{f_p}{f_u} \right) \left(\frac{\lambda}{\lambda_p} \right)^{\frac{2f_p}{f_u - f_p}} \right\} \qquad (6.31)$$

where f_u is the crushing strength of timber. In a later study, Yoshihara et al. modified Newlin and Gahagan's equation by inducing a secant modulus E_{sec} at the crushing stress point to reflect the plastic nature of stress-strain relationships:[59]

$$f_{cr} = f_u \left[1 - \left(1 - \frac{f_p}{f_u} \right) \left(\frac{\lambda}{\lambda_p} \right)^{\frac{2E_{sec}}{E}} \right] \qquad (6.32)$$

Yoshihara et al. compared their equation favorably with some small-scale specimens tested under axial loading.

Ylinen[60,61] proposed a critical load based on tangent modulus theory and an essentially empirical stress-strain relationship given in the following equation:

$$\mu = \frac{1}{E}\left[cf - (1-c)f_p\ln\left(1-\frac{f}{f_p}\right)\right]$$

(6.33)

in which, the coefficient c reflects the material properties, and $c = 1.0$ stands for the relationship of linear elastic materials. Differentiating Eq.6.33 in terms of stress f, the tangent modulus can be expressed as the function of the stress, f,

$$E_t = \frac{df}{d\varepsilon} = E\frac{f_p - f}{f_p - cf}$$

(6.34)

Let $f = f_{cr}$, and substitute Eq.6.34 into Eq.6.29, noting that $f_{cr} = N_{cr}/A$, we can obtain,

$$f_{cr} = f_E\frac{f_p - f_{cr}}{f_p - cf_{cr}}$$

(6.35)

This is a quadratic equation about f_{cr}, and the meaningful solution is

$$f_{cr} = f_p\left[\frac{1 + f_E/f_p}{2c} - \sqrt{\left(\frac{1 + f_E/f_p}{2c}\right)^2 - \frac{f_E/f_p}{c}}\right]$$

(6.36)

The bracket in Eq.6.36 is essentially the stability factor in the U.S. National Design Specification.[2] These equations developed for timber columns under axial loading are used to examine the axial load capacities of glubam members.

6.6.3 Behaviors of Glubam Columns and Analytical Capacities

Experimental studies on bamboo or engineered bamboo members under compression are very limited. Several researchers have studied the compressive behaviors of short bamboo culms,[62–65] whereas few have investigated buckling of bamboo culms.[66,67] Yu et al.[66] tested bamboo poles and developed a buckling strength based on tapered hollow section columns, intended for the design of bamboo scaffolding for building construction in Hong Kong. Harris et al. also used a numerical procedure to estimate the buckling strength for bamboo poles with different species, based on tapered columns with varying wall thickness.

For engineered bamboo columns, Li et al.[68,69] studied laminated bamboo lumber (LBL) columns, which are essentially similar to the thick-strip glubam. They first examined the stress-strain features of short columns and presented a tri-linear model;[68] they then examined buckling load and suggested an empirical stability factor based on numerical analysis and test results.[69] Li et al. also conducted eccentric compression tests on LBL columns.[70,71] Though not covered in this book, it is worthwhile

Table 6.4 Testing matrix of glubam columns

Specimen	Material	Length (mm)	Section (mm x mm)	Slenderness ratio	Repetition
G1B1-180	Glubam type-I	180	60x60	10	10
G1B1-400		400		23	3
G1B1-900		900		52	3
G1B1-1300		1300		80	3
G1B1-1700		1700		102	3
G2B1-400	Glubam type-II	400	56x56	25	3
G2B1-600		600		37	3
G2B1-1000		1000		65	3
G2B1-1200		1200		77	3
G2B1-1600		1600		102	3

Figure 6.27 Loading condition for axial compressed column.

mentioning that several researchers have studied compressive behaviors of columns made of bamboo scrimber and parallel bamboo strand.[72,73]

In the following discussion, the experimental studies of the author's group on axially compressed glubam members[74,75] and the evaluation of compressive loading capacities are summarized. Table 6.4 shows the testing matrix for the column specimens made of two types of glubam, the thick-strip (Glubam type-I) and the thin-strip (Glubam type-II).

As shown in Figure 6.27, all specimens were tested to simulate a loading condition as depicted in Figure 6.24. This is realized by placing a specially designed pin at the top and bottom of the column specimen. During testing, the strains of the column at the mid-height section and the deflections were measured using strain gauges and LVDTs, respectively. Test results are summarized in Table 6.5.

Figure 6.28 (a) and (b) compares the test results with various design codes. The suffix "c" refers to the results calculated based on the mean value of the compressive strength (shown as solid lines) while "cd" is based on the design values of the glubam materials suggested in Chapter 4 (dotted lines). The circle and triangular points represent the experimental values. The results show that the approaches based

Table 6.5 Test results of axially loaded glubam columns

Specimen	Ultimate Load		Deflection (mm)	Failure modes
	P_{cr} (kN)	Deviation		
G1B1-180	164	6.6%	-	S
G1B1-400	141	7.0%	3.8	S
G1B1-900	108	6.8%	5.8	B
G1B1-1300	67	4.5%	5.8	B
G1B1-1700	35	8.7%	6.3	B
G2B1-400	137	0.3%	3.8	S
G2B1-600	122	8.9%	3.5	S
G2B1-1000	76	0.0%	6.9	B
G2B1-1200	59	2.1%	9.8	B
G2B1-1600	36	10.3%	16.7	B

on Eurocode, adopting the mean value of the compressive strength, can effectively describe the capacity of the columns under axial load, whether short columns or long ones. On the other hand, the calculated results using the design compressive strength (dotted lines) are much lower than the test results, providing a necessary conservatism. This can be explained by the fact that the design compressive strength adopted considers the safety factor, which leads to a more conservative design. Therefore, it is feasible to design the glubam columns under axial loads based on the present design method for timber-based glulam.

Zhou et al.[76] compared the test results with various analytical models. They showed that the Euler's equation can capture the column load carrying capacity for slender columns (λ_{rel} > 1) made of both types of glubam. For GB1 specimens, the Newlin–Gahagan equation overestimates the experimental results for GB1 columns; Li et al.'s study,[69] however, provides a reasonable prediction of the results of the GB2 tests. Predictions based on the Yoshihara approach provide a different trend to the test results. The analytical curves using the tangent modulus theory with the proposed stress-strain model in Chapter 3 (Eq.3.9), and the method based on Ylinen's equation are close to each other and can provide a slightly conservative prediction of the test results for both GB1 and GB2 specimens for the entire range of the slenderness ratio.

6.7 Future Research Needs for Engineered Bamboo Components

As discussed in this chapter, considerable research has been carried out on various engineered bamboo components, particularly glubam, providing a basic experimental and analytical background for design and construction of engineered bamboo structures. However, on comparison with research on other well-established materials, such as timber, the research is still not sufficient. In particular, experimental data on large- to full-scale members need to be accumulated and established for calibrating various design equations. More research on long-term effects and time- dependent behaviors is urgently needed.

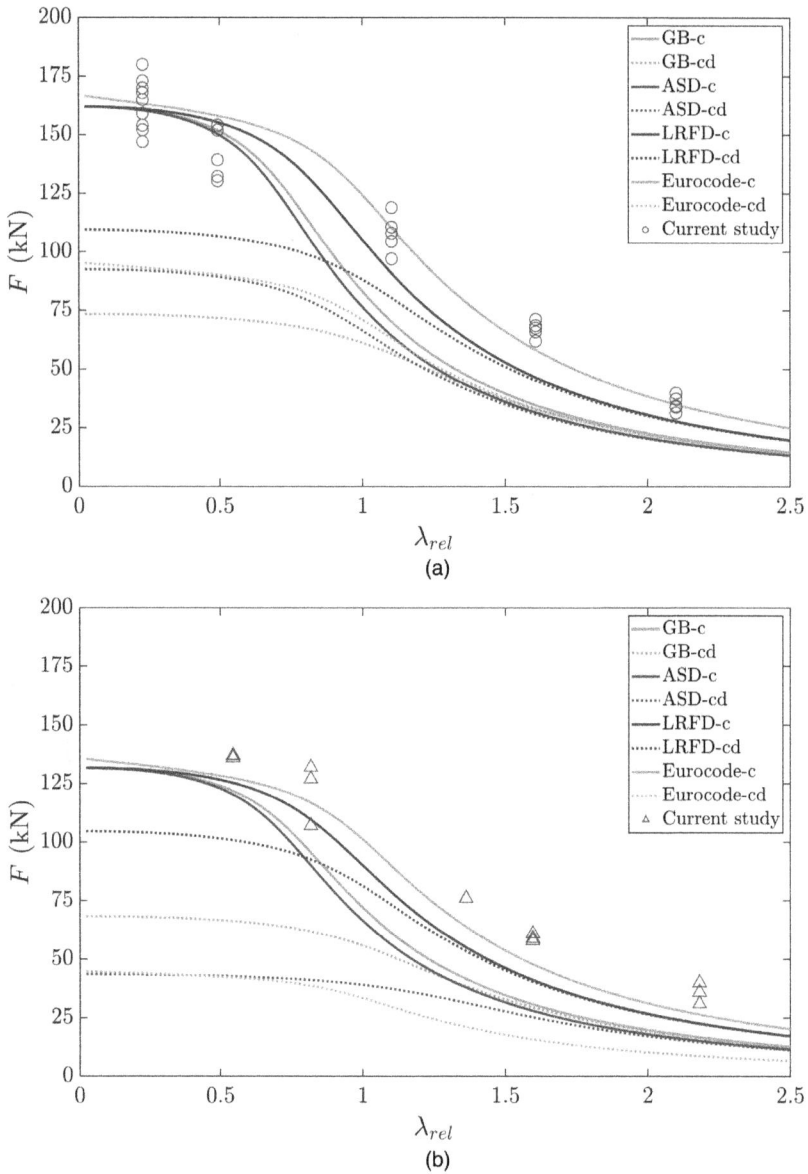

Figure 6.28 Experimental and analytical critical buckling load: (a) for thick-strip glubam specimens; (b) for thin-strip glubam specimens. (The relative slenderness ratio λ_{rel} is defined by λ / λ_p).

References

[1] Ministry of Housing and Urban–Rural Development of the People's Republic of China. (2017). GB 50005-2017: Code for design of timber structures (in Chinese).China Architecture & Building Press, Beijing.

[2] American Wood Council. (2018a). *National design specification for wood construction.* American Wood Council, Leesburg, VA.

[3] American Wood Council. (2018b). *NDS supplement: National design specification: Design values for wood construction.* American Wood Council, Leesburg, VA.

[4] Xiao, Y., Shan, B., Chen, G., Zhou, Q., & She, L.Y. (2008). Development of a new type of glulam – glubam. In Y. Xiao, M. Inoue & S.K. Paudel (Eds.), *Modern bamboo structures: Proceedings of First International Conference on Modern Bamboo Structures (ICBS-2007), Changsha, China, 28–30 October 2007.* CRC Press/Balkema, Leiden, The Netherlands.

[5] Xiao, Y., Zhou, Q., & Shan, B. (2010). Design and construction of modern bamboo bridges. *ASCE Journal of Bridge Engineering,* 15(5), 533–541.

[6] APA. (2017). *Glulam-product guide.* Report X440. APA – The Engineered Wood Association, Tacoma, WA.

[7] Zhou, Z., Shan, B., & Xiao, Y. (2008). Design and construction of a modern bamboo pedestrian bridge. In Y. Xiao, M. Inoue & S.K. Paudel (Eds.), *Modern bamboo structures: Proceedings of First International Conference on Modern Bamboo Structures (ICBS-2007), Changsha, China, 28–30 October 2007.* CRC Press/Balkema, Leiden, The Netherlands.

[8] Shan, B., Zhou, Q., & Xiao, Y. (2008). Construction of world's first truck-safe modern bamboo bridge. In Y. Xiao, M. Inoue & S.K. Paudel (Eds.), *Modern bamboo structures: Proceedings of First International Conference on Modern Bamboo Structures (ICBS-2007), Changsha, China, 28–30 October 2007.* CRC Press/Balkema, Leiden, The Netherlands.

[9] Yang, G.S. (2018). Experimental study on flexural property of glubam beams. Master of Engineering Thesis, supervised by Y. Xiao, Nanjing Tech University.

[10] Li, Z., Yang, G.S., Zhou, Q., Shan, B., & Xiao, Y. (2018). Bending performance of glubam beams made with different processes. *International Journal of Advances in Structural Engineering,* 22(2), 535–546.

[11] Zhou, Q. (2012). Experimental study and engineering application of glue laminated bamboo beams. Doctor of Engineering Thesis, supervised by Y. Xiao, Hunan University.

[12] Ministry of Housing and Urban–Rural Development of the People's Republic of China. (2017). GB/T 51226-2017: Technical standard for multi-story and and high-rise timber buildings (in Chinese). China Architecture & Building Press, Beijing.

[13] Li, H.T., Su, J.W., Zhang, Q.S., & Chen, G. (2015). Experimental study on mechanical performance of side pressure laminated bamboo beam (in Chinese). *Journal of Building Structures,* 36(03), 121–126.

[14] Nowak, T. (2016). Strength enhancement of timber beams using steel plates – Review and experimental tests. *Drewno,* 59, 75–90, DOI:10.12841/wood.1644-3985.150.06.

[15] Dagher, H.J. (2005). Current state of reinforced wood technology: New products, codes and specifications. In *Proceedings of the 3rd International Conference on Advanced Engineered Wood Composites, July 10–14, Bar Harbor, ME, USA.*

[16] Davalos, J.F., Qiao, P., & Trimble, B.S. (2000a). Fiber-reinforced composite and wood bonded interface, Part 1. Durability and shear strength. *Journal of Composites Technology and Research,* 22(4), 224–231.

[17] Davalos, J.F., Qiao, P., & Trimble B.S. (2000b). Fiber-reinforced composite and wood bonded interface, Part 2. Fracture. *Journal of Composites Technology and Research,* 22(4), 232–240.

[18] Davids, W.G., Nagy, E., & Richie, M.C. (2008). Fatigue behavior of composite-reinforced glulam bridge girders. *ASCE Journal of Bridge Engineering,* 13(2), 183–191.

[19] Fiorellia, J., & Diasb, A.A. (2006). Fiberglass-reinforced glulam beams: Mechanical properties and theoretical model. *Materials Research*, 9(3), 263–269.

[20] Weaver, C.A., Davids, W.G., & Dagher, H.J. (2004). Testing and analysis of partially composite fiber-reinforced polymer-glulam-concrete bridge girders. *ASCE Journal of Bridge Engineering*, 9(4), 316–325.

[21] Holzer, S.M., Loferski, J.R., & Dillard, A.D. (1989). A review of creep in wood: Concepts relevant to develop long-term behavior predictions for wood structures. *Wood and Fiber Science*, 21(4), 376–392.

[22] Granello, G., & Palermo, A. (2019). Creep in timber: Research overview and comparison between code provisions. *New Zealand Timber Design Journal*, 27(1).

[23] Xiao, Y., Li, L., & Yang, R.Z. (2014). Long-term loading behavior of a full-scale glubam bridge model. *ASCE Journal of Bridge Engineering*, 19(9).

[24] JTG D62. (2004). Design guidelines for reinforced and prestressed concrete highway bridges (in Chinese). People's Transportation Press, Beijing.

[25] Li, L., & Xiao, Y. (2016). Creep behavior of glubam and CFRP-enhanced glubam beams. *ASCE Journal of Composites for Construction*, 20 (1).

[26] Lewis, W.C. (1961). Design considerations for fatigue in timber structures. *Transactions of the American Society of Civil Engineers*, 126(2), 821–828.

[27] Smith, I., Landis, E., Gong, M. (2003). *Fracture and fatigue in wood*. Wiley, Chichester, UK.

[28] Kyanka, G.H. (1980). Fatigue properties of wood and wood composites. *International Journal of Fracture*, 16, 609–616, https://doi.org/10.1007/BF02265220

[29] Sasaki, Y., & Yamasaki, M. (2002). Fatigue strength of wood under pulsating tension-torsion combined loading. *Wood and Fiber Science*, 34(4), 508–515.

[30] Sasaki, Y., Oya, A., & Yamasaki, M. (2014). Energetic investigation of the fatigue of wood. *Holzforschung*, 68(7).

[31] Yildirim, M.N., Uysal, B., Ozcifci, A., & Ertas, A.H. (2015). Determination of fatigue and static strength of scots pine and beech wood. *Wood Research*, 60(4), 679–686.

[32] Köhler, J., & Faber, M. (2003). A probabilistic creep and fatigue model for timber materials. In A. der Kiureghian, S. Madanat, & J.M. Pestan (Eds.), *Conference: International Conference on Applications of Statistics and Probability in Civil Engineering ICASP9, at San Francisco, USA: Volume 2*, Millpress, Rotterdam, The Netherlands, pp. 1141–1148.

[33] Davids, W.G., Nagy, E., & Richie, M.C. (2008). Fatigue behavior of composite-reinforced glulam bridge girders. *Journal of Bridge Engineering*, 13(2), 183–191.

[34] Gao, L., Zhang, Z., Zeng, D., He, G., & Cai, J. (2016). Experimental research on fatigue behavior of larch glulam beams (in Chinese). *Journal of Building Structures*, 37(10), 27–35.

[35] Bhkari, M.N, Ahmad, Z., Afidah, A.A., & Tahir, P.M. (2016). Post-fatigue behaviour of Kekatong glued laminated timber railway sleepers. In *Proceedings of the International Civil and Infrastructure Engineering Conference*, pp. 819–831, DOI:10.1007/978-981-10-0155-0_69.

[36] Keogh, L., O'Hanlon, P., O'Reilly, P., & Taylor, D. (2015). Fatigue in bamboo. *International Journal of Fatigue*, 75, 51–56.

[37] Song, J., Surjadi, J.U., Hu, D., & Lu, Y. (2017). Fatigue characterization of structural bamboo materials under flexural bending. *International Journal of Fatigue*, 100(1), 126–135.

[38] Ali, A., Rassiah, K., Othman, F., Lee, H.P., Tay, T.E., Hazin, M.S., & Megat Ahmad, M.M.H. (2016). Fatigue and fracture properties of laminated bamboo strips from Gigantochloa scortechinii polyester composites. *BioResources*, 11(4), 9142–9153.

[39] Platts, M. (2014). Strength, fatigue strength and stiffness of high-tech bamboo/epoxy composites. *Agricultural Sciences*, 05, 1281–1290, DOI:10.4236/as.2014.513136.

[40] Scherer, J.F., Bom, R.P., & Barbieri, R. (2020). Torsional fatigue in bamboo fibers reinforced epoxy resin composites. *Engineering Research Express*, 2(1).

[41] Leichti, R.J., Falk, R.H., & Laufenberg, T.L. (2007). Prefabricated wood composite I-beams: A literature review. *Wood Fiber Science*, 22(1), 62–79.

[42] Harte, A.M., & Baylor, G. (2011). Structural evaluation of castellated timber I-joists. *Engineering Structures*, 33(12), 3748–3754.

[43] Allen, E., & Iano, J. (2013). *Fundamentals of building construction: Materials and methods*. John Wiley & Sons, Hoboken, NJ.

[44] Tang, Z., Shan, B., Li, W.G., Peng, Q., & Xiao, Y. (2019). Structural behavior of glubam I-joists. *Construction and Building Materials*, 224, 292–305.

[45] Racher, P., Bocquet, J., & Bouchair, A. (2007). Effect of web stiffness on the bending behaviour of timber composite I-beams. *Materials & Design*, 28(3), 844–849.

[46] APA. (2001). *Performance of rated I-joists*. APA –The Engineered Wood Association, Tacoma, WA.

[47] General Administration of Quality Supervision, Inspection and Quarantine, and Standardization Administration of the People's Republic of China. (2012). GB/T 28985-2012: Wood I-joist for building structures. Chinese Standards Press, Beijing.

[48] Canadian Standards Association. (2009). CSA O86-01: Engineering design in wood. Canadian Standards Association, Toronto.

[49] Chen, G., Wu, J., Jiang, H., Zhou, T., Li, X., & Yu, Y. (2020). Evaluation of OSB webbed laminated bamboo lumber box-shaped joists with a circular web hole. *Journal of Building Engineering*, 29, 101129, https://doi.org/10.1016/j.jobe.2019.101129

[50] Chen, G., Li, H.T., Zhou, T., Li, C.L., Song, Y.Q., & Xu, R. (2015). Experimental evaluation on mechanical performance of OSB webbed parallel strand bamboo I-joist with holes in the web. *Construction and Building Materials*, 101, 91–98.

[51] Yeoh, D., Fragiacomo, M., & Franceschi, M.D. (2011). State of the art on timber-concrete composite structures: Literature review, *ASCE Journal of Structural Engineering*, 10, 1085–1095.

[52] Santos, P.G.G., Martins, C.E.J., Skinner, J., Harris, K., Dias, A.M.P.G., & Godinho, L.M.C. (2015). Modal frequencies of a reinforced timber-concrete composite floor: Testing and modeling. *Journal of Structural Engineering*, 141(11), 04015029.

[53] Meena, R., Schollmayer, M., & Tannert, T. (2014). Experimental and numerical investigations on fire resistance of novel timber-concrete-composite decks, *Journal of Performance of Constructed Facilities*, 28(6), A4014009.

[54] Shan, B., Xiao, Y., Zhang, W.L., & Liu, B. (2017). Mechanical behavior of connections for glubam-concrete composite beams. *Journal of Construction and Building Materials*, 143, 158–168.

[55] Shan, B., Wang, Z.Y., Li, T.Y., & Xiao, Y. (2020). Experimental and analytical investigations on short-term behavior of glubam-concrete composite beams. *ASCE Journal of Structural Engineering*, 146(3).

[56] Wu, Y., & Xiao, Y. (2018). Steel and glubam hybrid space truss. *Engineering Structures*, 171, 140–151.

[57] Shanley, F.R. (1946). The column paradox. *Journal of the Aeronautical Sciences*, 13(12), 678.

[58] Newlin, J.A., & Gahagan, J.M. (1930). Tests of large timber columns and presentation of the forest products laboratory column formula. U.S. Department of Agriculture Technical Bulletin No. 167. U.S. Department of Agriculture, Washington, DC.

[59] Yoshihara, H., Ohta, M., & Kubojima, Y. (1998). Prediction of the buckling stress of intermediate wooden columns using the secant modulus. *Journal of Wood Science*, 44, 69–72.

[60] Ylinen, A. (1956). A method of determining the buckling stress and the required cross-sectional area for centrally loaded straight columns in elastic and inelastic range. *Memoires Association International des Ponts et Charpentes*, 16, 529–550.

[61] Zahn, J.J. (1992). Re-examination of Ylinen and other column equations. *ASCE Journal of Structural Engineering*, 118(10), 2716–2729.

[62] Li, W.T., Long, Y.L., Huang, J., & Lin, Y. (2017). Axial load behavior of structural bamboo filled with concrete and cement mortar. *Construction and Building Materials*, 148, 273–287.

[63] Mahzuz, H.M.A., Ahmed, M., Ashrafuzzaman, M., Karim, R., & Ahmed, R. (2011). Performance evaluation of bamboo with morter and concrete. *Journal of Engineering and Technology Research*, 3(12), 342–350.

[64] Okhio, C., Waning, J., & Mekonnen, Y.T. (2011). An experimental investigation of the mechanical properties of bamboo and cane. . *Journal of Selected Areas in Bioengineering* (JSAB), 11, 7–14.

[65] Deng, J.C., Chen, F.M., Wang, G., & Zhang, W.F. (2016). Variation of parallel-to-grain compression and shearing properties in moso bamboo culm (Phyllostachys pubescens). *BioResources*, 11(1), 1784–1795.

[66] Yu, W.K., Chung, K.F., & Chan, S.L. (2003). Column buckling of structural bamboo. *Engineering Structures*, 25, 755–768.

[67] Harries, K.A., Bumstead, J., Richard, M., & Trujillo, D. (2017). Geometric and material effects on bamboo buckling behavior. *Proceedings of the Institution of Civil Engineers – Structures and Buildings*, 170(4), 236–249.

[68] Li, H.T., Zhang, Q.S., Huang, D.S., & Deeks, A.J. (2013). Compressive performance of laminated bamboo. *Composites: Part B*, 54, 319–328.

[69] Li, H.T., Su, J.W., Zhang, Q.S., Deeks, A.J., & Hui, D. (2015). Mechanical performance of laminated bamboo column under axial compression. *Composites Part B*, 79, 374–382.

[70] Li, H.T., Wu, G., Zhang, Q.S., & Su, J.W. (2016). Mechanical evaluation for laminated bamboo lumber along two eccentric compression directions. *Journal of Wood Science*, 62, 503–517.

[71] Li, H.T., Chen, G., Zhang, Q.S., Ashraf, M., Xu, B., & Li, Y.J. (2016). Mechanical properties of laminated bamboo lumber column under radial eccentric compression. *Construction and Building Materials*, 121, 644–652.

[72] Li, Z., He, M.J., Tao, D., & Li, M.L. (2016). Experimental buckling performance of scrimber composite columns under axial compression. *Composites Part B*, 86, 203–213.

[73] Tan, C., Li, H.T., Wei, D.D., Lorenzo, R., & Yuan, C.G. (2020). Mechanical performance of parallel bamboo strand lumber columns under axial compression: Experimental and numerical investigation. *Construction and Building Materials*, 231, 117168.

[74] Lv, X.H. (2011). Axial compressive loading tests and FE analysis of glubam columns. MEng thesis supervised by Y. Xiao, Hunan University.

[75] Chu, F.Z. (2019). Experimental study on axial compression of glubam columns. MEng thesis supervised by Y. Xiao, Nanjing Tech University.

[76] Zhou, S.C., Chu, F.Z., Lv, X.H., & Xiao, Y. (2021). Experimental studies on glubam columns under axial compression. Journal of Building Engineering, 103453.

Chapter 7

Glubam Trusses

As a part of the building envelope, the bearing capacity of roof trusses is related to the safety performance of the whole structure. The light weight and high strength of a bamboo roof frame make it suitable for building large-span roof systems. Besides supporting roof systems, similar to timber structures, engineered bamboo trusses can be made for other purposes, such as floor structures, bridges, etc. Bamboo-based structural trusses have the advantages of low cost, fast construction speed, strong and flexible design, etc. Therefore, bamboo trusses can be designed for wide usage for various commercial, school and other public buildings, as well as residential buildings. This chapter discusses the design and analysis of glubam trusses combined with experimental research and practical project applications.

7.1 Roof Truss Design

Trusses are typically made of a basic triangular shape with three pinned stick-type elements, as shown in Figure 7.1. Although in reality, the basic stick-type elements are not pinned together, instead being welded or bolted together to be capable of moment transfer, they can be treated as connected by pins. Also, the upper and lower chords can be analyzed as individual elements connected between joints, though they need to be checked for bending under a transverse load.

Figure 7.2 shows several typical symmetric trusses used in timber structures, primarily for building roofs. The overall triangular shapes of the trusses can be made as single slope, unsymmetric, or trapezoidal shapes, to suit various structural and architectural purposes. Combinations of simple trusses can also be designed to form more complicated truss systems, or even a building system. For example, a single slope triangular shape truss can be designed and used as a basic module to build a frame system for a building,[1] as conceptually depicted in Figure 7.3. The author's research group worked with Indonesian architect Eco Prawoto on a low-cost residential building system, adopting modular glubam truss systems for the roof and walls (Figure 7.4).

Research on round bamboo culms in trusses is an old but important topic; [2,3,4] however, this chapter focuses mainly on trusses made of engineered bamboo.

7.2 Simple Triangular Glubam Trusses

As detailed in Table 7.1, Chen et al.[5,6] carried out static tests on two types of glubam trusses with simple triangular shapes (single triangle and kingpost types). The model

DOI: 10.1201/9781003204497-7

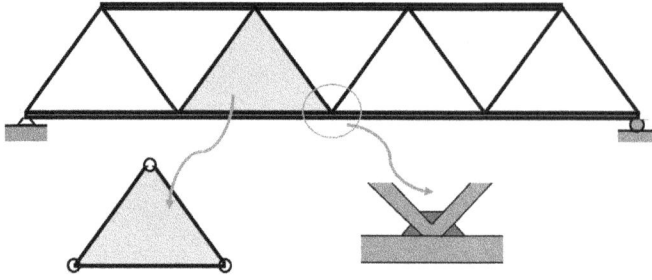

Figure 7.1 A structural truss and its triangular element and joint.

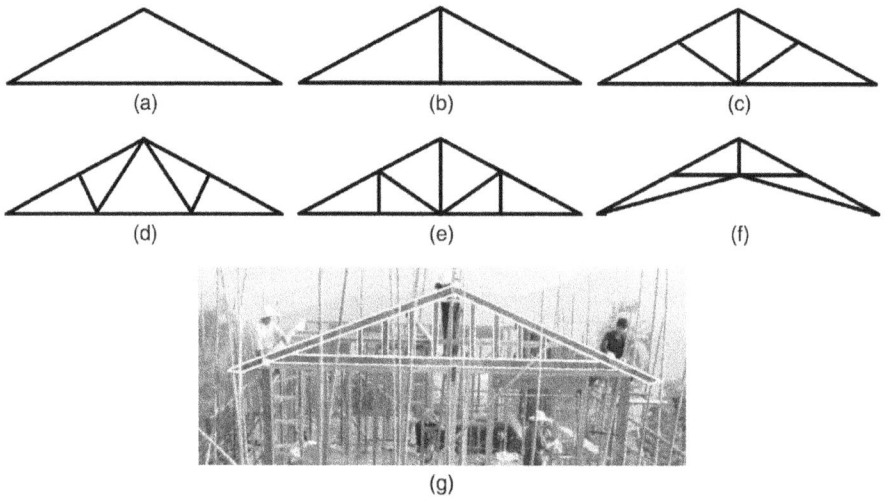

Figure 7.2 Typical types of structural truss: (a) simple triangle; (b) kingpost; (c) queen; (d) fink; (e) Howe; (f) scissor; (g) gable.

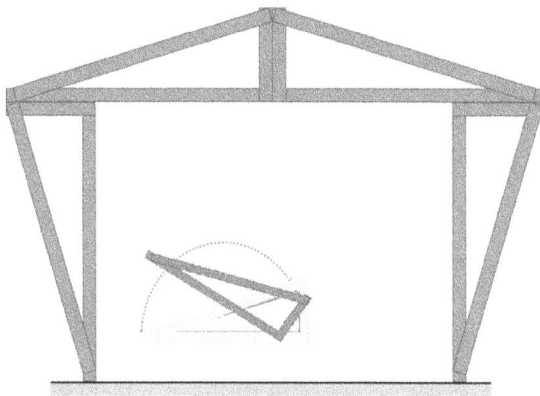

Figure 7.3 Formation of a building frame using simple triangular trusses.

Figure 7.4 Building example with modular trusses to support roof and walls.

Table 7.1 Roof truss size (unit: mm)

Roof truss number	Upper chord (mm × mm)	Lower chord (mm × mm)	Web (mm × mm)	Span (mm)	Slope*
BT11, BT12, BT13	28×150	28×150	-	5000	3:12
BT21, BT22, BT23	56×140	56×120	56×90	6000	6:12

Note: *3:12 means 3 units rise for every 12 horizontal units.

roof trusses were divided into groups BT1 and BT2, with three identical specimens in each testing group. The BT1 roof truss was a form used in the construction of the earthquake relief buildings in Guangyuan City after the Wenchuan Earthquake on May 12, 2008. The BT2 truss was the typical form designed for the construction of the world's first modern bamboo villa building at Hunan University, and other glubam buildings. Detailed dimensions of the two types of glubam truss specimens are shown in Figure 7.5 (a) and (b), respectively. The material used in the specimens was thin-strip glubam.

During testing, the load was applied for each upper joint using a lever arm to amplify the gravity load by steel blocks, as shown in Figure 7.6. To ensure the stability of the truss specimen, steel frames were used to prevent the premature out-of-plane deformation of the truss. During the testing, the following procedures were followed:

1 A standard load level, P_n, was first determined based on the prototype structure.
2 Preloading at an increment of 0.25 P_n until P_n was reached; the intervals between each level of loading was 30 min.
3 Loading to P_n similar to step 2, and holding the load for 12–24 hours, before unloading. The unloading was executed in two stages at 0.5 P_n decrement, with intervals of 30 min.

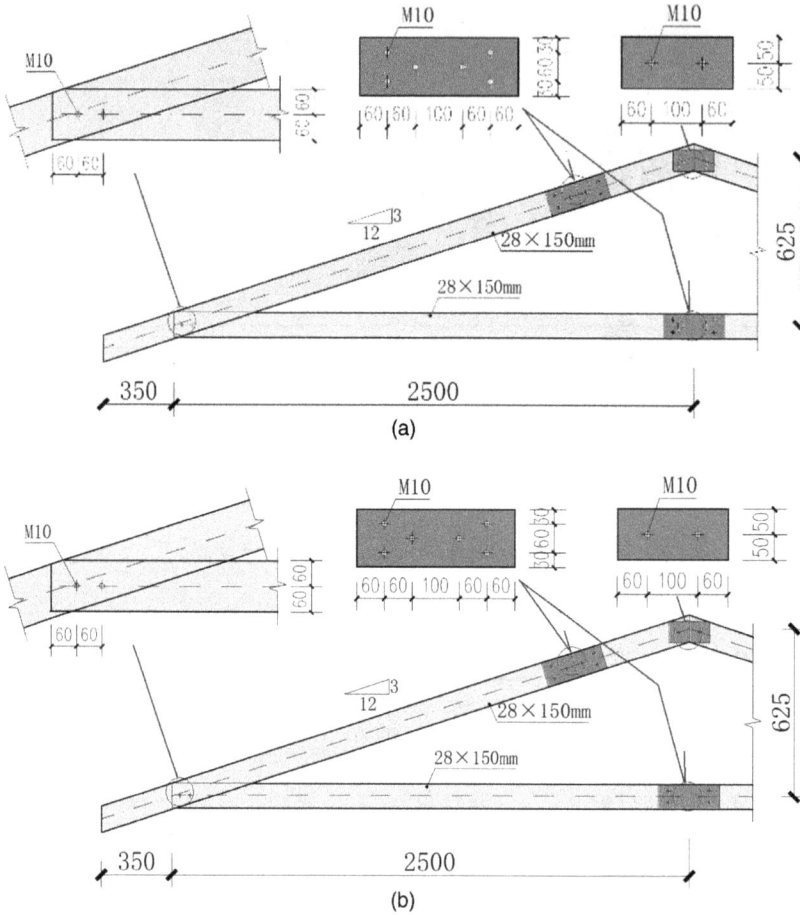

Figure 7.5 Details of truss specimens: (a) BT1; and (b) BT2.

4   Final loading to failure was executed first to the level of P_n, similar to step 2, but with an interval of 2 hours, then load to failure at an increment of 0.1 P_n, with intervals of 30 min. for each step.

During testing, the roof truss model was considered a failure if any of the following conditions occurred: loss of capacity in any rod or joint plate in the truss; a sudden and significant increase in truss deflection; the bearing enlargement at the bolted joints exceeding 8 mm; relative slip at tension joints exceeding 20 mm.

The applied total load and lower chord mid-span deflection relationships for the specimens are shown in Figure 7.7 for the two types of trusses, BT1 and BT2, respectively. The load–mid-span deflection curves of the roof trusses remain linear until they are loaded to the level of the design load, which was determined based on the prototype design. Nonlinear behavior becomes clearer when the applied load approaches to a level

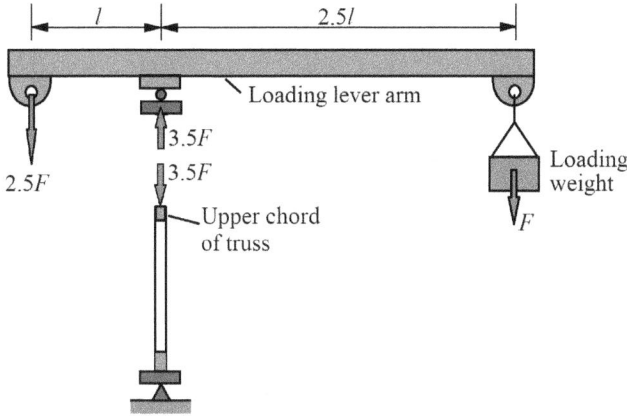

Figure 7.6 Roof truss gravity loading method.

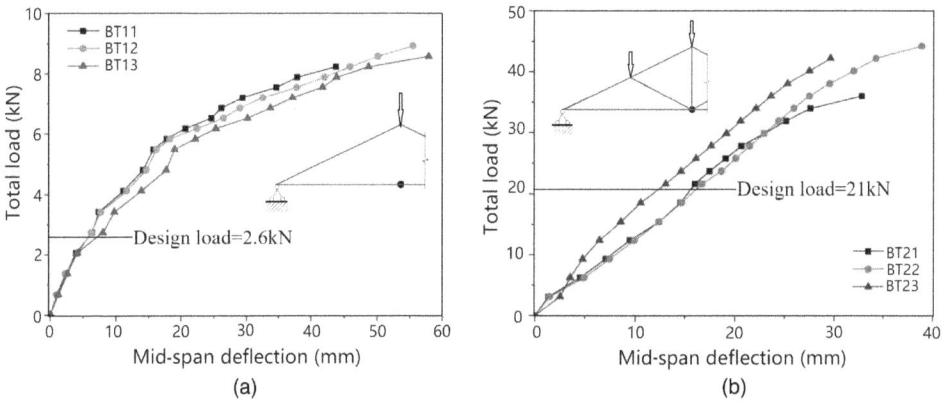

Figure 7.7 Recorded load and deflection curves: (a) BT1 specimens; (b) BT2 specimens.

that is twice the design load. Similar behaviors are quite common with timber trusses.[7,8] With the accelerated increase of the mid-span deflection, the specimens failed, due primarily to the out-of-plane buckling of the upper chords between the lateral supports. The main testing characteristic values are given in Table 7.2. Judging by the relative deflection corresponding to the design load, both trusses have adequate stiffness.

7.3 Analysis of Glubam Trusses

Modeling of timber trusses can be somewhat complicated; however, simple and practical methods can also be utilized.[9–11] For simple analysis of trusses, the main issue is the adoption of different joint models. For the glubam trusses tested, the joints were bolted with steel bolts and glubam cover plates. In reality, the connection can sustain a certain

Table 7.2 Summary of roof truss test results

Specimens	$P_u(N)$	P_u/P_d	Δ_d (mm)	Δ_d/l	Δ_u (mm)
Average BT1	8.75×10^3	3.4	6.44	1/776	52.37
Average BT2	40.8×10^3	2.0	15.53	1/392	29.79

Note: P_u is the ultimate bearing capacity; P_d is the design load; Δ_d is the mid-span deflection of the lower chord corresponding to P_d; Δ_u is the mid-span deflection of the truss under the ultimate load; l is the span length between the two supports.

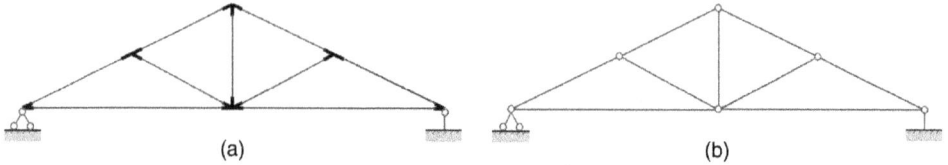

Figure 7.8 Truss models with: (a) rigid joints; and (b) pinned joints.

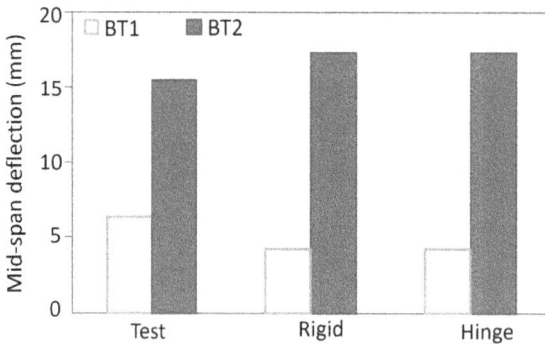

Figure 7.9 Comparison of deformation calculation results and experimental results of different node models.

amount of rotational resistance; however, for simplicity, the joints are modeled as either rigid or pinned, as shown in Figure 7.8 (a) and (b), respectively. The analytical results are compared with the test results in Figure 7.9. As articulated in Figure 7.9, both of the simple models provide similar analytical results of the mid-span lower chord joint deflections; these analytical results are reasonably close to the test results, though they underestimated the results of BT1, whereas they overestimated the deflection of BT2. Needless to say, both models can easily be built using any of the modern structural analysis software, and of course, the pinned joint model is far easier.

For the analysis of loading capacity of the truss models, the rigid joint model provides significant overestimation of the test results, while the pinned joint model largely underestimates the test results. Chen developed a semi-rigid joint model combined with an axially loaded member model.[2] Figure 7.10 shows the comparisons of analytical

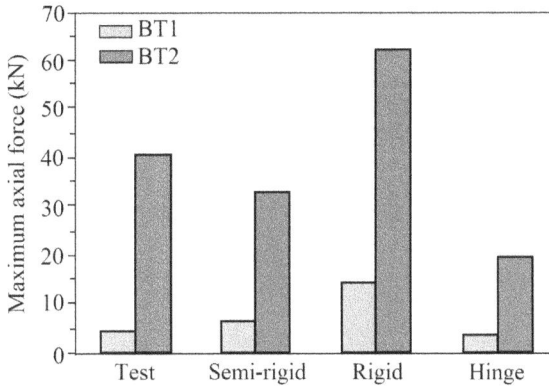

Figure 7.10 Comparison of bearing capacity calculation based on different nodal models and results from tests.

simulations using the three different modeling techniques against the testing results. The semi-rigid model gives overall a better prediction of the test results. However, the pinned model yields a conservative prediction of the capacities of both trusses.

7.4 Toothed Metal-Plate Connected Glubam Trusses

In light timber structures, the toothed metal-plate connections are widely used for making trusses and other components (for example, Figure 7.11), due to their cost effectiveness.[12,13] The toothed metal-plate connector (MPC) was first proposed in America in 1952, and it caused a revolution in the timber truss industry according to Rakesh et al.[12] Their literature review indicates that the number of research studies on this topic are relatively few compared with the widespread usage of the MPC. Several researchers carried out experimental research and analysis, mainly concentrating on the static behavior of MPC timber joints, including tension, shear, and bending.[14–19] Kent et al. studied MPC timber joints under dynamic, cyclic loading.[20] Limited research has also been carried out on the long-term performance of MPC joints.[21,22] Several specifications related to the testing and design of MPC timber joints are also available now to guide practice.[23–25]

As discussed in Chapter 5, cost-effective MPC was attempted for glubam joints with a considerable number of experimental tests, carried out by Wu and Peng in their thesis studies.[26,27] Peng further tested MPC glubam truss specimens with two different sizes (3.0 m long and 5.6 m long), as shown in Figure 7.12 (a) and (b), respectively. In Figure 7.12 (c), a typical joint with the toothed metal is exhibited. Due to the fact that the original glubam board had a limited length of 2.4 m, the upper and lower chords of the specimens had to be extended using butted connections, as shown in Figure 7.12 (d).

For all the specimens, the section dimensions of the upper and lower chords and web elements are all 30 mm thick and 65 mm high. Four specimens were tested repetitively for each size. The four-point loading configuration was used to simply but relatively closely simulate the uniformly distributed load as indicated by the arrows in Figure 7.12

Figure 7.11 Metal-plate connected trusses.

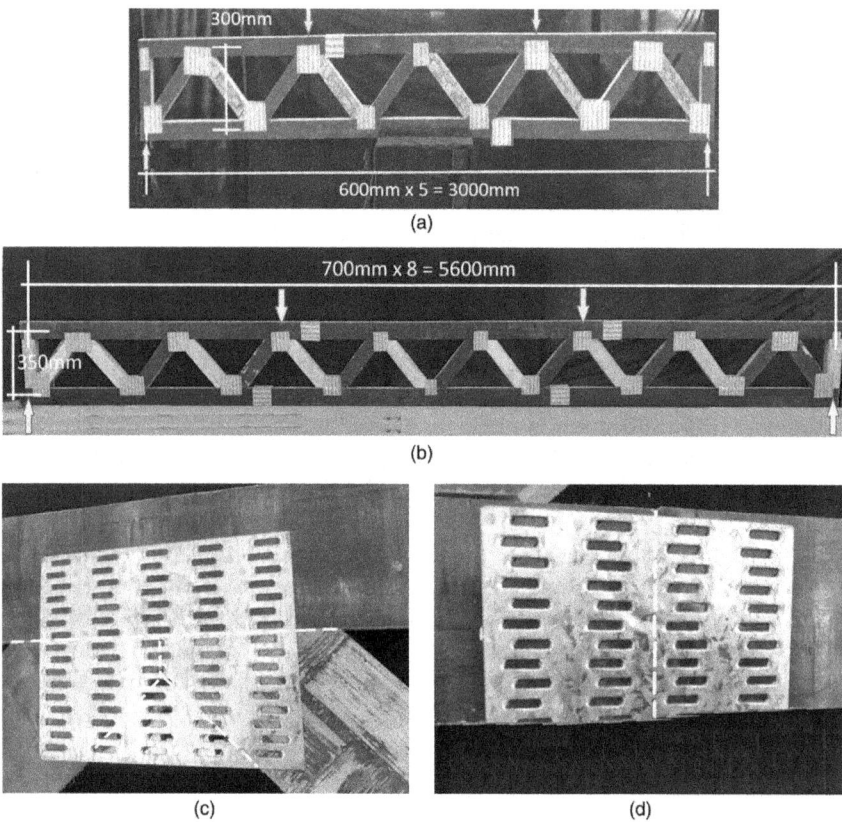

Figure 7.12 Parallel trusses of 3000 mm long (a); and 5600 mm long (b); with toothed metal-plate connections for joining chord and web elements (c); and extension of chords (d).

Figure 7.13 Test setup used in testing MPC truss specimens.

(a) and (b). The test setup including oil jack, supports, and load distribution beam is photographed in Figure 7.13. Based on potential prototype structure design, the standard design loads were estimated as P_k = 5.25 kN and P_k = 3.85 kN, respectively.

The testing method specified in Chinese standard GB/T 50329 (2012)[28] was followed, with loading procedures including trial loading to 0.6 P_k, loading to design load P_k, and final loading to failure. For each loading procedure, the increment of applied load was 0.1–0.2 P_k and the load was maintained for 30 min. for each loading step. The truss specimen was judged as a failure for any of the four conditions:

1   loss of capacity of any element or joints;
2   sharp and rapid increase in deflection;
3   split cracking of glubam at the joints;
4   apparent failure of the metal plates at joints, including the pull out of the metal teeth, rupture of plate or teeth, etc.

The two types of metal-plated trusses performed in an elasto-brittle manner with failure occurring in the joints. For the 3 m long trusses, the metal teeth were bent and pulled out along with some glubam that was cut off, as shown in Figure 7.14 (a). For the 5.6 m long trusses, the failure was caused by the rupture of the toothed metal plate for the butted connections of the lower chord, as shown in Figure 7.14 (b).

The results of the load and mid-span displacement relationships for the four specimens in each testing group are reasonably close, as depicted in Figure 7.15 (a) and (b), respectively. The vertical axis in Figure 7.15 has been normalized by the design load P_k, with its value provided. The analytical simulation using the general structural analysis software SAP2000 by Peng[29] is also shown in Figure 7.15. The MPC truss specimens developed load carrying capacities exceeding 3 times their design load. The behavior is close to linear until they reach 2 times their design standard load P_k. The trusses also have adequate elastic stiffness with their displacements corresponding to

(a)   (b)

Figure 7.14 Failure patterns: (a) pull out of metal teeth; (b) rupture of metal plates.

the design load of less than 1/1000 of the truss span, much lower than the code-specified limit of 1/300 of the span. The elastic modeling provides a reasonable simulation of the initial behaviors of the MPC glubam truss specimens.

7.5 Other Types of Glubam Trusses

7.5.1 Comparison of Conventional and String-beam Truss Systems

Different shapes of large-span roof trusses may have distinct performance and cost effectiveness. Based on an actual project, two 20 m long full-scale glubam trusses with two types of configurations were studied by Xie and Xiao,[30] in order to assess the mechanical performance and compare the cost. Figure 7.16 shows the details of the two types of roof trusses made of thin-strip glubam.

Model truss GBT-1 was a double Howe truss with total glubam volume of about 2.6 m³, and 427 kg of steel parts (Figure 7.16 (a)). The steel cable string and glubam beam model GBT-2 (Figure 7.16 (b)) had a total glubam volume of about 1.9 m³, and 288 kg of steel parts. In terms of manufacturing labor involvement, the string and beam type glubam truss GBT-2 was much more cost effective. The prototype roof truss spans 20.25 m, with a height of 3.66 m, and spacing of 4.05 m. The roof design has a constant gravity load of 0.875 kN/m² or a line load of 3.54 kN/m, and a roof live load of 0.5 kN/m² or a line load of 2.03 kN/m. In the testing, the four-point loading scheme with equal segments was adopted and the equivalent design load was calculated as 76.2 kN.

The glubam trusses were tested to failure under a gradually increased load. The experimentally obtained load and displacement relationships for the two types of large-span glubam truss models are shown in Figure 7.17.

Through the tests, the load-deformation curves and failure modes were obtained, and the mechanical behaviors evaluated and compared. Experimental results show that the failure mode of the two glubam trusses is the out-of-plane buckling of the upper compression chords between the lateral supports. Such a failure can be delayed or prevented in the actual structure if the lintel beams are densely arranged. The performance of the

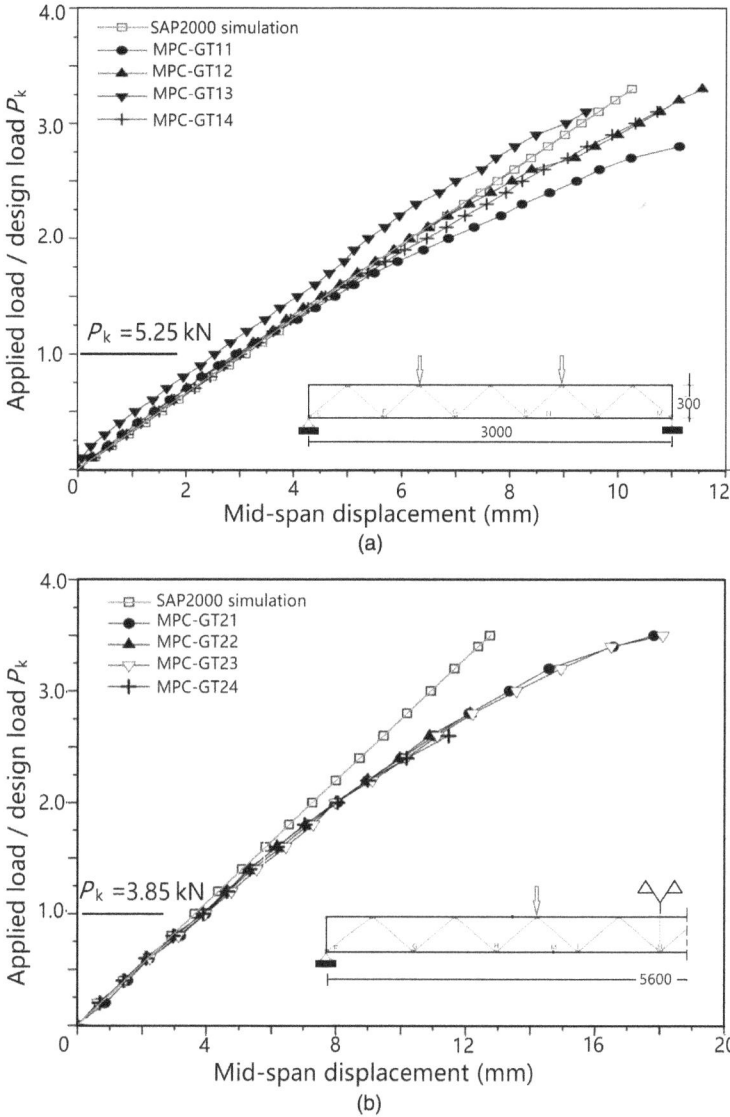

Figure 7.15 Load mid-span displacement results: (a) 3 m long truss; (b) 5.6 m long truss.

truss system with glubam elements as upper compression chords and the high-strength steel bars as tension chords is superior to the glubam truss with conventional config-uration; thus it is considered more suitable for actual construction. The finite element analysis (FEA) results show good agreement with the test results. The experimental and analytical capacities based on the current Chinese timber structures code are reason-ably close.

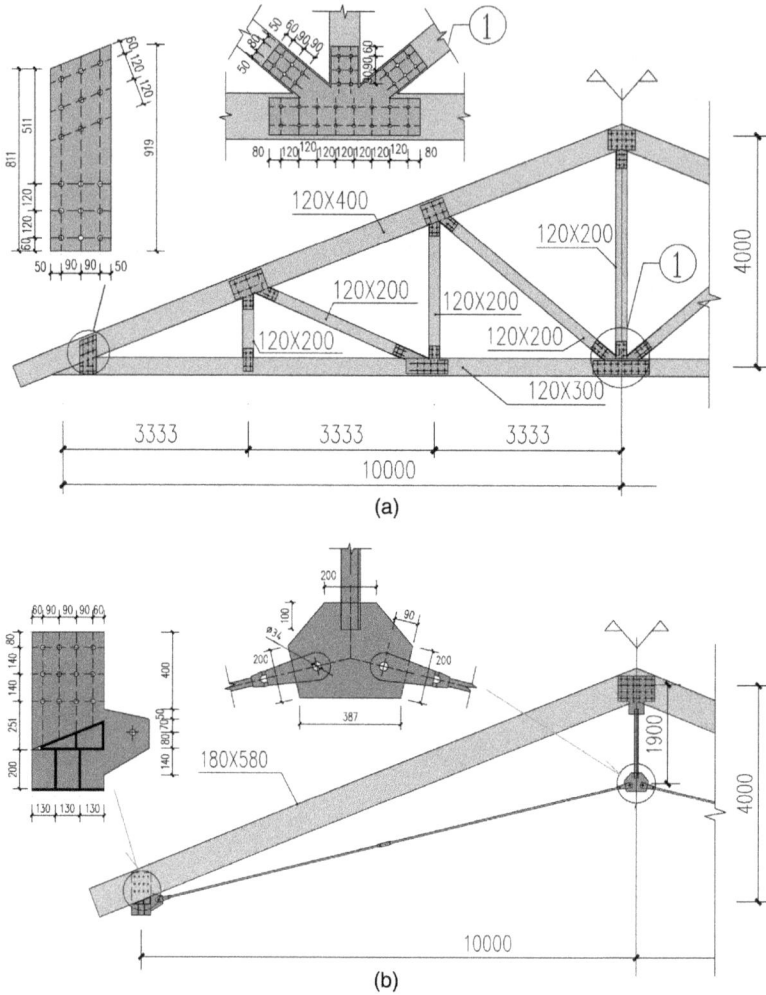

Figure 7.16 Large-span truss models: (a) Howe type truss GBT-1; (b) steel cable string and glubam beam truss GBT-2.

7.5.2 Steel–Glubam Hybrid Trusses

In order to efficiently utilize steel and glubam by combining their advantageous properties while avoiding disadvantages, a hybrid truss system was invented by Xiao and Wu for large-span roof structures.[31,32] As shown in Figure 7.18 (a) and (b), in this special truss system, the upper and web members are made of glubam elements, whereas the lower chords are made with steel elements, such as steel pipes. Figure 7.18 (c) exhibits a relatively large-span roof truss (in total, 20 m long, 10 m wide) under construction using the hybrid system. The three-dimensional spatial truss system can be easily converted into a plane truss system.[33]

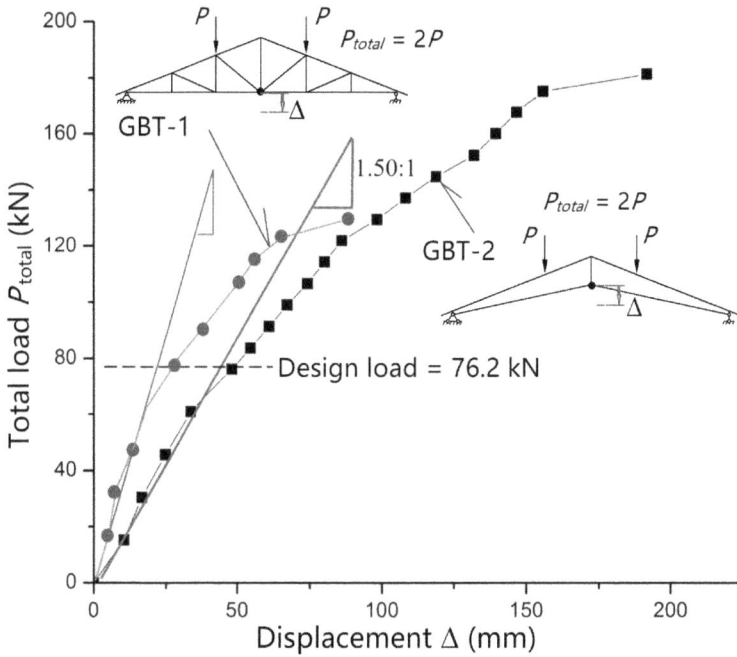

Figure 7.17 Load and displacement relationships.

Wu carried out a static loading test program on a full-scale hybrid steel and glubam truss, which was modeled from a canopy for an office building entrance.[32] As shown in Figure 7.19 (a), the truss had an overall dimension of 9.6 m×2.4 m and a height of 0.85 m. The model truss was composed of 16 (or 2×8) orthogonal spatial square pyramid modules in an upside-down position, with the length of each element equal to 1.2 m, and the covering area of each module being 1.2 m×1.2 m. The loading was applied to each of the upper joints by the use of hanging bags filled gradually with steel blocks, as exhibited in Figure 7.19 (b). The loads applied to the joints in the middle were loaded through a pulley system.

Different failure modes, such as buckling of web chords, shear fracture of bolted connecting joints, and yield of bottom steel pipes, were realized at different testing stages. The experimental results validate that the space truss system has an excellent load carrying capacity. If yielding of the lower chord steel pipes occurs, the system can also experience a large plastic deformation under the ultimate loading condition, as shown in Figure 7.20. It was shown that some selected steel lower chords can be selected to have a reduced section area, so they can be used as a fuse to yield earlier, therefore preventing brittle failure of other elements.

Analyses of the structural behavior under static load are compared with the experimental results, showing a good agreement. Quaranta et al. also conducted dynamic characterization tests using the same model tested by Wu and Xiao.[34] The results obtained from dynamic experimental tests include different modes and frequencies

(a)

(b)

(c)

Figure 7.18 An example of a hybrid steel and glubam spatial truss: (a) upper chord joint; (b) lower chord joint; (c) lifted truss.

from ambient vibration and free-decay vibration. Enhanced frequency domain decomposition (EFDD) and stochastic subspace identification (SSI) were adopted for output-only modal parameter estimation. Design values for the viscous damping ratio for glubam truss structures with steel bolted connections are finally recommended.[34] The numerical assessment of the human-induced vibration serviceability conditions for footbridges built by means of this structural system was also carried out. The damping ratio for the first vibration mode is identified as approximately 1.5%.

7.6 Summary of Glubam Trusses

The studies to date on various types of glubam trusses clearly demonstrate the suitability of such bamboo-based material in building or bridge structural trusses for many

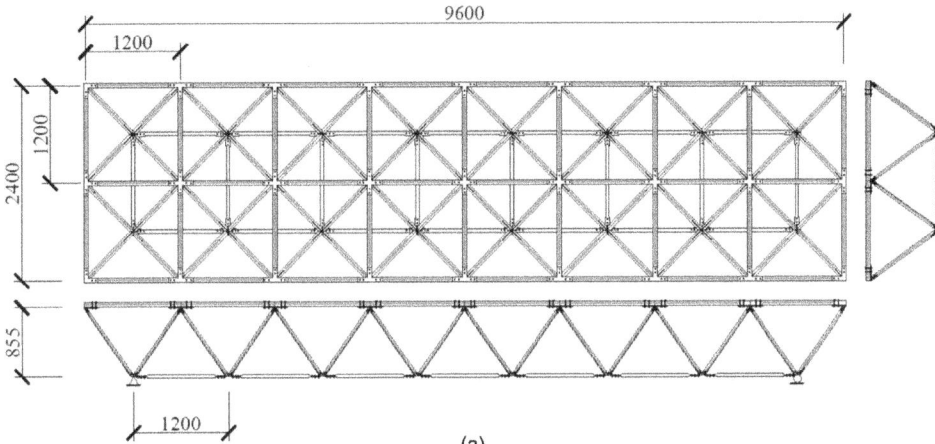

(a)

(b)

Figure 7.19 Static loading test on a hybrid steel and glubam truss model.
Source: Wu & Yiao (2018).[32]

utilizations. The glubam trusses designed following the conventional timber structure design appear to be adequate to offer satisfactory performance. The manufacturability of glubam trusses is essentially similar to the timber ones; however, special consider-ations are needed in properly choosing tools for nailing and pressing the toothed metal plates, as glubam is slightly harder than timber. Glubam can also be combined with steel or fiber reinforced composites (FRP) to be used in the design of a hybrid truss system, which can offer further advantages such as simplification of the connections for the lower tension chords. In the near future, the engineered bamboo trusses should be codified in terms of design and construction specifications for further promotion in

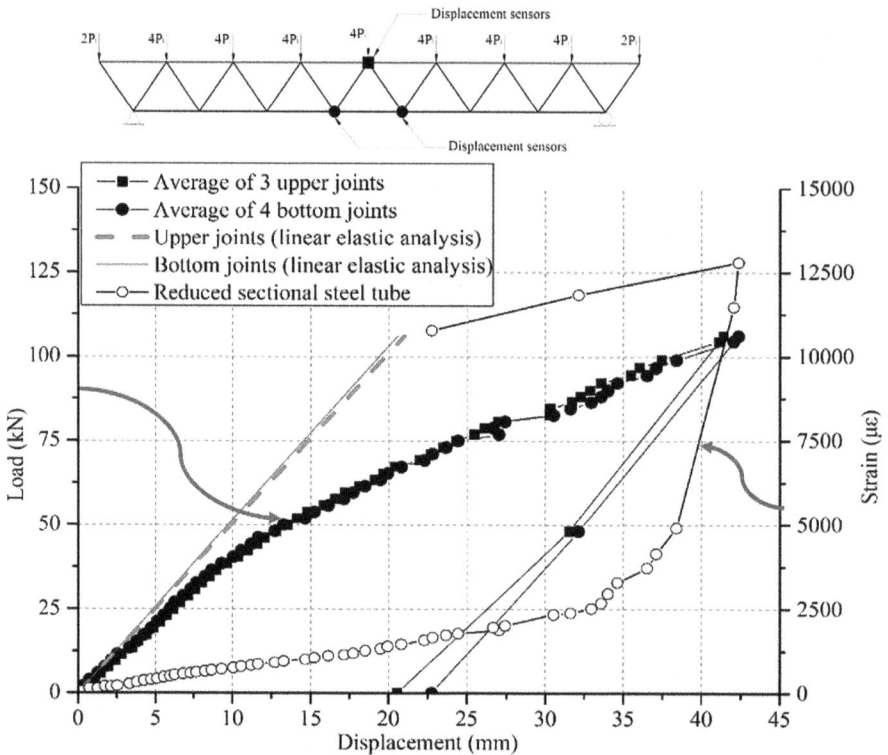

Figure 7.20 Load and displacement, strain and displacement responses.

construction applications. Standard trusses should also be developed for mass production and applications.

References

[1] Huang, X. (1959). *Roof bamboo structures* (in Chinese). Architectural Engineering Press, Beijing.
[2] Chen, Z.Y. (1958). Several questions about bamboo roofs (in Chinese). *Journal of Tsinghua University*, 4(2), 269–286.
[3] Albermani, F., Goh, G.Y., & Chan, S.L. (2007). Lightweight bamboo double layer grid system. *Engineering Structures*, 29, 1499–1506.
[4] Shan, B., Gao, L., Li, Z., & Xiao, Y. (2015). Experimental research and construction of prefabricated bamboo pole demonstration house (in Chinese). *Industrial Building*, 45(04), 33–41.
[5] Chen, G. (2011). Experimental research and engineering application of light frame bamboo structure. Doctoral Dissertation, supervised by Y. Xiao, Hunan University.
[6] Xiao, Y., Chen, G., & Feng, L. (2014). Experimental studies on roof trusses made of glubam. *Materials and Structures*, 47(11), 1879–1890.
[7] Wolfe, R.W. (1986). *Strength and stiffness of light-frame sloped trusses*. Report FPL-RP-471. U.S. Department of Agriculture, Forest Products Lab., Madison, WI.

[8] Wolfe, R.W., & LaBissoniere, T.G. (1991). *Structural performance of light-frame roof assemblies. II. Conventional truss assemblies.* Report FPL-RP-499. U.S. Department of Agriculture, Forest Products Lab., Madison, WI.

[9] Riley, G.J., & Gebremedhin, K.G. (1999). Axial and rotational stiffness model of metal plate connected wood truss joints. *Transactions of the ASAE*, 42(3), 761–770.

[10] Gupta, R., Gebremedhin, K.G., & Cooke, R.J. (1992). Analysis of metal-plate-connected wood trusses with semi-rigid joints. *Transactions of the ASAE*, 35(3), 1011–1018.

[11] Li, Z. (1998). Practical approach to modeling of wood truss roof assemblies. *Practice Periodical on Structural Design and Construction*, 3(3), 119–124.

[12] Rakesh, G., Milan, V., & Thomas, H.M. (1996). *Metal-plate-connected wood joints: A literature review.* Forest Research Laboratory, Oregon State University, Corvallis, OR.

[13] Mccarthy, M., & Wolfe, R.W. (1987). Assessment of truss plate performance model applied to southern pine truss joints. Research paper FPL-RP. U.S. Department of Agriculture, Washington, DC.

[14] Gupta, R., & Gebremedhin, K.G. (1990). Destructive testing of metalplate connected wood truss joints. *Journal of Structural Engineering*, 116(7), 1971–1982.

[15] Gebremedhin, K.G., Jorgensen, M.C., & Woelfel, C.B. (1992). Load-slip characteristics of metal plate connected wood joints tested in tension and shear. *Wood Fiber Science*, 24(2), 118–132.

[16] Groom, L., & Polensek, A. (1992). Nonlinear modeling of truss-plate joints. *Journal of Structural Engineering*, 118(9), 2514–2531.

[17] Gupta, R. (1994). Metal-plate connected tension joints under different loading conditions. *Wood Fiber Science*, 26(2), 212–222.

[18] Song, X.B., & Lam, F. (2009). Laterally braced wood beam-columns subjected to biaxial eccentric loading. *Computers & Structures*, 87(118), 1058–1066.

[19] Song, X.B., Lam, F., Huang, H., & He, M.J. (2010). Stability capacity of metal plate connected wood truss assemblies. *Journal of Structural Engineering*, 136(6), 723–730.

[20] Kent, S.M., Gupta, R., & Miller, T.H. (1997). Dynamic behavior of metal plate-connected wood truss joints. *Journal of Structural Engineering*, 123(8), 1031045.

[21] Percival, D.H., & Suddarth, S.K. (1989). Long-term tests of 4×2 parallel-chord metal-plate-connected wood trusses. Addendum to Research Reports Nos. 81-1 and 89-2. Small Homes Council–Building Research Council of the University of Illinois, Urbana-Champaign, IL.

[22] Donald, A.B., & Frank, E.W. (2011). Creep deflection in design of metal plate-connected wood trusses. *Practice Periodical on Structural Design and Construction*, 16(1), 10–14.

[23] American National Standards Institute (ANSI). (2014). ANSI/TPI 1-2014: National design standard for metal plate connected wood truss construction. Truss Plate Institute, Alexandria, VA.

[24] Canadian Standards Association (CSA). (2014). CSA S347: Method of test for evaluation of truss plate used in lumber joints. CSA, Toronto, ON.

[25] Chinese Standards. (2012). JGJ/T 265-2012: Technical code for light wood trusses. China Architecture and Building Press, Beijing.

[26] Wu, J.M., Xiao, Y., & Shea, E. (2016). Ultimate strength of metal-plate-connected glubam joints (in Chinese). *Industrial Construction*, 46(7), 118–123.

[27] Peng, Q., Xiao, Y., & Wu, J.M. (2018). Strength of metal-plate-connected glubam joints (in Chinese). *Building Structure*, 048(019), 91–96.

[28] Ministry of Housing and Urban–Rural Development and General Administration of Quality Supervision, Inspection and Quarantine of the People's Republic of China. (2012). GB/T 50329-2012: Standard for test methods of timber structures (in Chinese). China Architecture and Building Press, Beijing.

[29] Peng, Q. (2017). Research on mechanical behavior of metal-plate-connected GluBam floor trusses (in Chinese). Master of Engineering Thesis under supervision of Y. Xiao,Hunan University.

[30] Xie, Q.J., & Xiao, Y. (2016). An experimental study of long-span glubam roof trusses. *Journal of Building Structures*, 37(4).

[31] Xiao, Y., Wu, Y., & Yang, Q.W. (2016). A type of hybrid spatial truss. China Innovation Patent ZL 201610621173.4. Public announcement No. CN 106088468, on May 31, 2019.

[32] Wu, Y., & Xiao, Y. (2018). Steel and bamboo hybrid trusses. *Engineering Structures*, 171, 140–153.

[33] Li, Z., Li, T., Wang, C., He, X.Z., & Xiao, Y. (2020). Experimental study of an unsymmetrical prefabricated hybrid steel-bamboo roof truss. *Engineering Structures*, 201, 109781.1–109781.13.

[34] Quaranta, G., Demartino, C., & Xiao, Y. (2019). Experimental dynamic characterization of a new composite glubam-steel truss structure. *Journal of Building Engineering*, 25, 100773, DOI:10.1016/j.jobe.2019.100773.

Chapter 8

Engineered Bamboo Structural Walls

Structural walls are not only used as a vertical load bearing system, but most importantly are designed for resisting lateral force actions from loading such as wind and earthquake. Besides serving for load resistance, walls are also responsible for providing the external envelope to a building for maintaining a built environment inside. Of course, a building can be further divided by walls to form more units of built environment. This chapter discusses research findings, primarily from the author's group, on seismic behaviors and the design of glubam frame walls and their performance related to fire and thermal insulation.

8.1 Types of Lightweight Frame Structural Walls Involving Bamboo

In North America, timber-framed houses generally use a lightweight woodframe system, in which the main anti-lateral force members are so-called 2 × 4 walls; that is, the studs of the standard materials are connected by nails into a frame, plus plywood or oriented strand board (OSB) sheathing constitutes a structural wall. The inner surface is generally nailed with a fireproof gypsum board, and the intermediate cavity is filled with insulating materials, such as rock wool. The lightweight woodframe wall system was most likely created by practitioners;[1] however, recently it has been studied systematically by scholars in order to provide sound scientific backup and for developing methods for design.[2–5] The lightweight woodframe wall system is now becoming accepted in Asian countries including China. This type of structure is now included in the updated version of Chinese design codes for timber structures,[6] thanks to the contributions of many scholars,[7] and probably also the push by the Canadian Association of Timber Industry in China.

The first attempt to use engineered bamboo board or plybamboo in structural walls appeared to be led by Prof. Janssen of Netherland's Eindhoven University and documented in a doctoral thesis by Gonzalez-Beltran.[8] The author's research group started to devise various types of structural walls involving the usage of engineered bamboo from 2006. Figure 8.1 exhibits one example of the author's work – a single-story glubam building under construction, with lightweight glubam frame, which would be sheathed with plybamboo panels. The author's research group took the initiative in developing the following four types of lightweight frame wall systems with bamboo-based materials to partially or fully substitute the timbers.

DOI: 10.1201/9781003204497-8

Figure 8.1 Construction of a single-story building with glubam heavy frames and lightweight frame walls.

8.1.1 Bamboo Panel Sheathed Lightweight Woodframe Walls

As shown in Figure 8.2, this type of wall system is essentially the same as the conventional full timber walls except for the use of glue laminated bamboo panels as sheathing materials.[9] The sheathing panels have a basic modular dimension: 2440 mm tall and 1220 mm wide (8 ft. by 4 ft. based on the North American custom units). Longer walls can be assembled by using 1.2m wide (or long) modular wall panel units or be made as one piece. The process of assembling a prefabricated wood studs wall unit sheathed with thin-strip plybamboo is shown in Figure 8.3.

The advantage of this wall system is its possible immediate smooth implementation into existing timber design practice. The author strongly believes this should be given the highest priority for the code developing agencies in considering the adoption of bamboo for construction.

8.1.2 Bamboo Panel Sheathed Cold-formed Light-gauge Steel Frame Walls

Light gauge steel structures have been developed successfully for many years, and in North America,[10–12] they have been developed and pushed by some professional organizations with the intention of penetrating the huge timber housing market. The steel frames are typically made with galvanized cold-formed steel shapes. Conventionally, the light gauge steel frame walls are sheathed with oriented strand boards (OSB), plywood panels, etc. The author's group has shown it is promising to use plybamboo panels in light gauge steel frame shear walls.[13] Figure 8.4 exhibits a glubam building which was constructed partially with plybamboo-sheathed light gauge steel frames. This type of structural wall should also be very promising for inclusion in the well-developed codes of lightweight steel buildings.

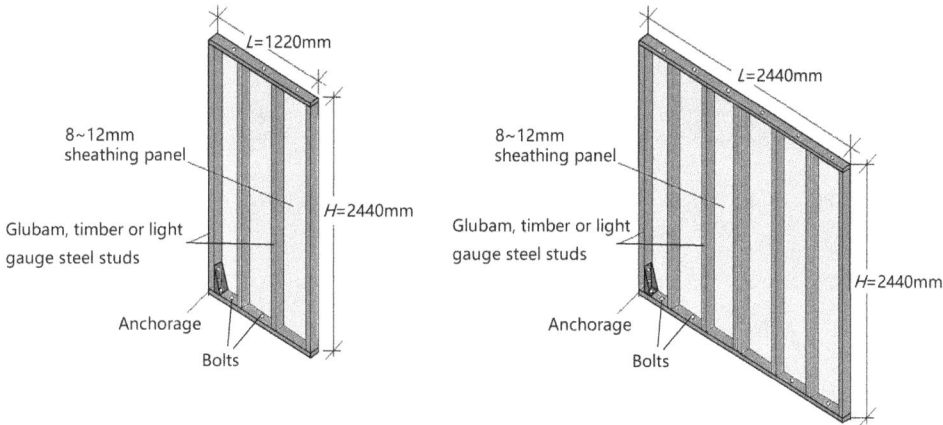

Figure 8.2 Lightweight bamboo-frame walls with modular dimensions.

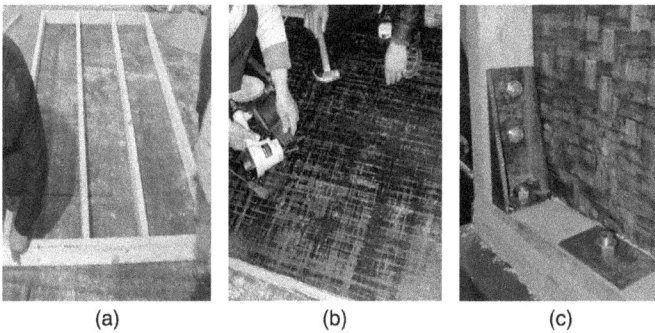

| (a) | (b) | (c) |

Figure 8.3 Fabrication of walls: (a) assembling frame skeleton; (b) nailing wall sheathing panels; (c) anchoring corner.

8.1.3 Glubam Lightweight Frame Walls

This is a type of structural wall system with a full bamboo solution; however, it maintains the similar geometrical features and intended functions of lightweight woodframe walls. The author's research group has studied primarily the thin-strip glubam walls,[14] whereas researchers from Colombia have tested laminated thick-strip Guadua bamboo (which can be considered as Guadua bamboo-based glubam) walls.[15]

8.1.4 Round Bamboo-culm Frame Walls

Another bamboo wall system the authors studied is that using lightweight frame walls with round bamboo culms as studs and sheathed with plybamboos.[16] The advantage of this wall system is the reduced manufacturing process for the studs, with the direct use

Figure 8.4 Plybamboo-sheathed light gauge steel frame walls.

Figure 8.5 Round bamboo wall buildings.

of bamboo culms involving only minor shaping work, and reduced labor and manufacturing cost. Figure 8.5 shows a building designed and constructed by the author's group using plybamboo sheathed around bamboo frame walls.[16]

The author has recently conceived the system of cross laminated bamboo and timber (CLBT or CLTB), in the light of the recent growing success of wood- based CLT.[17]

8.2 Seismic Behavior of Lightweight Bamboo Walls

To date, the author's research group has carried out a quite significant amount of shear tests to investigate the seismic behavior of structural walls with various types of lightweight frames (timber, cold-formed steel, glubam, and bamboo culms) sheathed with plybamboo panels, and other comparable conventional materials such as OSB and plywood.[9,13,14,16] Table 8.1 shows the details of four research programs testing different types of shear walls. The testing program for the woodframe shear wall is discussed in detail in this section, whereas the testing results for other types of shear walls can be found from the relevant references.

8.2.1 Plybamboo Sheathed Woodframe Walls

As shown in Figure 8.2, the first series of studies was for the model walls framed with grade IIIc SPF (spruce, pine, and fir) studs with a section size of 38 mm × 89 mm (or nominal 2 × 4). The sheathing panels were 1220 mm × 2440 mm plybamboo with 9 mm thickness. The details are shown in Table 8.2. The main testing parameters include loading cycles (monotonic and cyclic loading), wall sizes (one-unit or two-unit plybamboo sheathing panels), types of nails, and spacing configurations. The two types of nails were selected based on extensive trials of nailing to penetrate the relatively hard plybamboo properly. The wall lower corners are connected to the testing frame by the use of custom-made anchorage brackets connected by bolts, as also shown in Figure 8.2.

Table 8.1 Testing programs on bamboo walls with different combinations

Combination	Frame	Width (m) *	Comments
Wood + Bamboo	Timber studs	1.2–2.4	20 specimens; main parameters: nail types and spacing
Steel + Bamboo	CFS studs	1.2–2.4	16 specimens; main parameters: wall width, loading protocols
All bamboo-glubam	Glubam studs	1.2–2.4	24 glubam and 4 timber specimens; main parameters: plybamboo thickness, nail types
All bamboo-culms	Round bamboo culms	2.4	2 specimens were tested

Note: *All specimens had a height of 2.4 m.

Table 8.2 Parameters and test contents for woodframe walls with plybamboo sheathing

Specimen group	Size (m)	Nail connector type	Nail spacing (mm)	Loading mode
1	1.22×2.44	6d spiral nail	Edge: 150; field: 300	3 Monotonic and
2	1.22×2.44	51 mm staple nail	Edge: 150; field: 300	1 cyclic loading
3	1.22×2.44	51 mm staple nail	150 throughout	
4	2.44×2.44	51 mm staple nail	150 throughout	
5	2.44×2.44	6d spiral nail	Edge: 150; field: 300	
WF	2.44×2.44	45 mm T-shape nail	150 throughout	

Source: Xiao et al. (2015);[9] Wang et al. (2017).[14]

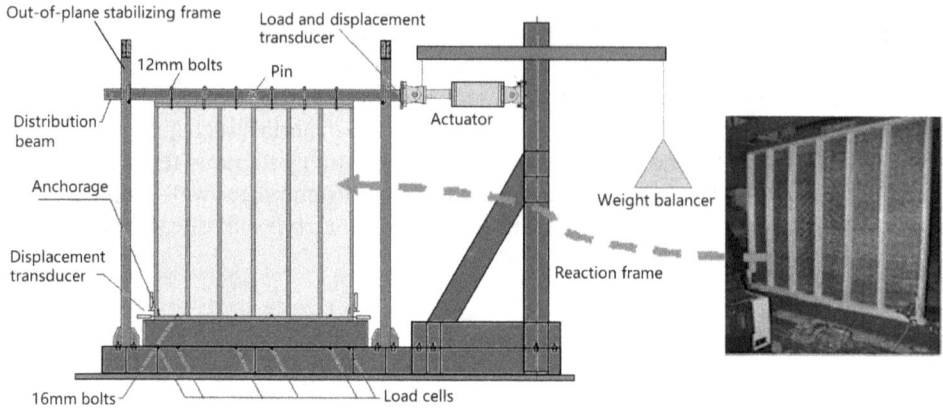

Figure 8.6 Wall test setup.

8.2.2 Testing Method

The seismic load simulation tests were carried out using the test setup shown in Figure 8.6. The lateral load is applied by a servo-controlled hydraulic actuator with a maximum stroke of 600 mm and a maximum loading force of 100 kN. The system is equipped with a corresponding displacement sensor and a force sensor, which are controlled by closed loop. For testing the two-unit plybamboo wall specimens, an upper distribution beam was added with a center pin to reduce the stiffness and to possibly obtain conservative testing results, avoiding overestimation of the capacity.[18]

For each group of specimens, three monotonically loaded tests were performed first to obtain the average lateral load and displacement curve; then the curve is used to determine the loading protocol for the remaining cyclic loading tests. Testing standard ASTM E2126[19] was adopted as the guideline for the monotonic and cyclic tests. From the three monotonic loading tests of the same testing group, the average ultimate displacement Δ_m can be found. The program for cyclic loading consisted of two displacement patterns, as follows: (1) five fully reversed single cycles at amplitudes of 1.25, 2.5, 5, 7.5, and 10% of the mean ultimate displacement Δ_m; and (2) phases including three fully reversed cycles of equal amplitude at peak displacements of 20, 40, 60, 80, 100, and 120% of the mean ultimate displacement Δ_m. The loading speed was 7.5 mm and 60 mm per min. for monotonic tests and cyclic tests, respectively.

8.2.3 Experimental Performance

For lightweight woodframe shear walls, there are mainly two damage/failure patterns; i.e., first, the damage of nail connections between the sheathing panel and the studs; second, the separation of the wall studs and the bottom seal plate. Such failure patterns are also recognized for bamboo-based shear walls. Figure 8.7 (a) and (b) shows the deformation of the plybamboo panels, resulting from the deformation or damage of the nails, exhibited in Figure 8.7 (c) and (d). The damage to the nails was mainly severe bending, as shown in Figure 8.7 (c) and (d); however, a limited number of steel staple nails were ruptured due to their relatively high brittleness, which is considered to be

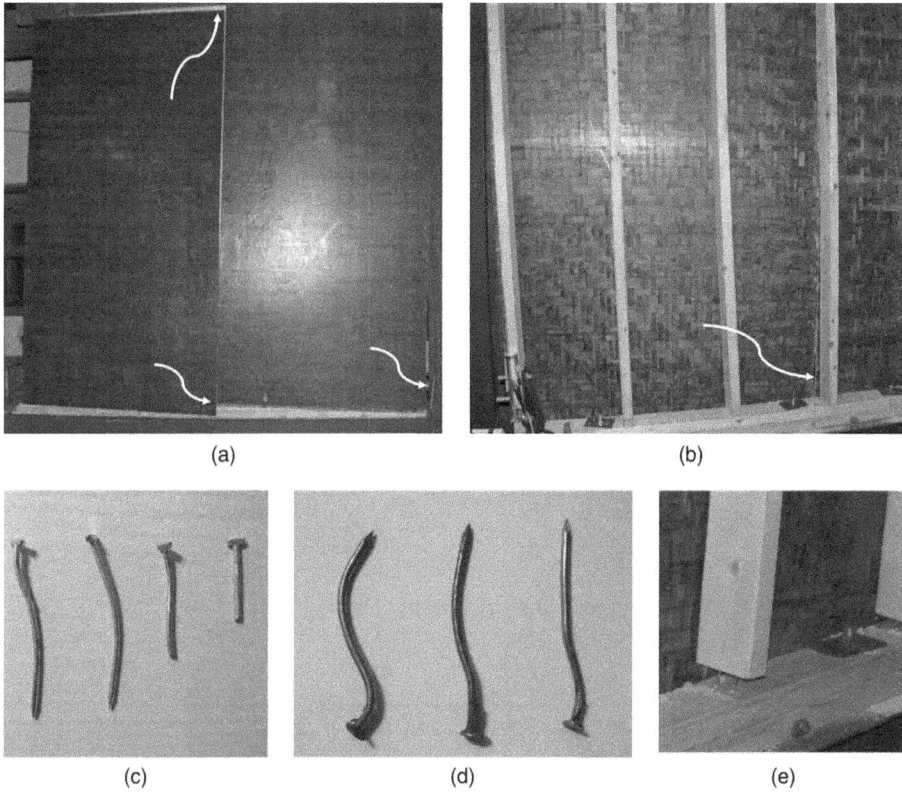

Figure 8.7 Damage to shear walls: (a) rotation of sheathing panels; (b) deflection of studs and separation from sheathing panel; (c) bending and rupture of steel staple nails; (d) bending of common nails; (e) pull out of stud from seal plate.

the sacrifice for pursuing the construction convenience of using them. On the other hand, the pull out or separation of the studs from the seal plate (Figure 8.7 (e)) was limited to the studs away from the wall end, which were not attached at the bottom with anchorage.

The load-displacement curves of each group of walls in monotonic loading and cyclic loading are shown in Figure 8.8. The monotonic loading curves shown in Figure 8.8 (a) are the average curves of three specimens with the same testing parameters. The hysteresis curves in Figure 8.8 (b)–(f) also show the relevant average curves from monotonic loading. Figure 8.8 (g) shows the result of specimen wall WF-1, which was a conventional timber wall model with SPF studs and OSB sheathing panels.

As apparently shown in Figure 8.8, the lateral loading capacity is dependent on the length of the walls, and can roughly double when the length changes from 1.22 m to 2.44 m. The envelopes of the cyclic loading hysteresis loops are close to the load and displacement behaviors up to the maximum loading capacities, then the cyclic degradation becomes apparent. The test results show that the types of fasteners and the spacing of

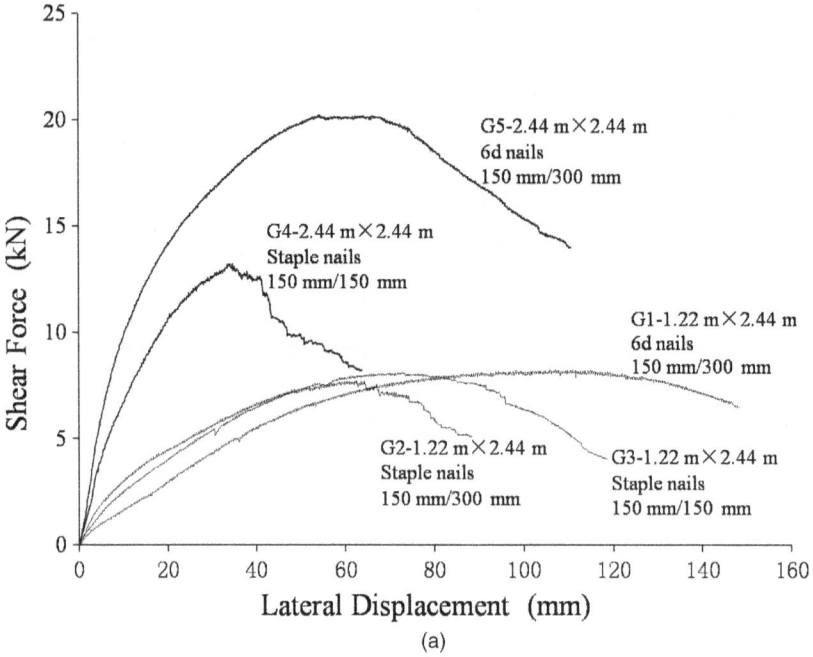

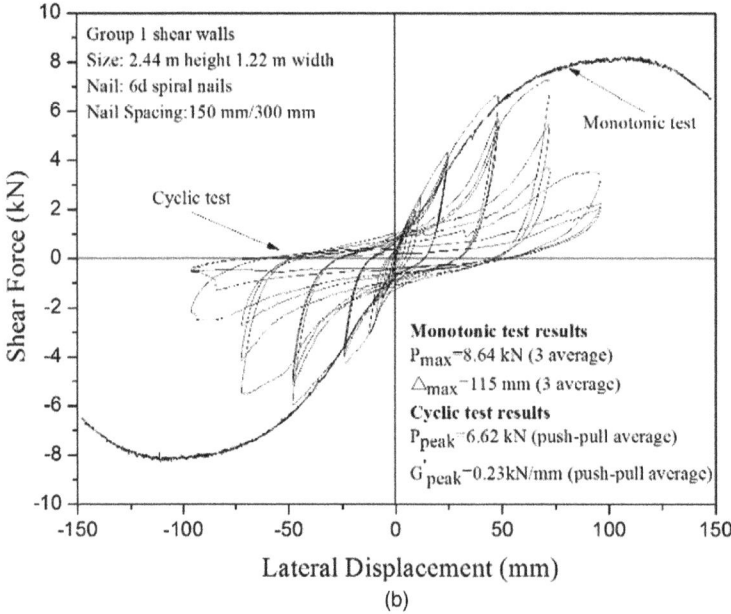

Figure 8.8 Load-deformation relationships of wall specimens: (a) monotonic curves (three average each); (b) 1.22 m × 2.44 m wall with 6d nails; (c) 1.22 m × 2.44 m wall with staple nails; (d) 1.22 m × 2.44 m wall with densified staple nails; (e) 2.44 m × 2.44 m wall with densified staple nails; (f) 2.44 m × 2.44 m wall with 6d round nails; (g) woodframe shear wall with 45 mm T-shape nails.

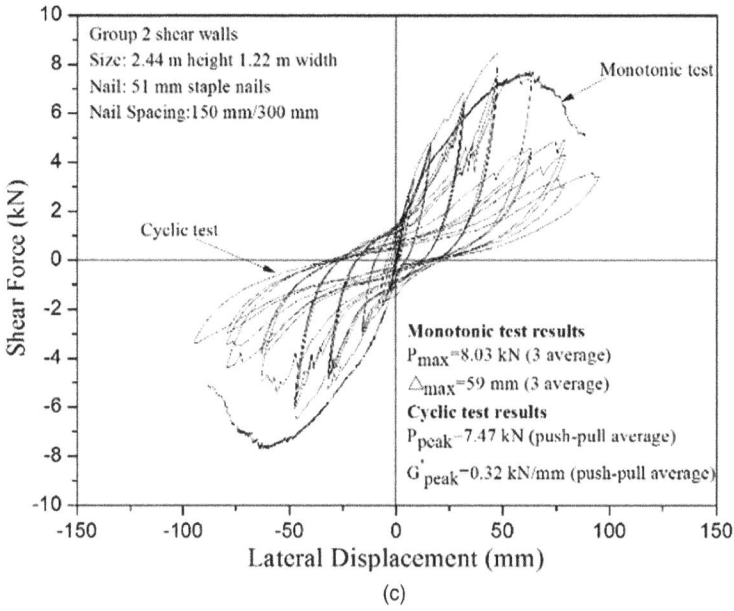

(c)

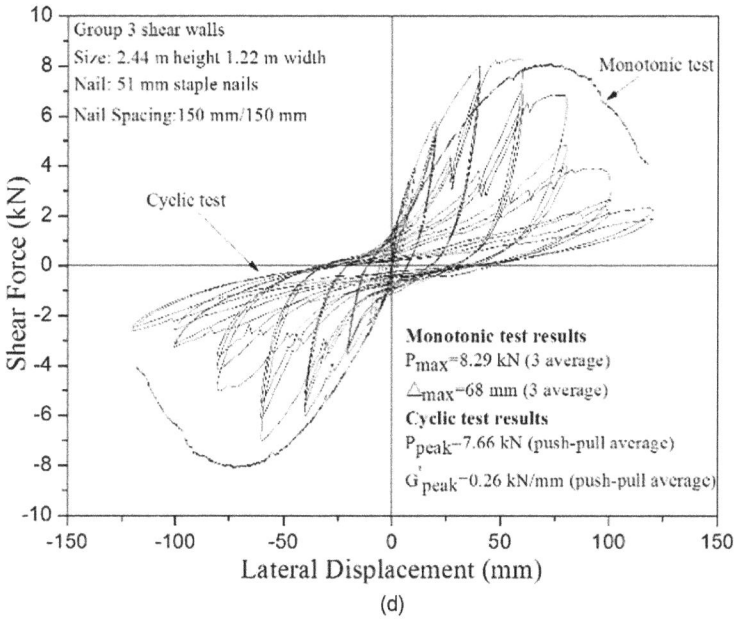

(d)

Figure 8.8 (Continued)

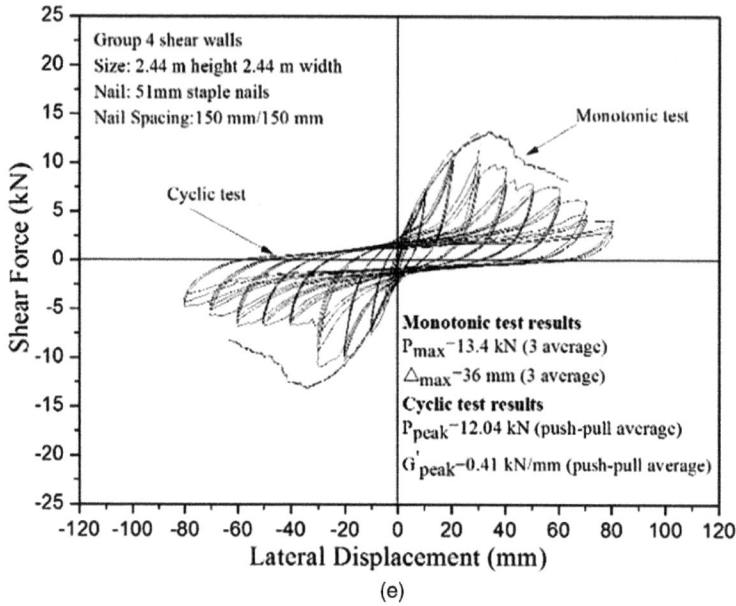

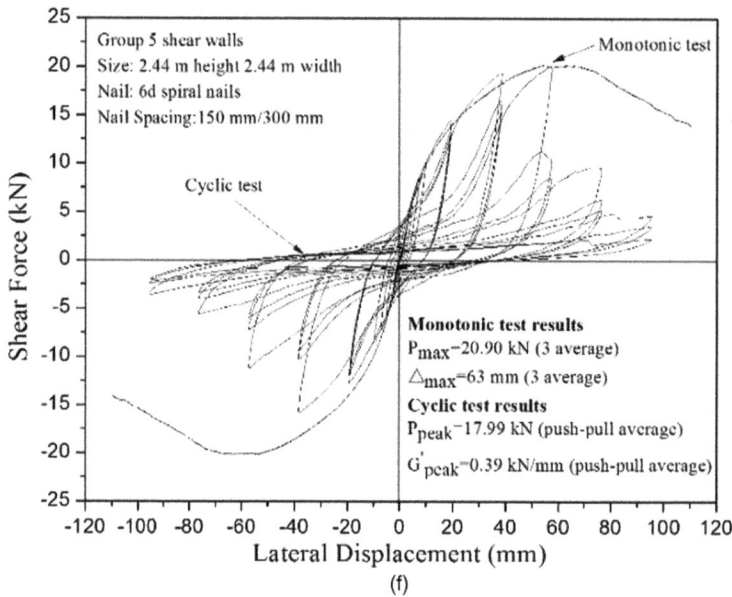

Figure 8.8 (Continued)

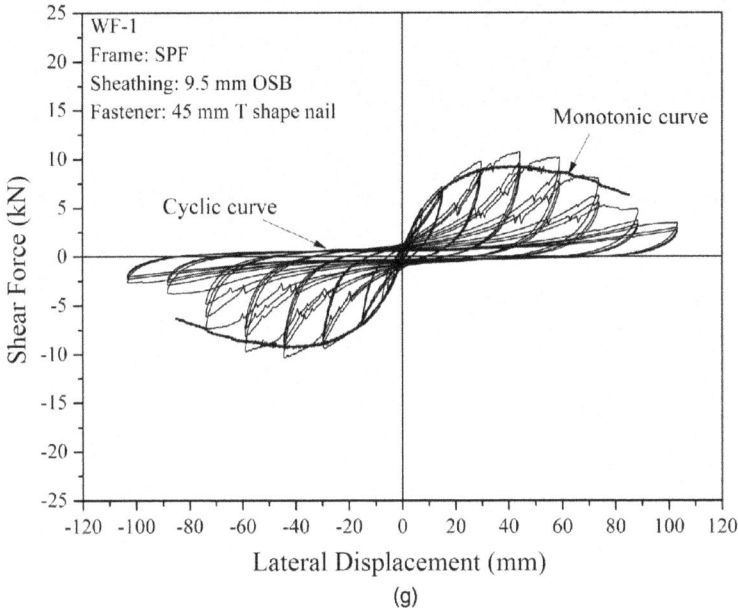

WF-1
Frame: SPF
Sheathing: 9.5 mm OSB
Fastener: 45 mm T shape nail

Monotonic curve

Cyclic curve

Shear Force (kN)

Lateral Displacement (mm)

(g)

Figure 8.8 (Continued)

fasteners significantly affect the behavior of the shear walls. Generally speaking, the walls with the 6d steel nails behaved better than those with the stapled nails; however, the stapled nails have significant construction efficiency.[9] Similar responses can be seen by comparing the results in Figure 8.8 (e) and (g) for the walls with similar nails but sheathed with plybamboo and OSB, respectively.

Based on ASTM E2126-11,[19] key features and parameters of the shear walls can be investigated. The wall shear strength v_{peak} is calculated according to the following equation,

$$v_{peak} = \frac{P_{peak}}{L} \ (kN/m)$$ (8.1)

where, P_{peak} represents the peak load value obtained by the hysteresis curve envelope diagram; L is the length of the test wall. The secant shear stiffness K is calculated from Eq.8.2:

$$K = \frac{P}{\Delta} \times \frac{H}{L} \ (kN/m)$$ (8.2)

where, P denotes the measured load value; Δ denotes the lateral displacement of the wall top; H is the height of the test piece, generally 2.44 m; L is the length of the wall of the test specimen, not including openings. Stiffness K can be defined for the load value at $0.4 P_{peak}$ and P_{peak}, respectively.

For evaluation of deformability, the displacement ductility factor μ can be defined using the ultimate displacement Δ_u and the yield deformation Δ_{yield},

$$\mu = \frac{\Delta_u}{\Delta_{yield}} \tag{8.3}$$

Further, the equivalent energy elastic-plastic (EEEP) can be assumed to model the behavior of the shear walls, following the similar procedure for lightweight woodframe shear walls.[19] The EEEP model is an elasto-plastic behavior which has the same dissipated energy as the original response with the same ultimate deformation. Based on such a definition, the equivalent yield force, P_{yield}, can be estimated,

$$P_{yield} = \left(\Delta_u - \sqrt{\Delta_u^2 - \frac{2A}{K_e}} \right) K_e \tag{8.4}$$

where A is the envelope area of the load-displacement curve till Δ_u; K_e is the secant modulus when the load reaches 0.4 P_{peak}, and is calculated as follows:

$$K_e = \frac{0.4 p_{peak}}{\Delta_e} \tag{8.5}$$

In Eq.8.5, Δ_e is the displacement corresponding to 0.4 P_{peak}.

The EEEP curves with normalized deformation calculated for the shear walls tested in previous experimental programs are shown in Figure 8.9. It can be seen that the average bearing strength of the T-shaped nail glubam shear wall with thin sheathing panels (6 mm) is 1.4 times higher than that of the wood one with the same nails to attach the sheathing panels, but with thicker OSB sheathings (9.5 mm). Increases in nail diameter can increase the bearing strength of shear walls until the failure mechanism changes from nail failure to pull-through failure of the plybamboo panel. Increases

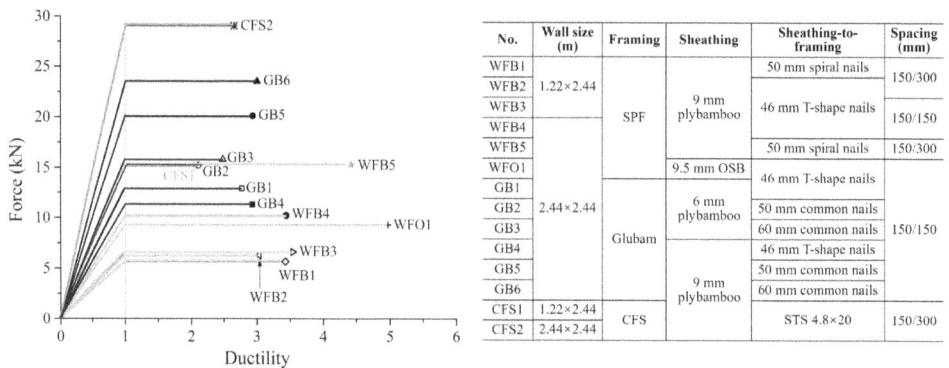

No.	Wall size (m)	Framing	Sheathing	Sheathing-to-framing	Spacing (mm)
WFB1				50 mm spiral nails	150/300
WFB2	1.22×2.44				
WFB3		SPF	9 mm plybamboo	46 mm T-shape nails	150/150
WFB4					
WFB5				50 mm spiral nails	150/300
WFO1			9.5 mm OSB		
GB1				46 mm T-shape nails	
GB2	2.44×2.44		6 mm plybamboo	50 mm common nails	
GB3		Glubam		60 mm common nails	150/150
GB4				46 mm T-shape nails	
GB5			9 mm plybamboo	50 mm common nails	
GB6				60 mm common nails	
CFS1	1.22×2.44	CFS		STS 4.8×20	150/300
CFS2	2.44×2.44				

Figure 8.9 Equivalent elasto-plastic curves of different types of walls.

in thickness of sheathing panels can also increase the bearing strength of shear walls corresponding to the change of failure mechanism from panel failure to nail yielding.

The calculation comparisons of the tested capacity of the engineered bamboo walls and those codified for timber walls show that the engineered bamboo walls have adequate design shear capacity.[9,14]

8.3 Seismic Analysis of Glubam Shear Walls

Over the last two decades, significant progress has been made in modeling and developing sophisticated analytical methods and tools for the seismic behaviors of timber structures.[20–23] An open-source computer program has also been developed under the U.S. National Science Foundation NEES project.[24] Taking advantage of the advancement in timber engineering, the author's research group attempted the modeling of shear walls with engineered bamboo. Chen[25] carried out numerical simulation analysis using the SAPWood program. As shown in Figure 8.10, the shear wall is regarded as an equivalent truss model, which is composed of four hinged rigid bar elements and two pairs of spring elements distributed along the diagonal. The lateral deformation resistance of the wall is realized by the spring element, while the rigid bar element bears only the axial force. The stiffness of the four rigid bar elements is much greater than that of the spring elements. The deformation of the wall comprises of the overall lateral displacement deformation of the equivalent model structure and the deformation of the spring element; the deformation of the rigid bar is not considered. In other words, the equivalent lateral stiffness of the wall is completely determined by the spring. The spring factor can be determined based on detailed FE analysis and calibrated against the experimental results.

Recently, Wang used the open-source structural analysis platform, OpenSees, to develop a simplified approach to predict the nonlinear response of lightweight glubam shear walls and structures based on nail connection testing data and modeling.[26–27] Based on the assumption that the characteristics of the sheathing-to-framing connections govern the performance of shear walls, the numerical model was developed based on the connection test information, as shown in Figure 8.11. A significant number of glubam nail connection tests were conducted considering different nail types and loading directions, and a performance database was assembled. The corresponding

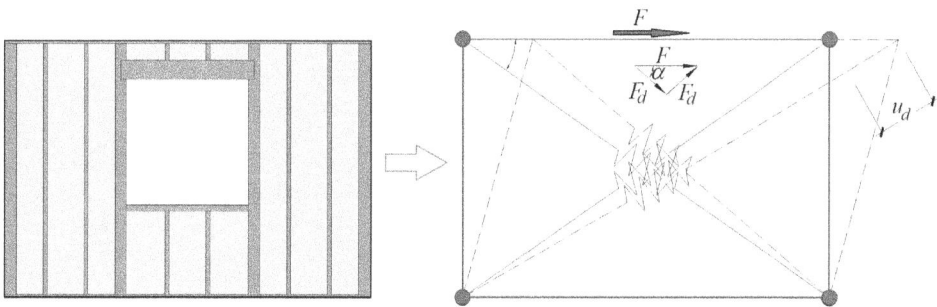

Figure 8.10 Equivalent truss model for shear wall.

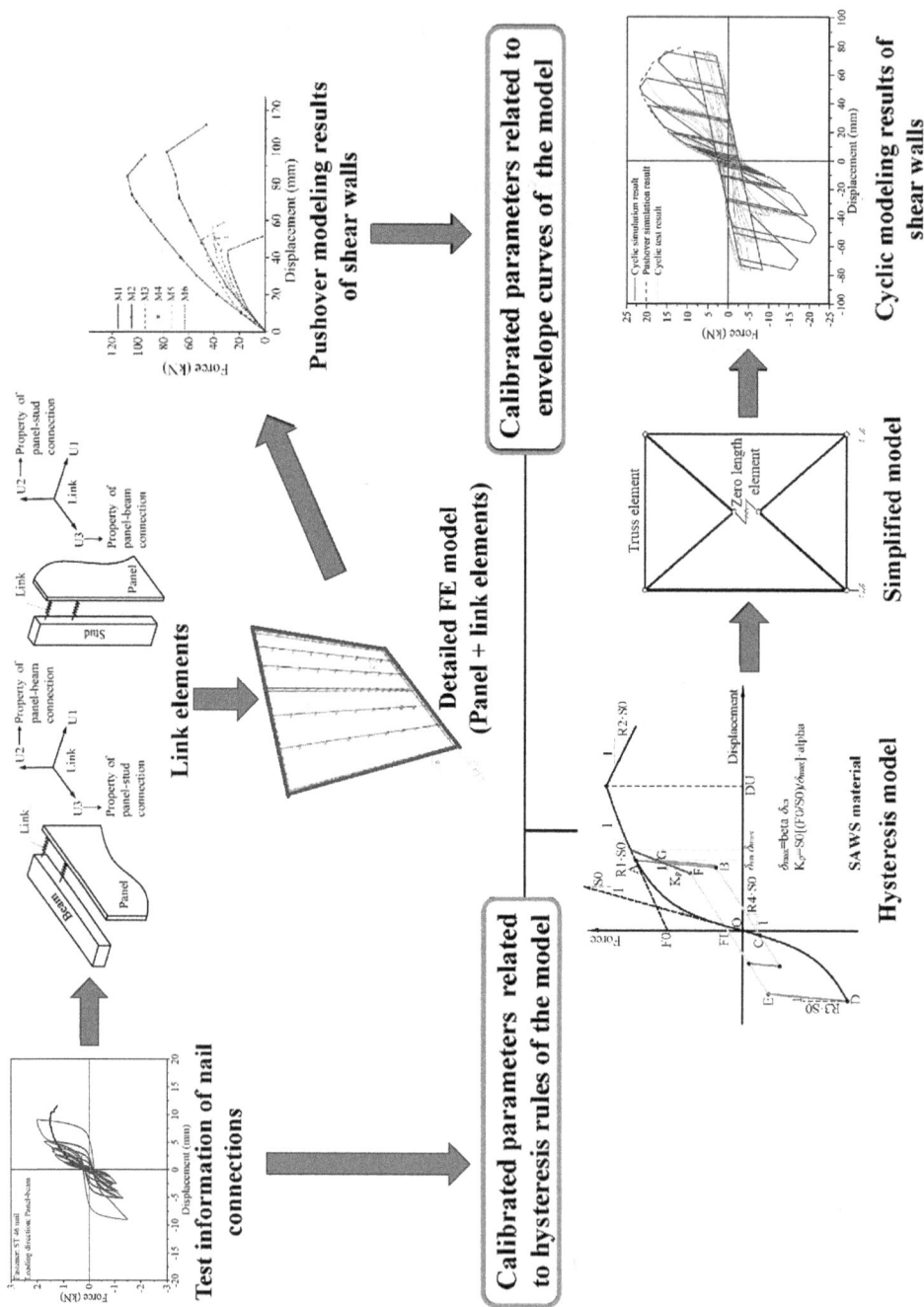

Figure 8.11 Modeling shear walls based on connection test data.

simplified hysteretic wall model is composed of two types of input parameters. First there are the parameters determining the wall's envelope curve under monotonic loading conditions; then there are the parameters that regulate the hysteresis rules.[28] The parameters related to the envelope curve are obtained on the pushover modeling results of shear walls; whereas those related to the hysteresis rules are estimated using the connection test data. The influences of different nails, wall sizes, and openings on the behavior of shear walls are evaluated based on the simplified model. As an example, the simulations for the cyclic loading and monotonic loading are compared in Figure 8.12 for the model glubam walls tested by Wang et al.[14] It is shown that the main features of the test results of the glubam walls can be reasonably described by the simulation using the simple modeling techniques.

8.4 Thermal Performance of Lightweight Frame Walls

As described in Chapter 3, thermal conductivity of glubam was studied, along with other materials, such as insulation and façade materials by several researchers[29–32] and the author's group.[33] Based on the material test results, Wang et al. also designed and fabricated four wall panel specimens for experimental study on their thermal insulation performance using guarded hot box equipment (GHB),[33] shown in Figure 8.13. In the GHB, testing is performed by establishing and maintaining a desired steady temperature difference across a test sample for a period of time that ensures constant heat flow and steady temperature, based on ISO 8990.[34]

The testing parameters of the four testing wall panel specimens are shown in Table 8.3., which also includes test results for thermal resistance R and transmittance U, defined as follows,

$$R_t = \frac{A(T_{si} - T_{se})}{\Phi}$$ (8.6)

where A is the area of the tested wall covered by the metering box; T_{si} and T_{se} are the temperatures of the inner (hot side) and outer (cold side) surface, respectively, and φ is the heat flow equal to the average electric power under steady-state conditions in the metering area heating circuit.

$$U = \frac{1}{R_{si} + R_t + R_{se}} = \frac{\Phi}{A(T_{ni} - T_{ne})}$$ (8.7)

where $R_{si} = A(T_{ni} - T_{si})/\Phi$, and $R_{se} = A(T_{ne} - T_{se})/\Phi$ are the inner and exterior surface resistances, respectively. Apparently, better insulation performances are related to larger values of R_t and lower values of U.

As shown in Table 8.3, the four case-study wall models were tested in the GHB apparatus to determine their thermal resistance and transmittance. These results are in agreement with the difference found using the hot plate test at the material level, described in Chapter 3. The full wood-based configuration (wall 1) has a thermal resistance slightly higher than the hybrid wood–bamboo-based configuration (wall 2) and

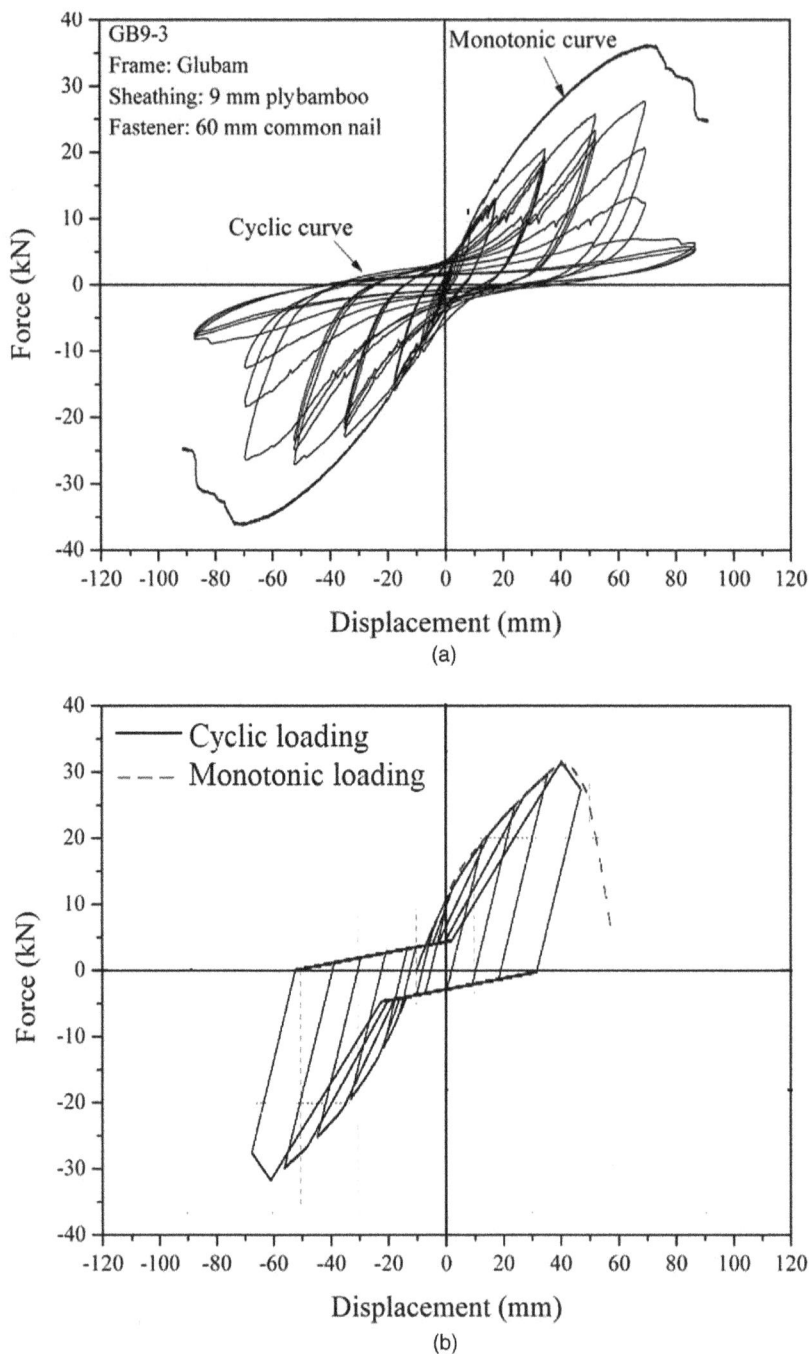

Figure 8.12 Lateral force and displacement hysteretic responses of glubam wall: (a) test by Wang et al.; (b) simulation.

Source: (a) Wang et al. (2017).[14]

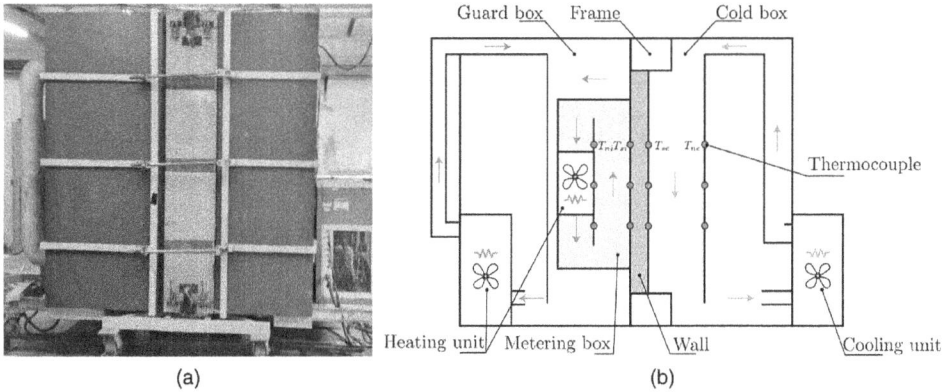

(a)   (b)

Figure 8.13 Photo (a) and schematic details (b) of hot-box equipment.

Table 8.3 GHN test parameters and results

	Wall 1	Wall 2	Wall 3	Wall 4
Sheathing panel	OSB	Plybamboo	Plybamboo	Plybamboo
Stud framing	SPF	SPF	Glubam	Glubam
R_t (m²K/W)	2.354	2.236	2.019	3.030
U (W/m²K)	0.395	0.402	0.457	0.308

the full bamboo-based configuration (wall 3). On the other hand, considering the full bamboo configurations, it is shown that increasing the stud thickness leads to a large increase in the thermal resistance. When the studs' thickness is changed from 89 mm (wall 3) to 140 mm (wall 4), the thermal resistance increases by +150% from 2.019 to 3.030 m²K/W. Compared with the other walls, wall 4 guarantees the best thermal insulation performance.

8.5 Fire Performance

Fire-resistant performance of timber and bamboo is one of the most important factors for assuring the safety of modern timber and bamboo buildings. However, there is still a lack of sufficient research in this field, thus hindering the wide acceptance of such types of bio-based materials in construction, particularly in China.

In order to provide experimental evidence for a previously executed demonstration project, a pilot study was carried out by the author's research group in testing a full-scale glubam wall panel, following the ISO standard temperature curve.[35,36] The overall size of the wall specimen was 3.7 m wide by 3.3 m tall, as shown in Figure 8.14 (a). The nominal size of the glubam studs was 30 mm wide by 90 mm deep, sheathed with 12 mm thick plybamboo panels on both sides. The sheathing panels were covered by 15 mm thick gypsum boards. Rock wool material was used to fill in the gaps of the sheathing panels for insulation, as shown in Figure 8.14 (b). A total of 12 thermal sensors were installed on the unexposed surface to detect the temperature increase. As a non-load

Figure 8.14 Fire test: (a) specimen details; (b) filling with rock wool insulation materials.

Figure 8.15 Test equipment (a); and final condition at 50 min. in testing process (b).

bearing wall, the required fire endurance time is 45 min. under the ISO fire standard, as required also by the Chinese code.[37]

The fire testing furnace used was 1.2 m deep and had a 3.0 m square opening, as shown in Figure 8.15. From the fire test, the temperature characteristics, fire-resistant influence factors, and failure mode were obtained. Temperature curves of each measuring point, temperature distribution, and fire-resistant limit were also recorded. The research results show that the highest temperature on the unexposed surface did not exceed 90°C, as shown in Figure 8.16, with a fire endurance duration exceeding 60 min., longer than the required 45 min. as per Chinese code GB/T 9978.[37] The test was also

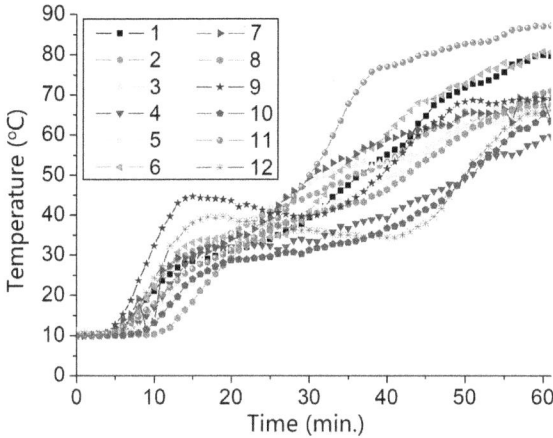

Figure 8.16 Temperature curves recorded for the unexposed surface.

simulated using fire dynamics simulation software FDS[38] and the results were satisfactory, with the temperature distribution of the unexposed surface reasonably well matching the instrumented temperatures.

8.6 Design Recommendations

The author believes that the design of some engineered bamboo structural components is now ready to be incorporated into the existing timber structure design specifications, reflecting the similar characteristics of bamboo and timber and the fact that the engineered bamboo structure is essentially developed as an alternative to or substitute for a timber structure. As the first step, the author suggests adding the engineered bamboo boards as one of the sheathing materials for woodframe shear walls, since it is relatively mature based on the experimental test results. For example, on page 97, clause 9.6.4 of the current Chinese timber structure design code GB 50005[6] states: "When using the wood-based structural boards as sheathing panels, the thickness shall not be less than 9.0 mm, for the maximum stud spacing at 410 mm; or 11.0 mm for the maximum stud spacing at 610 mm." It is suggested that "bamboo-based" be added to the statement and the clause be changed to: "When using the wood-based, or *bamboo-based* structural boards as sheathing panels, the thickness shall not be less than 9.0 mm, for the maximum stud spacing at 410 mm; or 11.0 mm for the maximum stud spacing at 610 mm." Following the timber design code, the shear resistance of a shear wall can be calculated based on its length l, and a design strength, f_d, as,

$$V = \Sigma f_d l \tag{8.8}$$

and the design strength f_d,

$$f_d = f_{vd} k_1 k_2 k_3 \tag{8.9}$$

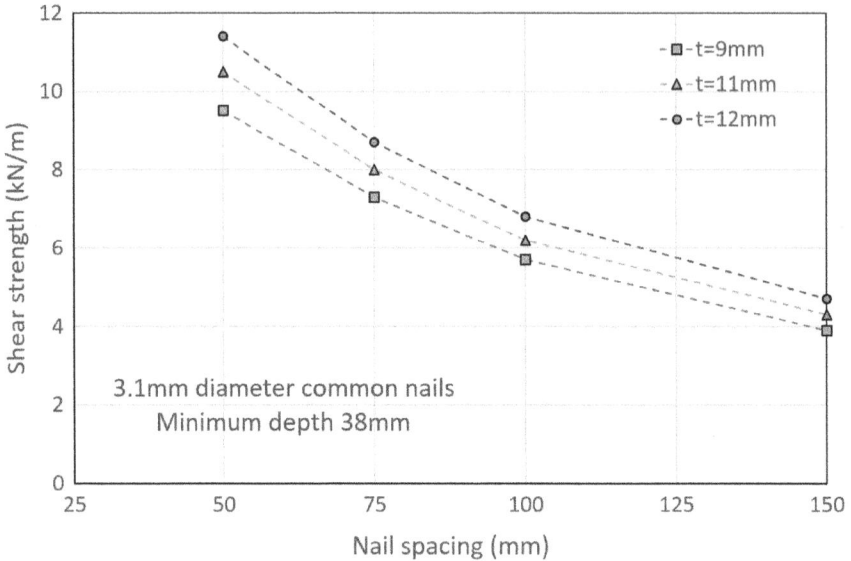

Figure 8.17 Basic design shear strength of woodframe shear walls.

where f_{vd} is the design shear strength of woodframe shear walls with wood-based sheathing panels and has been tabulated in the code; k_1 is the adjustment coefficient for the water absorption ratio, w (%), taken as 1.0 if $w<16\%$, and 0.8 if $16\%\leq w<19\%$; k_2 is the adjustment coefficient for different types of wood species for studs; for example, taken as 1.0 for Douglas fir, and 0.8 for SPF; and $k3$ is the strength adjustment coefficient for woodframe without the horizontal spacer elements. For shear walls with typical thickness of sheathing panels, $t = 9$–12 mm, and 3.1 mm diameter common nails, the shear strength, f_{vd} (kN/m) is depicted in Figure 8.17 for different nail spacing. Apparently, the closer the spacing is, the higher the shear strength.

Similar changes can be adopted for the design specifications of cold-formed light steel structural walls. Needless to say, the design specifications for the full engineered bamboo structures should be developed in the near future.

The walls should also be designed for thermal insulation. The previous study on thermal conductivity by the author's group can provide some guidance.[33] Using China as a regional example, the country is classified into five thermal regions distinguishing the thermal design requirements,[39] which includes the minimum thermal transmittance, U_{min}, for the thermal regions according to GB/T 50361[40] and GB/T 50189.[41]

Wang et al.[33] analyzed glubam shear walls with different configurations and concluded that all the engineered bamboo shear walls with a stud's cross-section size equal to $b \times w = 38 \times 140$ mm are suited for all the Chinese thermal regions (i.e., they satisfy the condition $U \leq U_{min}$). Furthermore, the bamboo shear walls with a stud's cross-section size equal to $b \times w = 38 \times 89$ mm can be adopted in the majority of the Chinese thermal regions, except in the "severe cold" areas, where the design uses stud spacing $s = 600$ mm and $s = 400$ mm; and in the "cold" areas, where the design uses

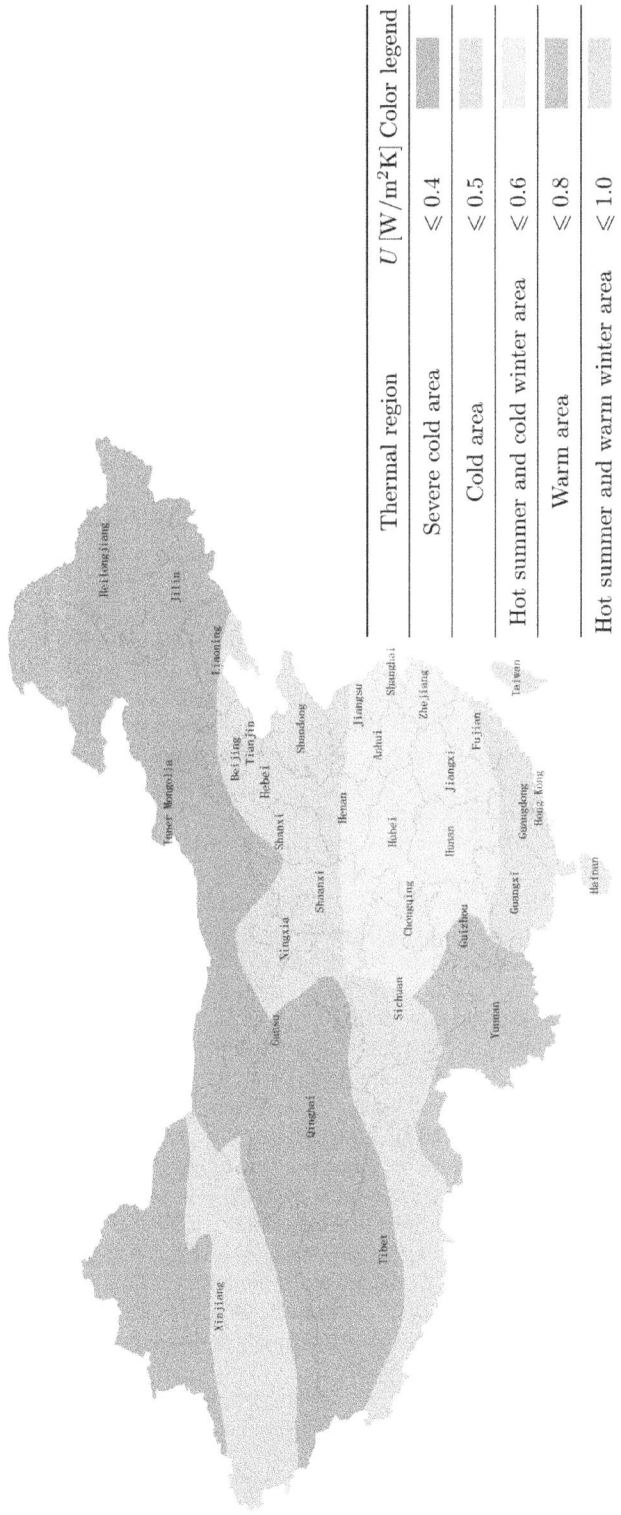

Thermal region	U [W/m²K]	Color legend
Severe cold area	$\leqslant 0.4$	
Cold area	$\leqslant 0.5$	
Hot summer and cold winter area	$\leqslant 0.6$	
Warm area	$\leqslant 0.8$	
Hot summer and warm winter area	$\leqslant 1.0$	

Figure 8.18 Map of thermal regions along with thermal requirements for buildings.

stud spacing $s = 300$ mm. This study can be considered as a useful reference for thermal design of engineered bamboo walls.

References

[1] Lienhard, J.H. (1997). Balloon frame houses. No. 779. The Engines of Our Ingenuity, https://uh.edu/engines/epi779.htm

[2] Lam, F., Prion, H.G.L., & He, M. (1997). Lateral resistance of wood shear walls with large sheathing panels. *Journal of Structural Engineering*, 123(12), 1666–1673.

[3] Pardoen, G.C., Waltman, A., Kazanjy, R.P., Freund, E., & Hamilton, C.H. (2002). *Testing and analysis of one-story and two-story shear walls under cyclic loading.* CUREE–Caltech Woodframe Project Report W-25. Consortium of Universities for Research in Earthquake Engineering (CUREE), Richmond, CA.

[4] van de Lindt, J.W. (2004). Evolution of wood shear wall testing, modeling, and reliability analysis: Bibliography. *Practice Periodical on Structural Design and Construction*, 9(1), 44–53.

[5] Li, M.H., Foschi, R.O., & Lam, F. (2012). Modeling hysteretic behavior of wood shear walls with a protocol-independent nail connection algorithm. *Journal of Structural Engineering*, 138(1), 99–108.

[6] Ministry of Housing and Urban–Rural Development of the People's Republic of China. (2017). GB 50005-2017: Code for design of timber structures (in Chinese). China Architecture & Building Press, Beijing.[7] Sun, X., He, M., & Li, Z. (2020). Novel engineered wood and bamboo composites for structural applications: State-of-art of manufacturing technology and mechanical performance evaluation. *Construction and Building Materials*, 249, 118751.

[8] Gonzalez-Beltran, G.E. (2003). Plybamboo wall-panels for housing: Structural design. PhD thesis, Eindhoven University of Technology, The Netherlands.

[9] Xiao, Y., Li, Z., & Wang, R. (2015). Lateral loading behaviors of lightweight woodframe shear walls with ply-bamboo sheathing panels. *ASCE Journal of Structural Engineering*, 141(3).

[10] American National Standards Institute (ANSI). (2013). *North American specification for the design of cold-formed steel structural members.* American Iron and Steel Institute, Washington, DC.

[11] Ministry of Housing and Urban–Rural Development of the People's Republic of China. (2017). GB 20018: Technical code of cold-formed thin-wall steel structures. Ministry of Housing and Urban–Rural Development of China, Beijing.

[12] Ministry of Housing and Urban–Rural Development of the People's Republic of China. (2010). JGJ 209: Technical specification for lightweight residential buildings of steel structure. Ministry of Housing and Urban–Rural Development of China, Beijing.

[13] Gao, W.C., & Xiao, Y. (2017). Seismic behavior of cold-formed steel frame shear walls sheathed with ply-bamboo panels. *Journal of Constructional Steel Research*, 132, 217–229.

[14] Wang, R., Xiao, Y., & Li, Z. (2017). Lateral loading performance of lightweight glubam shear walls. *ASCE Journal of Structural Engineering*, 143 (6).

[15] Correal, J.F., & Varela, S. (2012). Experimental study of glued laminated guadua bamboo panel as an alternative shear wall sheathing material. In Y. Xiao et al. (Eds.), *Novel and non-conventional materials and technologies for sustainability: Proceedings of NOCMAT-13-2011 Conference*, Trans Tech, Durnten, Zurich, Switzerland.

[16] Shan, B., Gao, L., Li, Z., & Xiao, Y. (2015). Experimental research and construction of prefabricated bamboo pole demonstration house (in Chinese). *Industrial Construction*, 45(4), 33–41.

[17] Xiao, Y., Cai, H., & Dong, S.Y. (2021). A pilot study on cross laminated bamboo and timber (CLBT) beams. *ASCE Journal of Structural Engineering*, 147(4), 06021002.

[18] Liu, Y., Ni, C., Lu, W.S., Lv, X.L., & Zhou, D. G. (2008). Effect of variable upper stiffness on the performance of wood shear walls (in Chinese). *China Civil Engineering Journal*, 41(11), 63–70.

[19] ASTM International. (2018). ASTM E2126-11: Standard test methods for cyclic (reversed) load test for shear resistance of vertical elements of the lateral force resisting systems for buildings. ASTM International, West Conshohocken, PA.

[20] Folz, B., & Filiatrault, A. (2004a). Seismic analysis of woodframe structures. I: Model formulation. *Journal of Structural Engineering*, 130(9), 1353–1360.

[21] Folz, B., & Filiatrault, A. (2004b). Seismic analysis of woodframe structures. II: Model implementation and verification. *Journal of Structural Engineering*, 130(9), 1361–1370.

[22] Pang, W.C., Rosowsky, D.V., Pei, S.L., & van de Lindt, J.W. (2007). Evolutionary parameter hysteretic model for wood shear walls. *ASCE Journal of Structural Engineering*, 133(8), 1118–1129.

[23] Folz, B., & Filiatrault, A. (2001). Cyclic analysis of wood shear walls. *Journal of Structural Engineering*, 127(4), 433–441.

[24] Pei, S.L., & van de Lindt, J.W. (2010). *User's manual for SAPWood for Windows – Seismic analysis package for woodframe structures, Version 2.0*. Colorado State University, Fort Collins, CO.

[25] Chen, G. (2011). Experimental research and engineering application of light frame bamboo structure. Doctoral Dissertation, supervised by Y. Xiao, Hunan University.

[26] Wang, R. (2019). Research on lightweight glubam frame structures. Doctoral Dissertation, supervised by Y. Xiao, Hunan University.

[27] Wang, R., Li, Z., & Xiao, Y. (2021). Fast modeling of lightweight glubam frame structures based on connection test information. *The Structural Design of Tall and Special Buildings*, e1903, https://doi.org/10.1002/tal.1903

[28] Folz, B., & Filiatrult, A. (2001). *Blind predictions of the seismic response of a two-story woodframed house: An international benchmark*. Report No. SSRP-2001/15. Division of Structural Engineering, University of California, San Diego, La Jolla, CA.

[29] Kiran, M., Nandanwar, A., Naidu, M.V., & Rajulu, K.C.V. (2012). Effect of density on thermal conductivity of bamboo mat board. *International Journal of Agriculture and Forestry*, 2(5), 257–261.

[30] Huang, P., Chang, W.S., Shea, A., Ansell, M.P., & Lawrence, M. (2014). Non-homogeneous thermal properties of bamboo. In S. Aicher, H.W. Reinhardt, & H. Garrecht (Eds.), *Materials and joints in timber structures: Recent developments of technology*. RILEM Bookseries Vol. 9, Springer, Dordrecht, The Netherlands, pp. 657–664.

[31] Huang, P., Zeidler, A., Chang, W.S., Ansell, M.P., Chew, Y.J., & Shea, A. (2016). Specific heat capacity measurement of Phyllostachys edulis (Moso bamboo) by differential scanning calorimetry. *Construction and Building Materials*, 125, 821–831.

[32] Shah, D.U., Bock, M.C., Mulligan, H., & Ramage, M.H. (2016). Thermal conductivity of engineered bamboo composites. *Journal of Materials Science*, 51(6), 2991–3002.

[33] Wang, J.S., Demartino, C., Xiao, Y., & Li, Y.Y. (2018). Thermal insulation performance of bamboo- and wood-based shear walls in light-frame buildings. *Energy and Buildings*, 168, 167–179.

[34] International Organization for Standardization (ISO). (1994). ISO 8990: Thermal insulation – Determination of steady-state thermal transmission properties – Calibrated and guarded hot box. ISO, Geneva, Switzerland, www.iso.org

[35] Tong, Y.J., Xiao, Y., Shan, B., & Chen, J. (2015). Experimental research on fire-resistant performance of Glubam structural wall (in Chinese). *Journal of Industrial Structures*, 45(4).

[36] International Organization for Standardization (ISO). (2019). ISO 834-2: Fire-resistance tests – Elements of building construction – Part 2: Requirements and recommendations for measuring furnace exposure on test samples. ISO, Geneva, Switzerland, www.iso.org

[37] General Administration of Quality Supervision, Inspection and Quarantine, and Standardization Administration of the People's Republic of China. (2008). GB/T 9978-2008: Fire-resistance tests – Elements of building construction (in Chinese). China Architecture and Building Press, Beijing.

[38] McGrattan, K., Hostikka, S., & Floyd, J. (2010). *Fire Dynamics Simulator (Version 5) User's guide*. National Institute of Standards and Technology (NIST) Press, Washington, DC.

[39] Ministry of Housing and Urban–Rural Development, General Administration of Quality Supervision, Inspection and Quarantine of People's Republic of China. (2016). GB/T 50176-2016: Code for thermal design of civil buildings (in Chinese). Construction Press, Beijing.

[40] Ministry of Housing and Urban–Rural Development, General Administration of Quality Supervision of People's Republic of China. (2018). GB/T 50361-2018: Technical code for infills orpartitions with timber framework (in Chinese). Construction Press, Beijing.

[41] Ministry of Housing and Urban–Rural Development, General Administration of Quality Supervision of People's Republic of China. (2015). GB/T 50189-2015: Design standard for energy efficiency of public buildings. Construction Press, Beijing.

Chapter 9

Design and Construction of Engineered Bamboo Structures

Like timber structures, engineered bamboo structures are prefabricated systems, suitable for buildings with various usages. As discussed in previous chapters, engineered bamboo structural elements can be designed and produced in modular forms for different dimensions of buildings, bridges, and other engineering facilities. When designing a structural system, such modular elements can be selected along with proper design and selection of connections. This chapter discusses the design issues for engineered bamboo structures, including the introduction of several typical demonstration projects completed by the author's research group.

9.1 Modular Mobile Buildings

Quickly installed and temporary buildings are widely used for disaster relief, office and utility facilities at constructions sites, mobile houses, etc. The service period of a temporary building can sometimes be very long, such as years, but it is not considered as permanent. In many cases, the building needs to be reused several times. In North America, most temporary buildings are made with timber structures, and are manufactured with good mobility, so they can be transported on wheels and set up at the site. In many cases, the temporary buildings can be very large. Figure 9.1 shows an example of a prefabricated mobile classroom in a quiet city in California. Based on the author's record, the building has been in service for at least 15 years since it was trucked in and set up at the site.

In China, most of the typical temporary buildings are made with a light-gauge steel structure,[1,2] though retired shipping containers are becoming more and more utilized, as shown in Figure 9.2.

Considering the stringent codification requirements for permanent building sectors, the temporary building market might be the first that engineered bamboo structures can enter. With such understanding and prediction, the author's research group successfully developed several temporary mobile building systems with glubam, including emergent houses for earthquake disaster relief, offices for construction sites, mountain cabins, municipal toilets, etc.[3]

9.1.1 Modular Panel Units

The basic module of the walls for modular temporary buildings is a glubam sandwich unit, shown in Figure 9.3, including glubam frame, insulation infills, plybamboo

DOI: 10.1201/9781003204497-9

Figure 9.1 Semi-permanent wooden structure classrooms.
Source: Photo courtesy of Connie Liao.

Figure 9.2 Light gauge steel building and container unit building.

board covers and interior panel. The panel unit can be assembled by hand or using an automated production line. The cover boards and panels are nailed or bolted to the glubam frame. Figure 9.4 exhibits modular units with different configurations for the window or door. The typical size of the unit is 1.2 m wide and 2.4 m tall, with the thickness depending on the insulation design for different areas with different thermal insulation requirements.

(a)

(b)

(c)

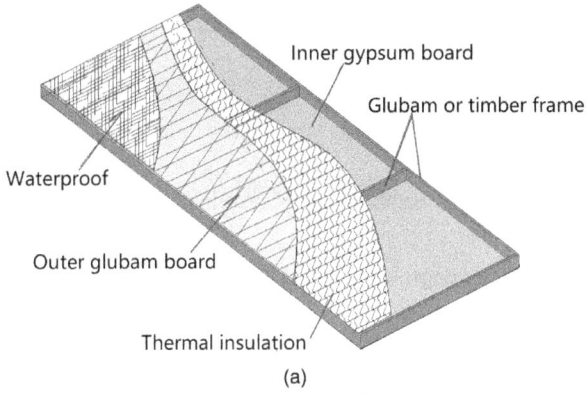

Figure 9.3 Wall panel element: (a) structural schematic diagram of panel element component; (b) manual process of infilling with insulation materials; (c) piles of manufactured wall panels.

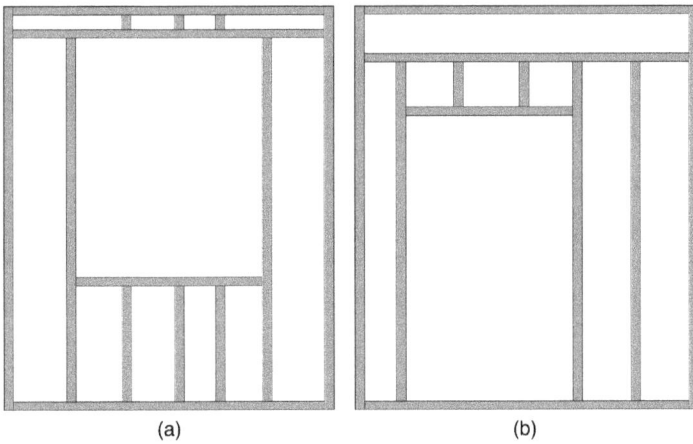

(a)

(b)

Figure 9.4 Wall frame patterns with: (a) window; (b) door.

(a)   (b)

Figure 9.5 Roof trusses (a) made of single layer of plybamboo; with bolted connection (b).

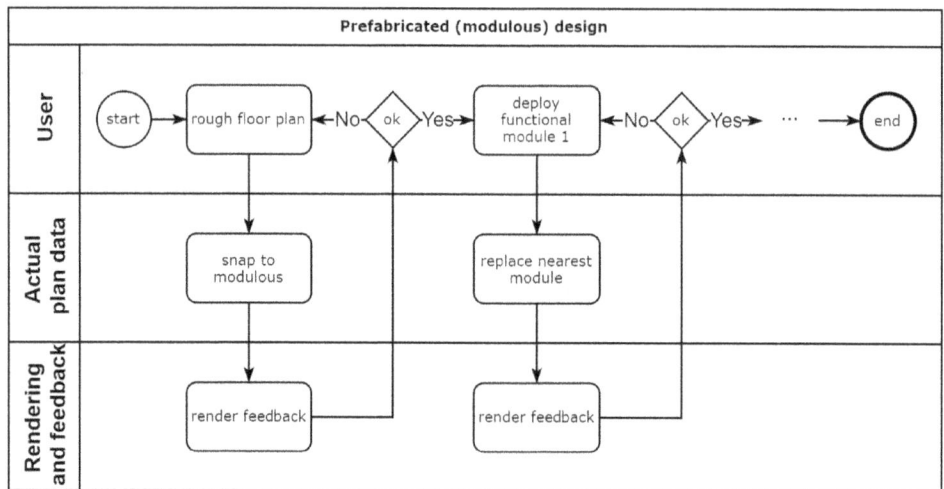

Figure 9.6 BIM flowchart.

For the roofs of relatively small mobile houses, simple trusses as shown in Figure 9.5 can be adopted, which are designed with upper, lower chords, and vertical web elements made of glubam. The elements are essentially linear members that can be transported with a relatively short length, but can be connected by bolts at the construction site.

9.1.2 Modular Mobile Building Design

Several typical modular mobile houses were developed by the author's group, in which all have the basic modular unit of a 1.2 m × 2.4 m panel. A modular mobile building can be designed using the building information modeling (BIM) approach. Figure 9.6 shows the flowchart of the design.

As shown in Figure 9.6, when an owner or user decides on the building area, the designer, assisted by a program, can quickly generate several possible floor plans for the owner to choose from. The program can automatically snap the length of each edge to the nearest possible value afforded by the given modules of glubam panel and makes sure that the plan is fully optimized for prefabricated production, while each step gives its own

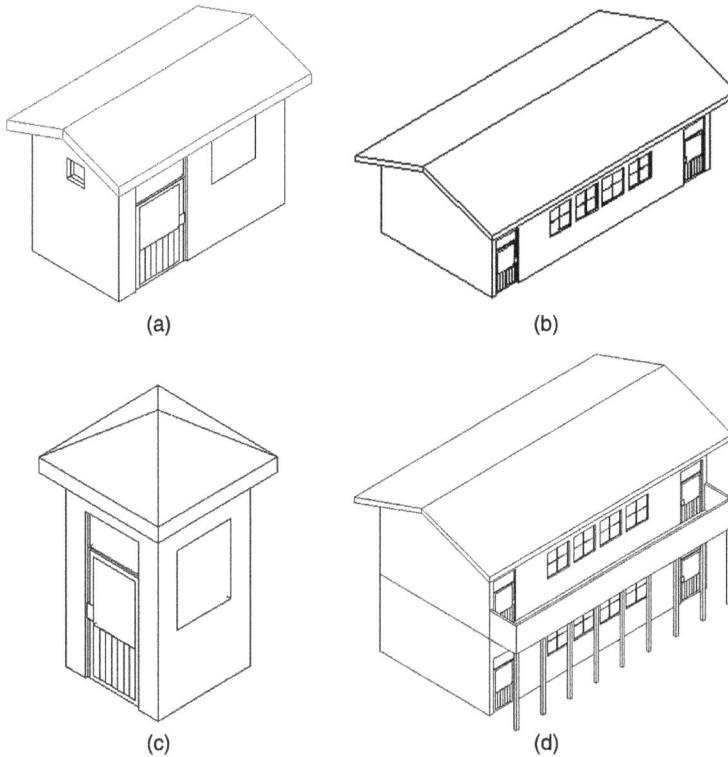

Figure 9.7 Modular building examples: (a) small single-story house unit; (b) single- story classroom; (c) sentry box; and (d) two-story office.

possible feedback; for example, a three-dimensional (3D) rendering to assist with further design. Once the floor plan is determined, the designer can provide an initial design with windows and doors. Then a preliminary building envelope can be established. If needed, a 3D model with rendering effects can also be worked out. Several typical glubam mobile buildings designed by the author are illustrated in Figure 9.7.

9.1.3 Construction of Prefabricated Glubam Houses

Depending on the mobility of a mobile building, the construction and installation vary. The whole building or a portion of the building can be prefabricated in the factory and trucked to the site to be set up on the foundation. Alternatively, the prefabricated panels can be transported to the site and the building erected on-site. For transportation of the full or partial building, the wall panels are connected to a steel foundation frame which can be made with light-gauge steel shapes, as exhibited in Figure 9.8.

At the construction site, the foundation should be compacted for setting up the mobile building. On-grade concrete flooring can be built. A brick or concrete strip foundation can be constructed to seat the steel foundation frame or the sill plates for

Figure 9.8 Modular unit on steel beams.

| (a) | (b) |

Figure 9.9 In-situ construction: (a) installation and connection of wall panels; (b) erection of roof trusses.

the prefabricated wall panels. Typically, the sill plates or steel foundation frame should be slightly higher than the ground to prevent rainwater coming in. Figure 9.9 shows the erection of wall panels and the installation of the roof trusses on the wall panels, during the construction of an earthquake disaster relief house.

9.2 Performance Evaluation of Glubam Mobile Buildings

9.2.1 Lateral Resistance of the Prefabricated Glubam Mobile House

It is important to design mobile buildings with sufficient lateral resistance against wind and earthquake actions. Utilizing a full-scale model, the author's group carried out

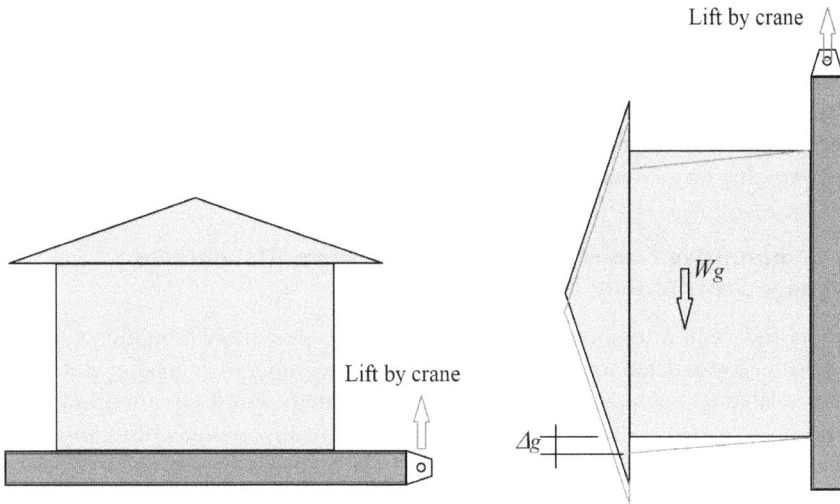

Figure 9.10 Concept of simple seismic loading test.

| (a) | (b) |

Figure 9.11 Full-scale glubam house unit in tilting tests along: (a) x-direction; (b) y-direction.

simple tests to study the lateral integrity of the prefabricated glubam mobile house under a lateral load approximately equal to its own weight.

The overall plan dimension of the testing model was 4.88 m by 3.66 m with a clear floor height of 2.44 m. The walls were made with prefabricated glubam panels. Triangular trusses were built on top of the walls and covered by glubam sheathing boards. A stiff steel frame made with I40a I-shape steel was specially fabricated to support the model for testing. The concept of the testing is depicted in Figure 9.10, and the actual tests were photographed and are shown in Figure 9.11 (a) and (b) for the two directions, respectively.

During testing, a crane was used to lift one side of the steel girder frame until the model was lifted completely and tilted 90 degrees, and then the relative deformation between the top and the bottom of the walls was measured. For each direction, the tests were repeated three times. The results show that the glubam model had good integrity against the lateral load equal to its self-weight and the relative deformation was extremely small and negligible. Essentially, the glubam model building behaved like a bamboo wooden box subjected to the tilting loads.

9.2.2 Comparative Fire-resistance Tests between Glubam and Light-gauge Steel Mobile Houses

Fire safety has been a major concern for light-gauge steel mobile buildings, which are often poorly constructed or maintained. There are numerous media stories about accidents related to light-gauge steel temporary buildings. However, good examples are also reported for earthquake relief compounds with a massive number of light-gauge steel residences after the Great Sichuan earthquake.[4] The main reason was due to efficient and strict fire control and management, sufficient building spacing, etc., rather than the fire-proof qualities of the light-gauge steel building itself. The type of light-gauge steel temporary buildings massively used in construction sites, disaster relief, temporary lodging facilities in China have not been covered in fire studies in previous relevant literature.[5]

During the execution of the project funded by the Blue Moon Fund for developing glubam mobile homes and classrooms in the aftermath of the Great Sichuan Earthquake in 2008, the author's group studied the fire performance of the typical light-gauge steel and glubam building models and conducted full-scale fire tests. Figure 9.12 (a) shows the plan design of the models, whereas Figure 9.12 (b) and (c) illustrate the light-gauge steel and the glubam testing model houses, respectively.

The light-gauge steel model house was contracted to be built by professionals specializing in such construction. The structure of the light-gauge steel model consisted of light steel frames with inserted panels made from two layers of thin steel-sheet skins with 50 mm thick polystyrene plastic foam in between for insulation. The glubam model had standard wall and roof panels, 2400 mm long and 1200 mm wide, with 10 mm thick plybamboo outside board and a 10 mm gypsum internal skin, with 30 mm thick rock wool insulation material sandwiched in between. For both models, commercial single-layer glass with plastic frames was used for the windows.

As shown in Figure 9.12 (a), wooden logs were piled at the center inside the testing model and ignited to generate the fire source. The total mass of the fire logs was about 100 kg, designed to generate similar heat to that of equivalent furniture, such as wood desks, chairs, beds, etc.

During the testing of the light-gauge steel model, the steel skin sheets fell down at about 9 min., and the foam in the panels started to generate thick smoke with a poisonous smell. At 12 min., the wall panels started to deform. The fire was extinguished at 30 min., and the final condition of the light-gauge steel model after cooling is exhibited in Figure 9.13 (a), which shows that the wall panels have become individual flexible and twisted sheets.

As shown in Figure 9.13 (b), for the prefabricated bamboo structure, the window glass cracked and the upper side of the door burned out at the 10th minute from ignition. At the 33rd minute, the wood crib collapsed; however, the main part of the house kept a good structural integrity, showing no signs of burning out.

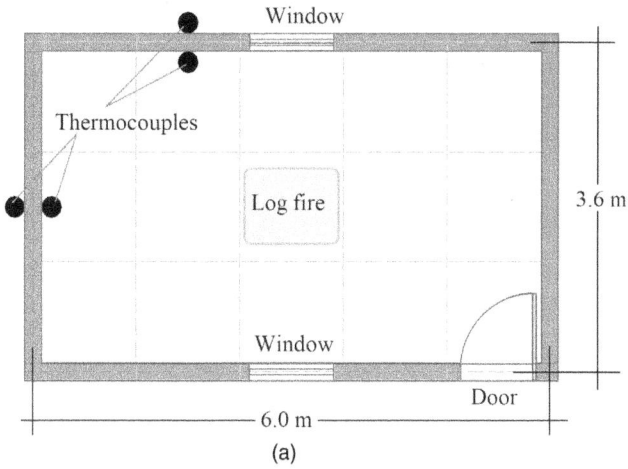

(a)

(b)

(c)

Figure 9.12 Fire tests of full-scale mobile building units: (a) plan view; (b) light-gauge steel unit; (c) glubam unit.

(a)

(b)

Figure 9.13 Final conditions of: (a) light-gauge steel unit; (b) glubam unit.

Based on the test results and data analysis, the following main conclusive recommendations can be made: (i) For light-gauge steel mobile buildings, it is recommended to add fire-proof layers to the internal surfaces of walls and ceiling; (ii) the mobile building consisting of 50 mm thick sandwich glubam panels along with internal gypsum board can withstand an in-room fire without serious problems for at least 30 min., which is considered sufficient for temporary buildings.

9.2.3 Wind Resistance and Water-tightness Study

For temporary buildings, particularly those for disaster relief, it is important to keep the buildings away from wind and rain damage and provide the residents with as much comfort as possible. A study was carried out by the author's research group on the wind resistance and water tightness of glubam mobile buildings for potential deployment in Southeast Asia, where cyclones with heavy rains are frequent.

The model glubam mobile building shown in Figure 9.14 was designed to withstand potential cyclones in Bangladesh. The architectural design was provided by Indonesian architect Eco Prawoto using truss modules similar to the truss module described by Huang.[6] The considered wind speed is 60 meters per second (m/s) which corresponds to

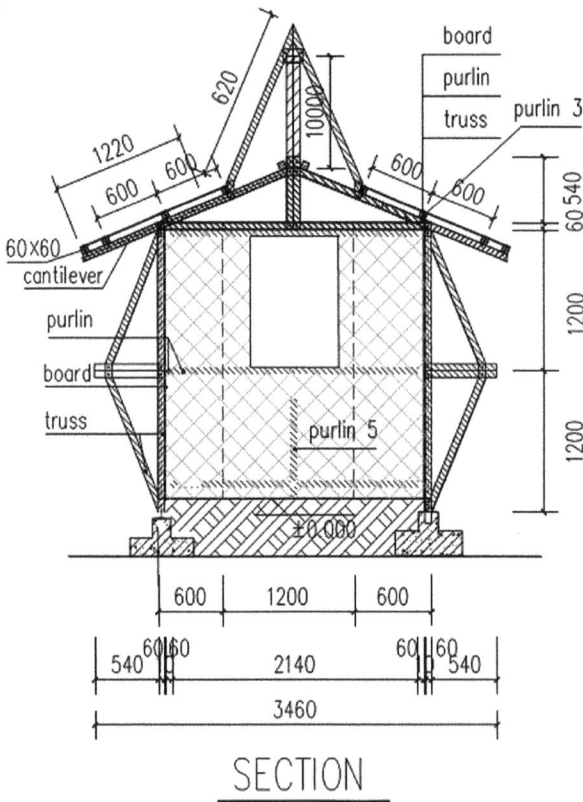

Figure 9.14 Wind and rain resistance testing model.

Figure 9.15 Test equipment setup for wind and rain resistance.

a most severe pressure of 1670 Pa on buildings located in the inland region B where the Bangladesh capital Dacca is situated.[7] For buildings lower than 5 m, the pressure can be reduced to 1242 Pa. The hold-down bolts for the foundation beams were designed to prevent overturning under the lateral pressure applied to the building. Glass was not installed in the windows and the door, but instead they were covered by glubam boards, simulating actual practice in extreme cyclone situations.

As exhibited in Figure 9.15, a propeller was used to generate wind pressure on the building whereas water sprays were arranged on steel frames around the testing model. The tests were conducted for the two main directions of the building model. For each direction, a wind pressure of 600 Pa along with water spray at 3.4 liters $(L)/m^2/min.$ was first applied for 15 min. The structural integrity and the water tightness were observed using cameras placed inside the building. After a visual investigation, the second stage of testing was continued for another 15 min. with a wind pressure of 1200 Pa and the same water spray.

During the testing, no obvious deformation was observed. However, some water leakages were found along some of the joint seams and around the holes of the nails and screws. Measures such as covering the potential leakage areas with water-proof tapes may be considered. In addition, while the model was securely bolted down to the foundation beam during the testing, in the real-life situation, adding temporary pulling cable supports to safeguard the buildings during extreme winds is recommended.

9.2.4 The Great Wenchuan Earthquake Relief Efforts

On May 12, 2008, the magnitude 8.2 Wenchuan Earthquake struck the Sichuan Province of China, resulting in close to 90,000 casualties and missing persons in total,

Figure 9.16 Prefabricated glubam mobile buildings for earthquake disaster relief: (a) classrooms; (b) medical center.

and significant damages to infrastructure. With the support of the Blue Moon Fund, the author's research group designed, manufactured, and deployed more than 2000 m² disaster-relief glubam mobile buildings for classrooms, temporary offices, and medical relief clinics. The modular buildings were all manufactured in Hunan Province and transported more than 1000 km to the city of Guangyuan near the epicenter. Figure 9.16 depicts some examples of the glubam relief buildings. The author's group also conducted an investigation three years after the deployment. It was found that a significant number of light-gauge steel relief buildings constructed by the local government generated waste that was difficult to reuse or recycle. However, many of the glubam relief buildings deployed by the author's research group were removed and their parts reassembled for more permanent buildings constructed in nearby villages.

9.3 Lightweight Glubam Frame Buildings

Lightweight woodframe structures are widely used in North America for residential and office buildings. The lightweight glubam frame building system developed by the author essentially follows the same construction procedures as the lightweight woodframe structures. The lightweight woodframe building system originated in the United States[8] to overcome the disadvantage of a shortage of skilled labor in building European-style heavy frame wooden buildings, and take advantage of the abundant forest resources in the new continent. Relatively slender lumbers are used to form closely spaced frames. Instead of using a mortise beam to connect columns, the members are nailed together and then covered by panels like skin. For this reason, the building system gained the nickname of balloon frame.

There are two types of lightweight woodframe construction. One is the original balloon type construction in which the column studs extend from the lower floor to the upper floor. The other is the so-called platform construction, in which the stud columns are constructed only for one floor height to support the upper floor slab, creating a working platform for the upper floor. This is a highly convenient method for constructing buildings, by using relatively short components which are thus easy to handle.

The structural system of a lightweight woodframe building is constructed mainly using relatively small-size lumbers of regulated dimension, such as lumbers with a nominal dimension of 2 in. × 4 in. (actual dimension of 38 mm × 89 mm), particularly for structural walls. For this reason, the system is also often referred to as "two by four" construction. In dealing with the widespread failure of the traditional timber residential buildings during the January 17, 1995 Hyogo-Ken Nanbu Earthquake, many Japanese companies introduced this two by four (in Japanese pronounced as "tsu bai hou") method, recognizing its good seismic resistance.

Based on extensive research and references of design practices for lightweight woodframe buildings, the author's research group established the lightweight glubam frame structures.[9,10]

9.3.1 Basic Design Features of Lightweight Glubam Frame Structures

The first lightweight glubam frame house was designed and constructed under the supervision of the author's research group, as shown in Figure 9.17 (a).[11,12] The two-story building had an approximately 260 m² area including five bedrooms, three bathrooms, kitchen, dining area, living room, fireplace, and two-car garage. The structure of the building was completed at the end of 2007, and the house was opened in the spring of 2009. The house was a milestone in modern bamboo structure development and attracted wide international public attention and media coverage. In October 2013, five years after its completion, the house was moved (Figure 9.17 (b)) and relocated to a new location (Figure 9.17 (c)) about 100 m away from its original location on the campus of Hunan University.

The structural system of a lightweight glubam frame building is shown in Figure 9.18, which mainly comprises the following:

Foundation: The walls and columns of a lightweight glubam frame building can be constructed directly on a concrete slab above a densified soil foundation, as shown in Figure 9.19 (a). Waterproof materials such as asphalt sheets should be used to isolate the sill plates of the walls from the foundation to prevent moisture. In most cases, an elevated foundation should be designed to provide ventilation to the first floor, as shown in Figure 9.19 (b). The elevated foundation for the first floor is typically formed with perimeter cripple walls and internal pedestals. The cripple walls and pedestals can be built with concrete or masonry blocks and bricks. The cripple walls can also be made in a similar way to shear walls.

Vertical elements – shear walls: Shear walls are the most important structural elements in lightweight frame buildings. Figure 9.20 (a) shows the wall studs before they are covered with sheathing panels. They are designed not only to carry vertical gravity loads, but also to resist lateral loads from wind and earthquakes. Similar to timber structures, a lightweight glubam frame shear wall consists of edge elements, studs, columns, and sheathing panels integrated together using nails or screws. The edge elements and studs are made with glubam lumbers

Figure 9.17 Engineered bamboo buildings: the world's first modern bamboo lightweight frame building completed at the end of 2007 (a); relocated in 2013 (b); and settled at current location on Hunan University campus (c); and a prefabricated glubam building built (assembled) at the Maseno University, Kisumu, Kenya in 2009 (d).

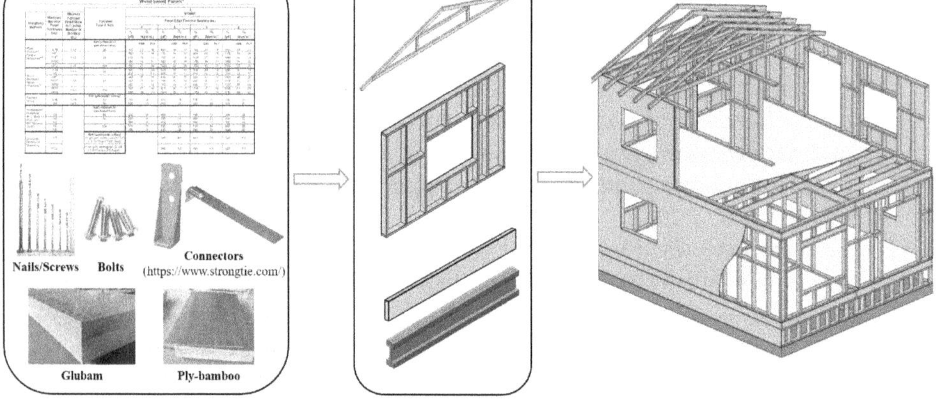

Figure 9.18 Lightweight glubam frame system in BIM.

Figure 9.19 Foundation for lightweight glubam frame buildings: (a) on-grade floor; (b) elevated first floor.

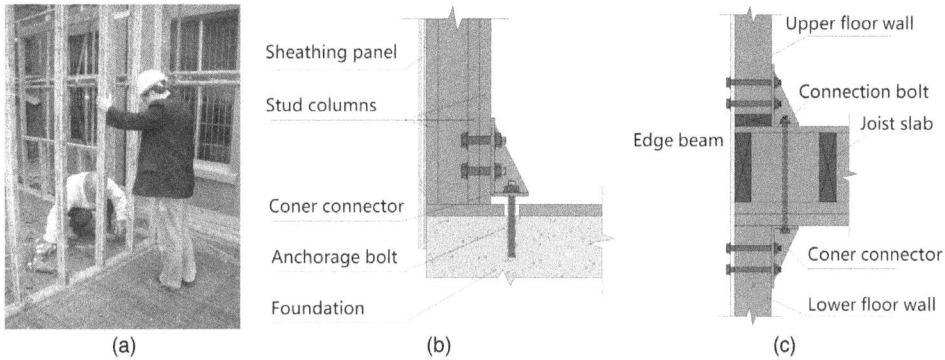

Figure 9.20 Shear walls: (a) installation of shear wall stud columns; (b) a possible connection with foundation; (c) connections between floors.

of standard dimension (30–40 mm wide and 90–120 mm deep). Plybamboo sheathing boards are nailed to the glubam studs on either one side or both sides of the wall, depending on the design. The connection of a shear wall with the first-floor slab is shown in Figure 9.20 (b), whereas the connections of the shear walls between the two stories and the slab are shown in Figure 9.20 (b). As discussed in Chapter 8, the studs can also be made of timber, or light-gauge steel.

Vertical elements – columns or posts: In a lightweight glubam frame building, with the lateral load resisted by shear walls, gravity load carrying columns can be used to provide a relatively large floor space. Examples of gravity load carrying columns or posts are shown in Figure 9.21, in which glubam square column and round bamboo columns are used to support the beam in the upper floor.

Joist slabs: Closely configured glubam beams (rectangular or I-shape) are used to form the support to the floor slab, as shown by the soffit view of an upper floor

Figure 9.21 Columns.

(a)

(b)

Figure 9.22 Joist slab: (a) soffit of a joist slab; (b) slab as the working platform for second floor construction.

slab in Figure 9.22 (a). Plywood boards are sheathed on top of the beams and then the flooring surface is covered. The slab of the second or higher floor is also used as the working platform for that floor to build walls and an upper floor or roof, as shown in Figure 9.22 (b). For a kitchen or bathroom, a cementitious mortar layer can be added on top of the plybamboo board with waterproof

Figure 9.23 Building roof construction: (a) tilt-up of a truss; (b) covering with sheathing panels.

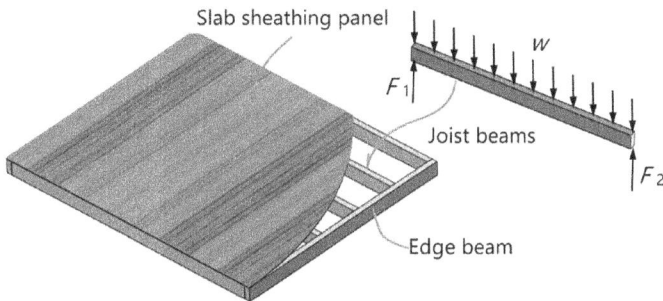

Figure 9.24 Vertical load on joist beams.

materials in between. For the joist slab design, it is important to maintain the continuity of the perimeters to form either rigid or flexible diaphragms.

Roof trusses: As discussed in Chapter 7, glubam trusses can be prefabricated using toothed metal connection plates or bolts. On top of the upper chords of the roof trusses, plybamboo sheathing panels are nailed to form a roof diaphragm. All the trusses should be placed on the top of edge beam elements to assure the proper transfer of the vertical load onto the walls or columns. In the orthogonal direction of the trusses, supports need to be provided to prevent the domino tilting and collapse of the roof trusses. Figure 9.23 shows the process of constructing the roof for a two-story glubam building.

9.3.2 Vertical Load Transfer Mechanisms

Due to the flexibility in slab sheathing and supporting joist beams, the distributed load from the slab surface should be distributed to the closest joist beam, which should be considered to be simply supported at its ends by walls or beams, as shown in Figure 9.24. If the beam is supported on a column, the column should be securely supported on the foundation. If a relatively large beam is carried by a shear wall, additional studs may be needed to bear the vertical load.

9.3.3 Lateral Load Transfer Mechanisms

Earthquake and wind are the main sources of lateral load that need to be considered for the design of lightweight timber or bamboo frame buildings. These loads are applied to the walls, slabs, and roofs and are then resisted and transferred to the foundation by shear walls. Thus, shear wall design for timber or glubam lightweight frame structures is most important. For regions with high seismicity, moment- resistant steel frames can be used to collect the earthquake forces and pass them on to the foundation.

The wind load values to be considered should follow the codified requirement based on the location of the building. The load path for wind in a lightweight glubam frame structure is shown in Figure 9.25. The wind is applied to the wall or roof surfaces and distributed by the sheathing panels to the studs (or rafters, trusses for roofs), which are supported at the top and bottom by slab diaphragms as simply supported beams. Then, the diaphragms transfer the force to the shear walls.

Under an earthquake attack, the earthquake force is applied, as inertia force, to the mass of the structure, or in other words, is distributed to all parts of a building. Therefore, the fundamental basics of design aim to prevent each part of the building from "flying away" under the inertia force application. As shown in Figure 9.26, taking the one-story simple box-type building as an example, the inertia forces applied to the slab and upper half of the walls need to be considered as being transferred to the foundation by the shear walls; whereas the inertia forces on the lower half of the walls can be considered as directly transferred to the foundation. Thus, the lateral earthquake force F is shared by the shear walls in the direction parallel to the direction under consideration.

As is clear from Figure 9.26, the slab diaphragm should be able to function as a "flat beam." Such a beam needs to possess adequate shear strength and stiffness provided by

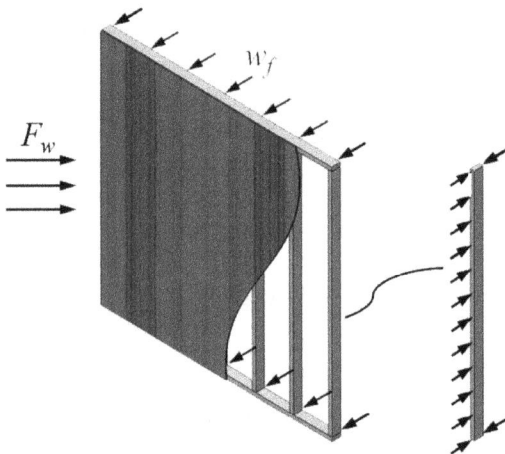

Figure 9.25 Wind load applied on an external shear wall.

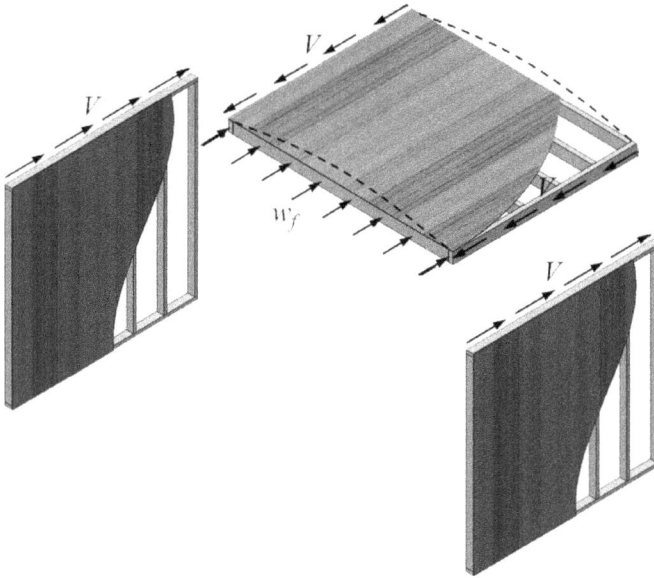

Figure 9.26 Earthquake loads applied on slab and shear walls.

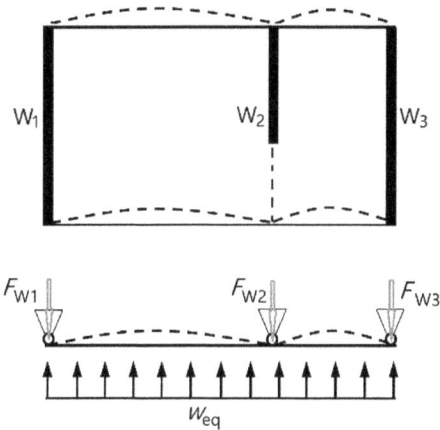

Figure 9.27 Flexible slab diaphragm model.

the joist beam and slab sheathing surface panels, similar to shear walls. On the other hand, flexural strength and stiffness are to be provided mainly by the adequate size and continuity of the edge beams.

For simplicity, the transfer of the slab diaphragm force to the shear walls is considered using two possible extreme mechanisms, the flexible diaphragm model and the rigid diaphragm model, as shown in Figure 9.27, and Figure 9.28, respectively.

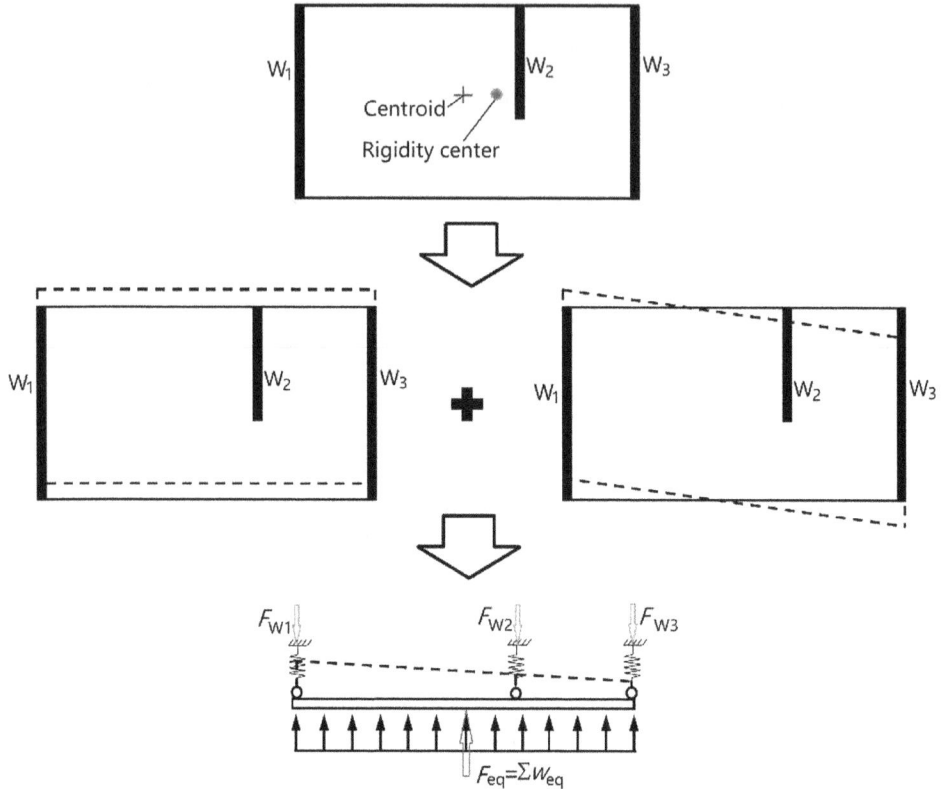

Figure 9.28 Rigid diaphragm model.

The flexible diaphragm model is widely used since it does not involve an estimation of the stiffness of the shear walls. The lateral force generated by wind or earthquake from the diaphragm is considered to be shared by the walls based on the tributary areas, as shown in Figure 9.27. In a rigid diaphragm model, the displacements of the walls are regulated by the rigid diaphragm based on translational and rotational modes and the stiffness of the walls. For simplicity, the stiffness of a wall can be replaced by the horizontal length of the wall, if all the walls can be considered to have the same stiffness per unit length.

9.4 Performance of Lightweight Glubam Frame Buildings

9.4.1 Seismic Behavior

To validate the seismic performance of lightweight glubam frame structures, a shake table test program was carried out by Chen.[12,13] The creation of the model is schematically shown in Figure 9.29, in which one of the corner rooms of the first bamboo house at Hunan University, built by the auhor's research group, was taken as the prototype.

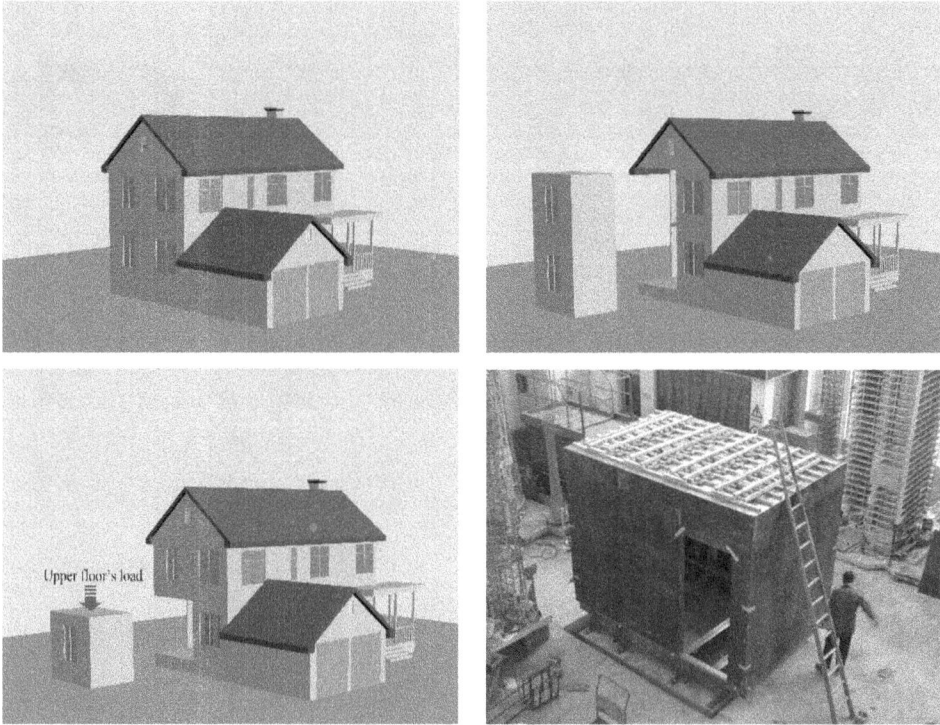

Figure 9.29 Schematics of model room creation.

The full-scale room model had a floor plan of 3.66 m long by 2.44 m wide, with a clear height of 2.6 m, as shown in Figure 9.30. A door and a window were designed for each of the longitudinal walls (*x*-direction in Figure 9.28a).

The walls of the model glubam unit were assembled from modular glubam panel units (1.2 m wide × 2.4 m tall). Each of the panel units was composed of "2 × 4" studs (actual sectional dimension: 40 mm x 84 mm) spaced every 410 mm on center, the bottom and top edge elements nail-integrated by plybamboo sheathing panels (1.2 m wide, 2.4 m long, 10 mm thick). Glubam beams with a section of 56 mm wide by 185 mm deep were arranged at 405 mm intervals, topped with 10 mm thick sheathing panels. In order to improve the lateral stability of the bamboo beams, shear braces were arranged between the bamboo beams. Anchor bolts (diameter 12 mm) were used for the connections between the first-floor wall and the foundation steel beam, and the spacing of the anchor bolts was 610 mm. Special metal hold-down anchors were arranged at the end of the shear wall and on both sides of the opening, as shown in Figure 9.31.

The shake table testing was carried out at the Earthquake Engineering Laboratory of Guangzhou University. The size of the shake table is 3 m × 3m × 1.2 m; weighing 6 tons, it is capable of testing a specimen with a mass of 10 to 25 tons. The working frequency is 0.1–50 hertz (Hz). The stiff steel table is driven by 8 actuators, 4 of which are in the horizontal direction and 4 in the vertical direction for simulating earthquake waves in three

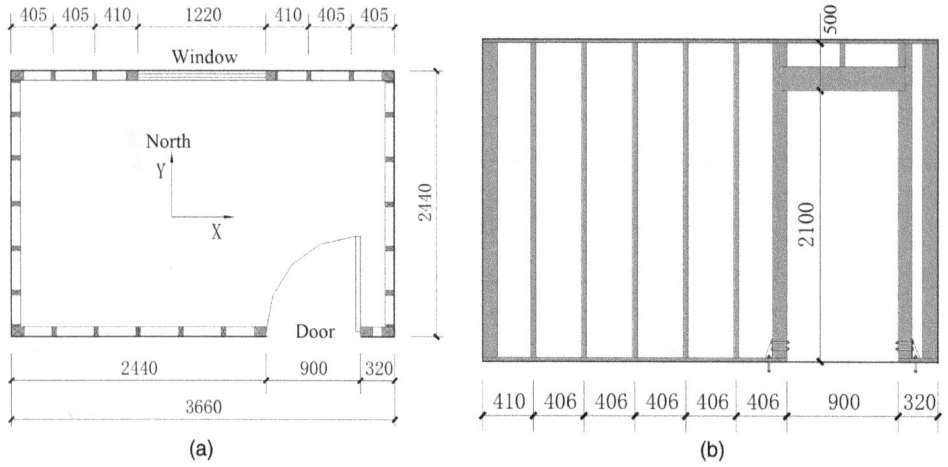

Figure 9.30 Full-scale glubam building room unit: (a) plan view; (b) elevation view of a wall.

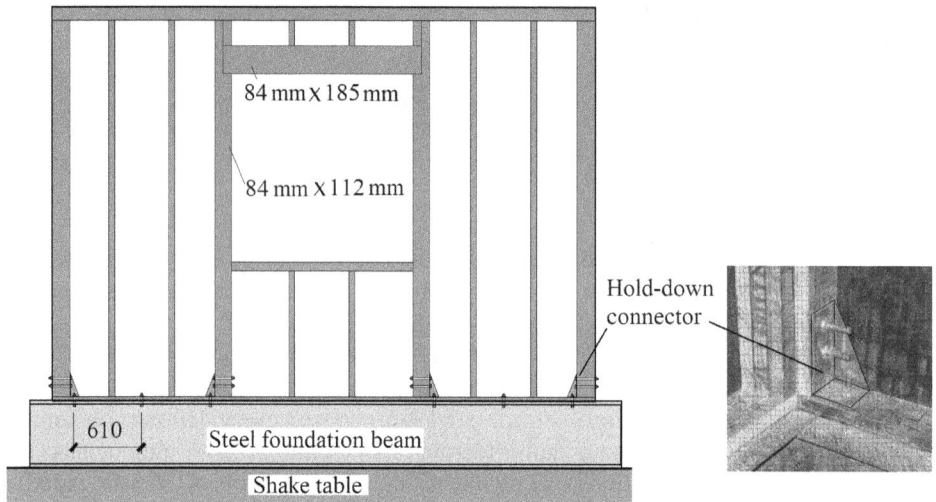

Figure 9.31 Connections between the glubam model walls and foundation beam.

dimensions. According to the requirements of the Chinese load code GB 50009,[14] the standard live load of the roof and the floor for residential buildings shall be 0.5 kN/m² and 2.0 kN/m², respectively. Based on the combination of the total loads on the second floor, an additional mass of the floor slab was assessed as 1920 kg, and the self-weight of the glubam structure model was estimated as 1600 kg. As shown in Figure 9.32, the additional mass was simulated using steel blocks placed on the second-floor slab.

Two representative actual earthquake records (El Centro and Taft) and one artificial seismic wave were selected as the input motion of the shake table. The peak values of

Figure 9.32 Simulated floor mass using steel blocks.

input seismic waves were adjusted according to the requirements of the test conditions, as shown in Table 9.1. Note that the peak ground accelerations 0.1 g, 0.2 g (and 0.3 g), and 0.4 g correspond to the seismic intensities VII, VIII, and IX, respectively, based on the Chinese seismic design code GB 50011.[15] Before and after each excitation, the model was scanned with white noise (0.5 Hz to 50 Hz) to obtain the natural frequency and damping of the model. The amplitude of white noise remained at 0.7 m/s^2 during the whole test.

During the testing, accelerometers were used to detect the actual accelerations of the table surface, and building roof slab. The acceleration records were integrated to obtain the displacements. Laser beam displacement sensors were affixed at selected locations at the bottom of the walls to monitor the potential uplift deformation. The hold-down anchor forces were monitored using center hole load cells.

During the testing, the glubam full-scale model room unit behaved exceptionally well. When the peak acceleration was below 0.3 g, the model structure was basically in an elastic state without visible damage. When the acceleration exceeded 0.4 g, the vibration of the model appeared to be more violent, and some nails pulled out at the corner of the wall and the openings of the door and window. In the 0.5 g test, slight torsion could be observed in the model indicating the accumulation of damage.

Before and after each different earthquake input level, the acceleration response time history curve was obtained by scanning the glubam model with 0.07 g amplitude white noise, and spectrum analysis was carried out, so as to obtain the amplitude frequency curve and phase frequency curve of the structure transfer function. Then, the dynamic characteristic parameters of the model structure were obtained. The changes in natural frequency and damping rations corresponding to the increase of the peak acceleration are shown in Figure 9.33 (a) and (b), respectively. As exhibited in Figure 9.33 (a), despite the accumulative shaking events, the change in frequency was small, and eventually degraded only about 2%. Since the structural stiffness K is directly related to the natural frequency f ($K = 4\pi^2 m f^2$), so the change in the natural frequency of the model

Table 9.1 Test working conditions

Test cases	Input wave	Direction	Peak ground acceleration (g)	
			Set value	Actual value
1	White noise	x	0.07	
2	El Centro	x	0.1	0.09
3	Taft	x	0.1	0.084
4	Artificial wave	x	0.1	0.087
5	White noise	y	0.07	
6	El Centro	y	0.1	0.118
7	Taft	y	0.1	0.1
8	Artificial wave	y	0.1	0.097
9	White noise	x	0.07	
10	El Centro	x	0.3	0.268
11	Taft	x	0.3	0.282
12	Artificial wave	x	0.3	0.338
13	White noise	y	0.07	
14	El Centro	y	0.3	0.346
15	Taft	y	0.3	0.296
16	Artificial wave	y	0.3	0.298
17	White noise	x	0.07	
18	El Centro	x	0.4	0.394
19	Taft	x	0.4	0.384
20	Artificial wave	x	0.4	0.368
21	White noise	y	0.07	
22	El Centro	y	0.4	0.368
23	Taft	y	0.4	0.412
24	Artificial wave	y	0.4	0.39
25	White noise	x	0.07	
26	El Centro	x	0.5	0.45
27	Taft	x	0.5	0.433
28	Artificial wave	x	0.5	0.466
29	White noise	x	0.07	
30	White noise	y	0.07	
31	El Centro	y	0.5	0.498
32	Taft	y	0.5	0.472
33	Artificial wave	y	0.5	0.426
34	White noise	y	0.07	

Note: x is direction of seismic direction along the longitudinal wall; y is direction of seismic direction is along the transverse walls.

structure reflects the change in terms of structural stiffness. Besides natural frequency, damping ratio is also an important characteristic index for structural design. As shown in Figure 9.33 (b), the damping ratio increased following the increase of seismic input levels. The damping ratio of the structure in the initial state was about 3.7%. After the 0.5 g earthquake test, the damping ratio increased by 20% in the x direction and 8.5% in the y direction. The damage in the x direction was more severe than that in the y direction, probably due to the existence of the window and door openings in the walls.

Figure 9.34 exhibits the time histories of the response acceleration at the second-floor slab, shown for earthquake input waves corresponding to peak acceleration of

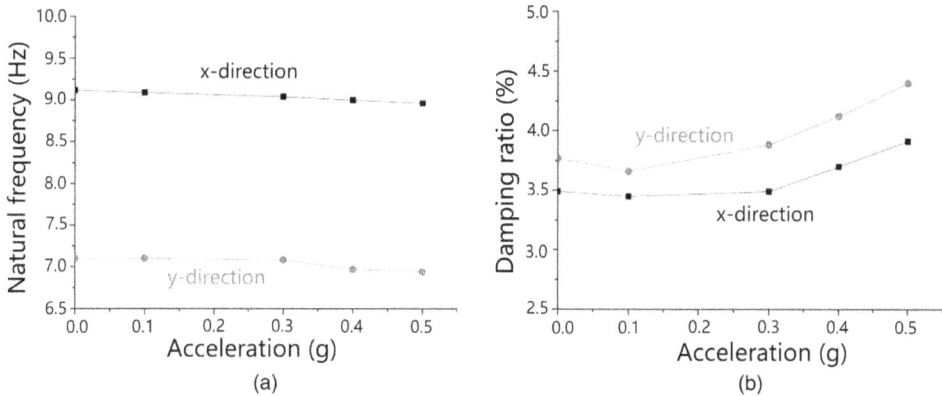

Figure 9.33 Vibration characteristic parameters: (a) natural frequency; (b) damping ratios.

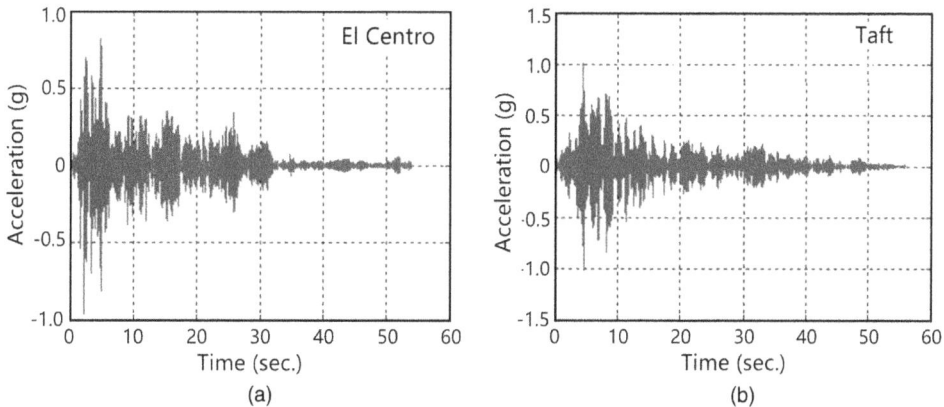

Figure 9.34 Floor accelerations: (a) El Centro wave at 0.5 g input; (b) Tafts wave at 0.5 g input.

0.5 g. Apparently, response to the input earthquake waves was amplification at the slab level. An amplification factor can be defined as the absolute ratio of the peak response acceleration and the peak input acceleration. Figure 9.35 shows the acceleration amplification factors for the x and y directions. The larger amplified acceleration in the y-direction was due to the relatively stiff configuration of the shear walls without openings.

After completing all the loading cases of the shake table tests, the full-scale room model was disassembled and the parts were shipped back to the laboratory at Hunan University. The model was then reassembled in a reaction frame for final push-over testing, as shown in Figure 9.36. The model behaved reasonably well without catastrophic failure. The damage was mainly concentrated at the wall panel seams with pull out of nails as shown in Figure 9.37 (a), and wall uplift at the corners, as shown in Figure 9.37 (b).

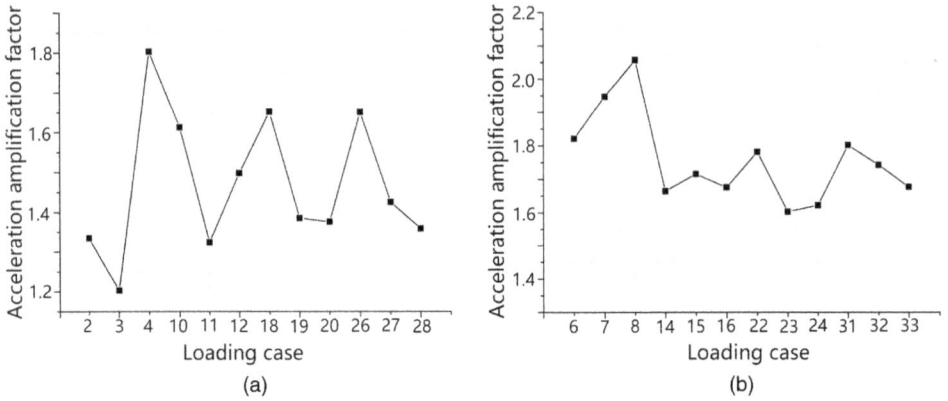

Figure 9.35 Acceleration amplification factors for loading cases in (a) x-direction; (b) y-direction.

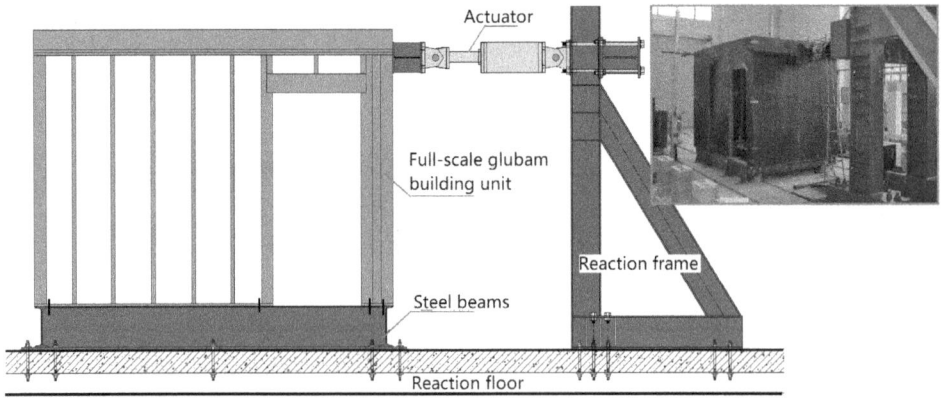

Figure 9.36 Test setup for push-over loading.

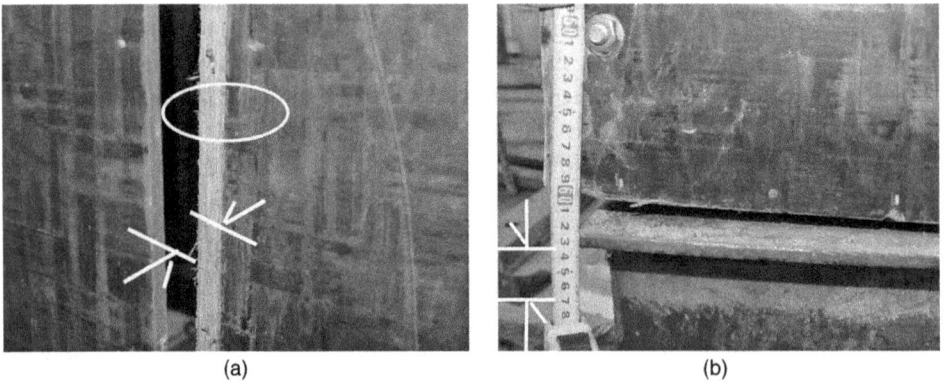

Figure 9.37 Damage in push-over test: (a) pull out of nails; (b) lift up of shear wall at the corner.

The lateral force and displacement relationship obtained from the push-over loading test is shown in Figure 9.38, along with the equivalent elasto-plastic curve. The elastic stiffness k_e is defined as the secant modulus corresponding to the load at 40% of the maximum capacity, F_{peak}. Taking the displacement corresponding to 0.8, the ductility factor μ_d can then be defined as,

$$\mu_d = \frac{\Delta_u}{\Delta_y} \tag{9.1}$$

where the yield displacement is defined as the yield force F_y divided by the stiffness k_e, and the yield force is determined using the concept of equivalent absorbed energy, as follows,

$$F_y = \left[\Delta_u - \sqrt{\Delta_u^2 - \frac{2A}{k_e}} \right] k_e \tag{9.2}$$

In the above equation, A is the enclosed area of the push-over curve up to the failure displacement Δ_u in Figure 9.38. The displacement ductility factor, calculated on the basis of Eq.9.1 was 3.1, which is similar but slightly larger than the ductility factors obtained by Wang et al. from the testing of glubam shear walls;[16] however, it is slightly smaller that that obtained for wood shear walls.[17]

Using the FEM software SAPWOOD, Chen analyzed the shake table test results of the full-scale glubam model unit.[12] The simulation results were reasonable with the damping ratio, frequency, and the response amplification factors all within 10% difference compared with the test results. Wang et al.[18] also performed the simulation of the shake table tests of the full-scale glubam unit, using a simple two-dimensional model developed in OpenSees.

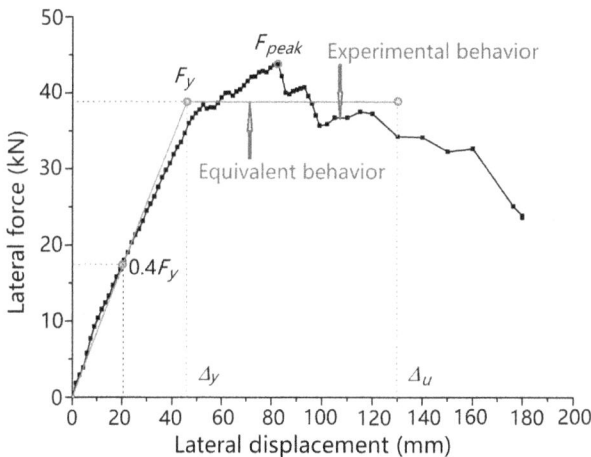

Figure 9.38 Lateral force and displacement relationships.

9.4.2 Fire Performance

One of the major hurdles hindering the wide development of timber and bamboo construction, particularly in China, is the misunderstanding and misconception about their fire resistance. Though timber and bamboo are combustible materials, the structures made with timber and bamboo do not have to be combustible and can be built with high quality in terms of fire resistance and endurance.

In order to clear out some of the doubts in people's minds, the author's group carried out a demonstration test to study the fire endurance of a full-scale glubam building room model.[19] The model took advantage of the leftover components and parts from the shake table testing and was built by reinstalling the walls and slabs along with additional insulation and interiors. Figure 9.39 exhibits the procedure for fabrication of the fire test model. A total of 210 steel blocks, 10 kg each, were arranged on the top of the second floor (roof for the model) to generate a distributed load of 2.6 kN/m^2, simulating the load on the second floor and the load transferred from the upper portion of the building.

After the reinstallation of the structural system, ordinary gypsum boards with details shown in Figure 9.40 (a) for fireproofing were stapled onto the studs of the walls and grid frames of the ceiling. The thickness of the gypsum boards was 10 mm for the walls, but 20 mm for the ceiling. As shown in Figure 9.39 (b) and (c), three different gypsum boards were installed in the middle part of the longitudinal walls (W1 and W2), which were nearest to the fire, in order to compare the fire-resistant performance of gypsum composite board (GCB), ordinary gypsum board (OGB), and fire-resistant gypsum board (FGB). The doors and windows were made with hardened plastics for the frames and glass for window panes.

(a) (b) (c)

(d) (e) (f)

Figure 9.39 Construction of fire-testing model: (a) install walls on masonry foundation; (b) ceiling frame; (c) filling of rock wool insulation materials; (d) gypsum board finishing; (e) placement of steel blocks on slab; (d) completed model.

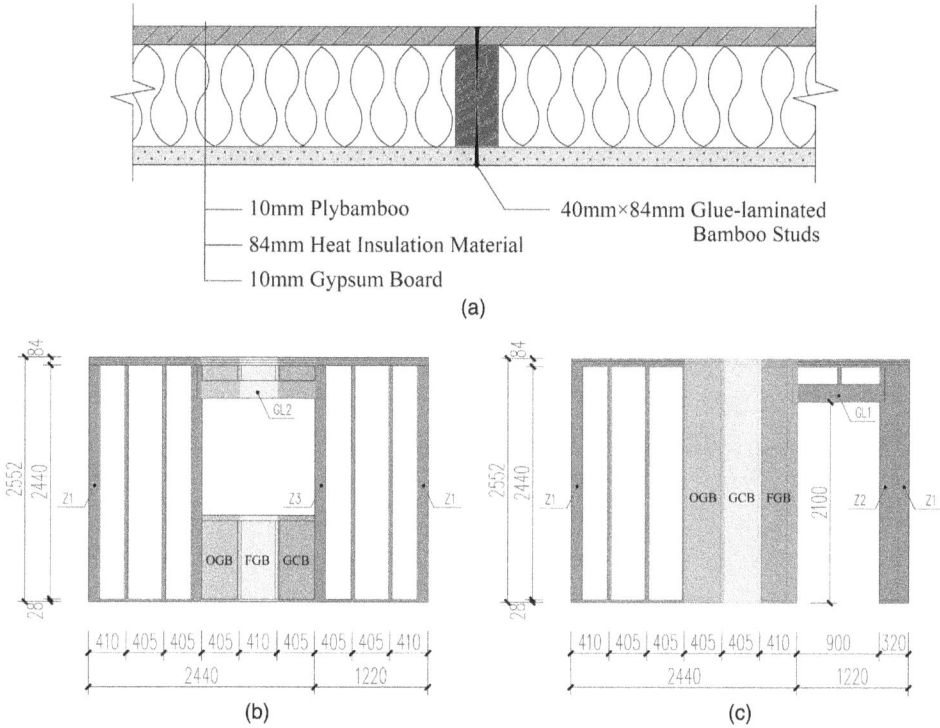

— 10mm Plybamboo
— 84mm Heat Insulation Material
— 10mm Gypsum Board

— 40mm×84mm Glue-laminated
Bamboo Studs

(a)

(b)

(c)

Figure 9.40 Installation details: (a) cross section of walls; (b) longitudinal wall with window; (c) longitudinal wall with door.

A wooden crib placed at the center of the model building was ignited to simulate a fire caused by the potential furniture inside the room. Wood sticks (occupying a total space of about 0.8 m³) and weighing 300 kg were piled to form a wood crib, which can generate about 6010 MJ heat load. Thermocouples were installed in the center of the slab and walls to detect and record the temperature change during the fire testing. The fire endurance testing and damage process of the model glubam building is shown in Figure 9.41.

At beginning of the experiment, the wood crib was sprayed with 500 ml 100% purity ethanol and was then ignited. The first noticeable sign was the cracking of glass on the door and window at the 9th min. from ignition, following by the twisting deformation of the plastic door and window frames. The glass fell out completely at about 16 min. into the burning, allowing the thorough combustion of the wood crib with whole house being fully ventilated. At about the 28th min., the wood crib collapsed, and around the same time, cracks could be seen on the gypsum boards through the burnt-out window; however, the gypsum composite board was still intact. With continued burning, some damaged parts of the gypsum boards started to spall. Based on observation, it is deemed that gypsum composite board works better than ordinary gypsum board and fire-resistant gypsum board. After burning of about 40 min., the combustion started to

Figure 9.41 Fire damage process: (a) window damage at 9 min.; (b) flash over at 13 min.; (c) maximum fire at 18 min.; (d) cracking of gypsum boards; (e) inside view at 60 min.; (f) glubam unit at 60 min.

Figure 9.42 Damage to the glubam beams after 12 hours of smoldering.

slow down as the wood burned out. The fire was extinguished by spraying water on the wood crib 60 min. after ignition.

It is known that due to the formation of a carbonized layer, timber has better fire endurance performance than unprotected steel.[20,21] The similar carbonization effects were also observed in the fire testing of the glubam model building. On the morning of the second day, or 12 h after the fire was extinguished, a thorough inspection was carried out. A smoldering zone near the upper portion of one of the walls was discovered. As shown in Figure 9.42, a small portion near the end of a beam was burnt for about a quarter of its depth due to the smoldering. These phenomena illustrate the good resistance of the glubam elements once charring is formed.

During the fire simulation test of the model glubam building, 16 thermocouples were used to record the wall and slab temperatures at selected locations. It was found that the highest temperature of an interior wall or ceiling surface was 686°C, while the highest temperature of an exterior wall surface was between 46°C and 84°C, lower than the ignition temperature of most materials. The typical temperature versus time histories for inner and outer surfaces at the center zone of one longitudinal wall and the slab are exhibited in Figure 9.43. As shown in Figure 9.43, the temperature of the interior of the wall or slab increased drastically, particularly after 15 min. of burning, and reached a peak about 30 min. after ignition. However, the temperature of the exterior of the wall stayed low with a very gentle increase. These test results demonstrated the excellent

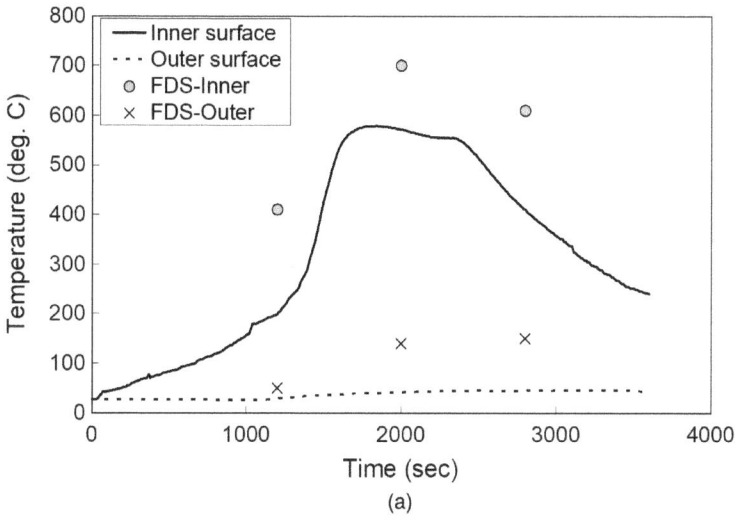

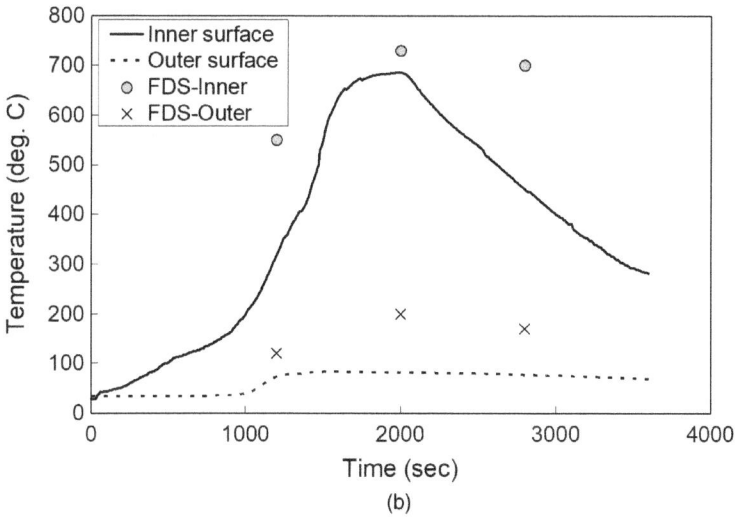

Figure 9.43 Temperature time histories for: (a) longitudinal wall center; (b) slab center.

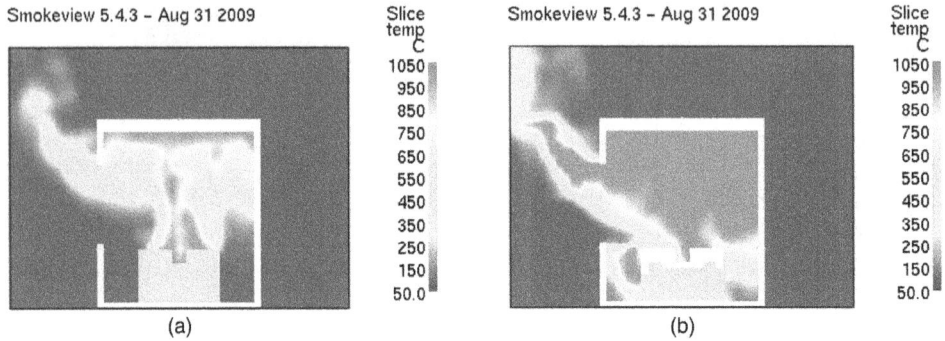

Figure 9.44 Flue gas temperature isosurface: (a) center section in narrow direction at 1200 s; (b) center section in narrow direction at 2000 s.

insulation effect of the bamboo wall system composed of a glubam exterior panel and a gypsum interior panel sandwiched with glubam structural grids filled by rock wool.

The results from the numerical simulation using the Fire Dynamics Simulator (FDS) software [22–24] are compared with experimental results, as shown in Figure 9.44. The difference between the experimental and analytical results for the highest temperature on the interior wall is within 10%; however, this difference was more than 100% for exterior surfaces. Overall, the FDS simulated results are conservative compared with the temperature histories from the test. On the other hand, the NIST-FDS simulation provides a good visual illustration and some insights into the fire combustion process, as exhibited by the Smokeview shots shown in Figure 9.44.

9.4.3 In-room Air Quality Evaluation

People today spend a significant amount of time in the indoor environment, breathing indoor air. Thus, the indoor air quality is closely related to the health of a building's occupants.[25–27] For the first modern bamboo house, an investigation was carried out by Xiao et al.[28] to inspect the indoor air quality according to the Chinese standard GB50325 (2010). This standard has been updated in 2020[29] with more stringent requirements for the limitation of harmful pollutants. In this section, the inspection results obtained by Xiao et al.[28] are summarized and compared with the new standard.

According to GB 50325 (2020),[29] if the room is less than 50 m2, the inspection can be allocated to only one location within the room. As shown in Figure 9.45, altogether, six rooms were inspected at a point in the center of the room. At the time of testing, the bamboo house had just been completed. The interiors were covered only by plybamboo boards, without final decoration of the interiors nor any furniture, curtains, etc. In other words, the tests were conducted on the glubam structural components in their worst (bare) condition.

During sampling, a Testo 8347 temperature and humidity anemometer was used to measure the on-the-spot humiture and air movement speed. The concentration of formaldehyde (HCHO, CH_2O) was detected and analyzed using phenol reagent colorimetry

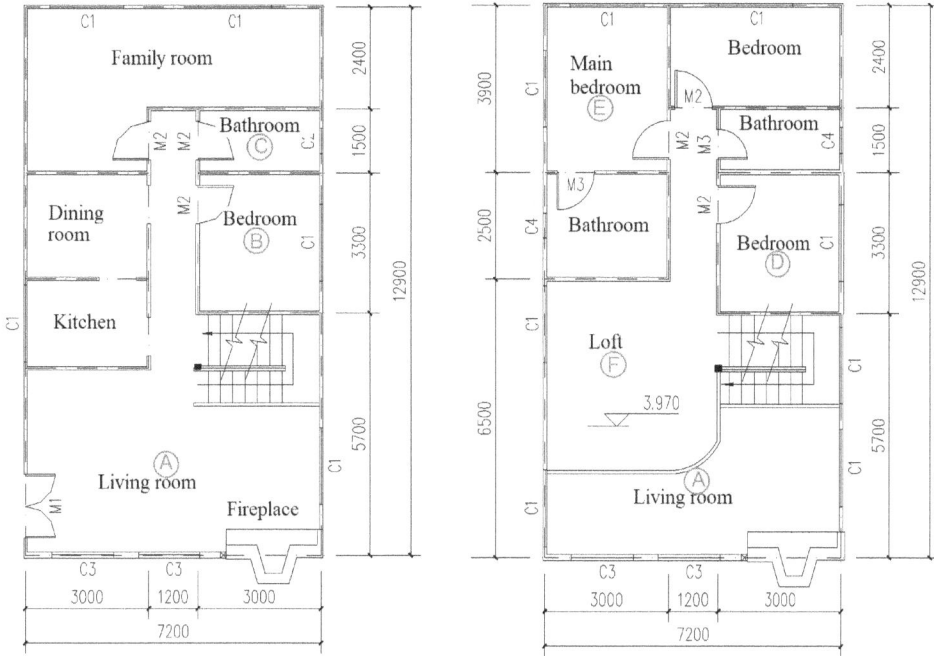

Figure 9.45 Bamboo structure villa layout: (a) ground floor; (b) second floor.

as specified in standard GB/T 18204.2.[30] Nessler's reagent spectrophotometry was used[30] to detect the concentration of ammonia, and the model 1027 radon apparatus was employed to detect radon contents. The total volatile organic compounds (TVOC) and other benzene volatile organic compounds (VOCs) were collected by activated carbon tube, then extracted by sulfur dioxide, and analyzed using a GC-9160 gas chromatograph hydrogen flame ionization device to determine the retention time and peak values. Before sampling, the test rooms were closed for more than 12 h. The doors and windows all remained closed during sampling. The temperature was about 26°C, and the relative humidity was above 75% for all the rooms during the testing.

The main results of the in-door air quality inspection are shown in Table 9.2, along with the limits specified by the GB 50325 (2020) standard.[29] The results show that the glubam house satisfies the current GB standards for indoor air quality.

9.5 Design Example of Lightweight Glubam Frame House

In this section, the bamboo house built at the famous Black Bamboo Park in Beijing is taken as an example to illustrate the design process for lightweight glubam frame buildings. The total area of the floor space is 110 m² for the 2-story portion, as shown in Figure 9.46. The platform balcony is ignored in the design. The height of the bamboo house is 6.4 m. A 50-year service life is considered, with a II waterproofing classification as an ordinary building.[31] Since the building is located in the capital city, Beijing,

Table 9.2 In-door air quality test results

Contents	Measurement points						Type-I building*
	A	B	C	D	E	F	-
Formaldehyde (mg/m³)	0.01	0.03	0.05	0.02	0.01	0.01	0.07
Benzene (μg/m³)	1.6	1.9	2.1	1.4	1.1	0.9	60
Toluene (μg/m³)	0.8	12.3	30.3	2.9	1.5	0.7	150
Paraxylene (μg/m³)	3.1	6.7	5.9	3.3	1.6	1.2	200
TVOC (mg/m³)	0.03	0.06	0.13	0.05	0.05	0.05	0.45

* Based on the GB 50325 (2020) standard.[29]

(a)

Figure 9.46 Glubam tea-house at Beijing Black Bamboo Park: (a) outside view; (b) ground floor plan; (c) second-story floor plan.

the design intensity is considered as VIII. Therefore, the corresponding designed basic earthquake acceleration is 0.2 g, as per the Chinese design code.[15] The site category is III.

Based on the Chinese loading code,[14] the basic wind pressure for a return period of 50 yearsis 0.45 kN/m²; basic snow pressure is 0.40 kN/m²; the ground roughness is classified as C. The dead loads for roof, floors, the interior walls, and the exterior walls are 1.91 kN/m²,1.24 kN/m², 0.39k N/m² and 0.78k N/m², respectively.

For simplicity, the bamboo house is assumed to be a double degree of freedom system and its natural period of vibration can be estimated according to the empirical formula. The floor slabs are assumed to be flexible diaphragms; thus the horizontal resistance of the shear walls can be estimated based on their tributary areas. The structural design is based on both the current Chinese design code for timber structures[32] and the U.S. AWC design specifications.[33]

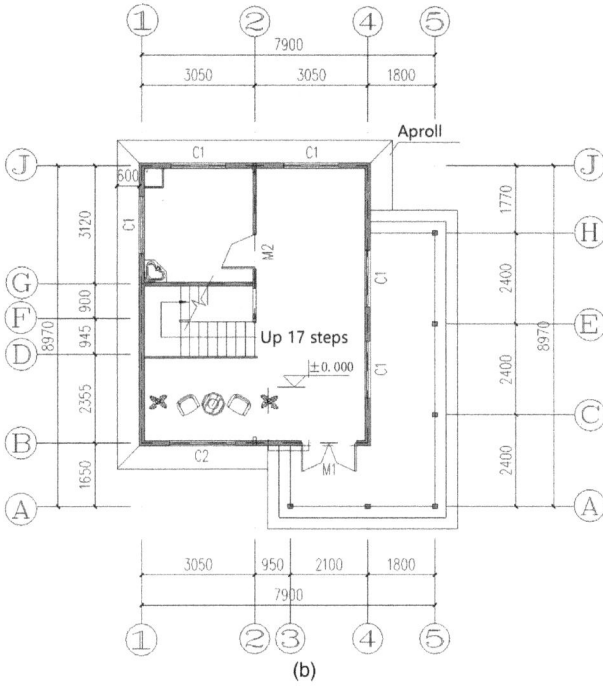

(b)

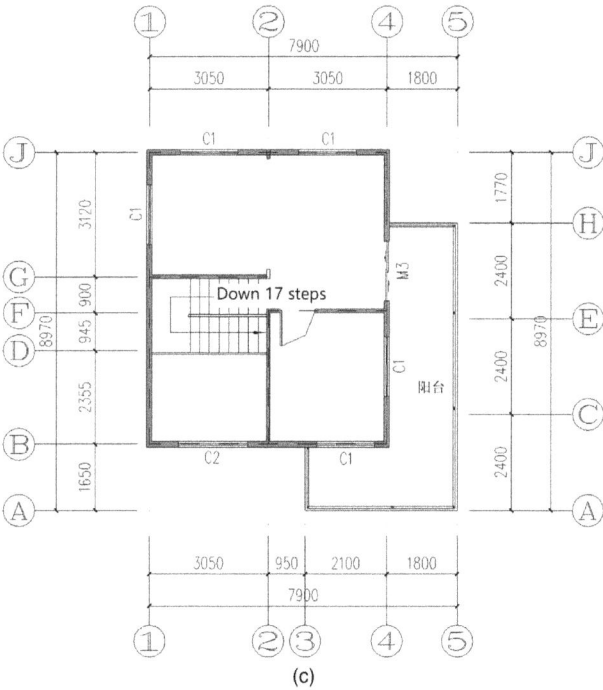

(c)

Figure 9.46 (Continued)

9.5.1 Computation of Loads

9.5.1.1 Wind Load

The wind load calculation is based on Figure 9.47.

1 Horizontal wind load on roof:

Transverse direction (perpendicular to the roof ridge):

$$F_{2-H} = \left[(0.27+0.17)\times 2.58/2 + (0.17-0.15)\times \sin 18.4^\circ \times 1.1\right]\times 7.46 = 4.29 kN$$

Longitudinal direction (parallel to the roof ridge):

$$F_{2-Z} = (0.27+0.17)\times\left(\frac{2.58}{2}+1.1\right)\times 6.24 = 6.56\,\text{kN}$$

2 Vertical wind load on roof:

$$F_{2-up} = (0.15+0.17)\times(3.12+0.54)\times\cos 18.4^\circ \times 7.46 = 8.29\,\text{kN}$$

3 Horizontal wind load on building

Transverse direction:

$$F_{1-H} = (0.27+0.17)\times 2.58\times 7.46 = 8.47\,\text{kN}$$

Longitudinal direction:

$$F_{1-Z} = (0.27+0.17)\times 2.58\times 6.24 = 7.08\,\text{kN}$$

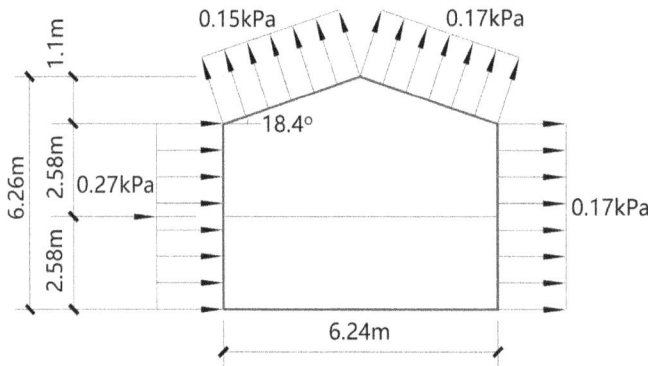

Figure 9.47 Standard wind load.

9.5.1.2 Earthquake Load

The natural period of vibration of the building T is,

$$T = 0.05(h_n)^{0.75} = 0.201s$$

where, h_n is the height in meters.

The horizontal force applied at each floor level is,

$$F_i = \frac{G_i H_i}{\sum_{j=1}^{n} G_j H_j} F_{EK}(1-\delta_n) = \frac{G_i H_i}{\sum_{j=1}^{n} G_j H_j} \alpha_1 G_{eq}(1-\delta_n)$$

where α_1 represents the horizontal seismic influence coefficient, and $\alpha_1 = \left(\frac{T_g}{T}\right)^{\gamma} \eta_2 \alpha_{max}$, in which, $T_g = 0.55$, $\eta_2 = 1.0$. $0.1 < T = 0.201 < T_g = 0.45$, so $\alpha_1 = \eta_2 \alpha_{max} = 0.16$. For low-rise buildings, the whiplash effect of the roof can be ignored, with the roof influence coefficient taken as $\delta_n = 0.0$; G_{eq} represents the equivalent structural gravity load. For the current building, the total gravity load can be taken as 85% of the total gravity load representative value. Therefore,

The dead load weight at the mass center of the roof is:

$$G_{2-eq} = 187.79 \, \text{kN}$$

The dead load weight at the second-floor level is:

$$G_{1-eq} = 182.21 \, \text{kN}$$

The equivalent total gravity load is:

$$G_{eq} = 0.85 \times (G_{1-eq} + G_{2-eq}) = 314.5 \, \text{kN}$$

Therefore, the story force is:

$$F_{2-eq} = \frac{187.79 \times 5.71}{(187.79 \times 5.71 + 182.21 \times 2.58)} \times 0.16 \times 314.5 = 34.98 \, \text{kN}, \text{ applied at a height of}$$

5.02 m to the top of the foundation.

$$F_{1-eq} = \frac{182.21 \times 2.58}{(187.79 \times 5.71 + 182.21 \times 2.58)} \times 0.16 \times 314.5 = 15.34 \, \text{kN}, \text{ applied at a height of}$$

2.58 m above the top of the foundation.

By comparison of the wind and the earthquake loads, it is clear that the earthquake effects govern the design.

9.5.2 Design of Structural Components

9.5.2.1 Roof Design

The roof system is composed with roof trusses patched with 10 mm thick sheathing panels. All upper chords have a sectional dimension of 56 mm wide and 140 mm deep.

Nails with a 2.8 mm diameter are used to connect the panels with the glubam members with spacings of 150 mm and 300 mm along the edges and inside the panels, respectively. The total length of the walls along the transverse direction is shorter, thus governing the design, for a uniformly distributed design load of,

$$w_f = 1.3 \times \frac{34.98}{7.32} = 6.21 \text{kN} / \text{m}$$

It is assumed that the roof diaphragms (two inclined planes) are merged with the diaphragms at the lower chord level, as one diaphragm. This load needs to be resisted by the walls along the axes B, G, and J, as shown in Figure 9.46 (b). The interface length of the wall along axis G is the shortest; i.e., 3.05 m. The interface of shear wall G and the diaphragm in the span of BG is subjected to the largest shear force.

1 SHEAR BEARING CAPACITY OF THE ROOF DIAPHRAGM

The designed shear bearing capacity of the roof is:[32]

$$V = f_d B = f_{vd} k_1 k_2 B$$

In this formula, $k_1 = k_2 = 1.0$, $B \approx 6.24$ m, $f_{vd} = 6.4 \text{kN} / \text{m}$
Thus, the designed shear bearing capacity along axis G is:

$$V = f_d \frac{B}{\gamma_{Eh}} = 6.4 \times 1.0 \times 1.0 \times \frac{3.05}{0.85} = 16.60 \text{kN} > 0.5 \times 6.21 \times 4.2 = 13.04 \text{kN}$$

2 BEARING CAPACITY CHECK FOR BOUNDARY ELEMENTS

The boundary members of the roof truss are composed with edge beams with a section of 56 mm × 140 mm. For simplicity, in the flexural analysis, the existence of the walls along the axes G and F is neglected. Thus, the tension force (same value for compression) in the boundary elements is:

$$N_f = \frac{M_1}{B_0} = \frac{w_f L_1^2}{8 B_0} = \frac{6.21 \times 7.32^2}{8 \times 6.1} = 6.82 \text{kN}$$

The axial bearing capacity of the boundary members of the roof diaphragm is controlled by its tensile capacity:

$$N_t = 2 \times 56 \times 140 \times 5.4 \times 10^{-3} = 84.67 \, kN > N_f = 6.82 \text{kN}$$

9.5.2.2 Second-floor Shear Wall Check

The shear wall is composed with the stud, the raising plate, the bottom plate, and the wall plate. The stud is the 40 mm × 120 mm glubam, spaced at 300 mm; the wall plate is 11 mm thick plybamboo board, sheathed for both sides; the pitch of the 3.25 diameter nails is 150 mm. The total lateral force to be designed is 34.98 × 1.3 = 45.47 kN.

I SHEAR CAPACITY

Flexible diaphragm assumption is used in assessing the shear input to the shear walls. Since the wall along axis F is only provided for the second floor, its resistance is ignored. Thus, the shear forces of the walls along the axes J, G, and B can be calculated based on the tributary floor areas.

Along axis J: $V_{SW1} = 45.47 \times \dfrac{1.56}{7.32} = 9.69\,\text{kN}$

Along axis G: $V_{SW2} = 45.47 \times \dfrac{3.68}{7.32} = 22.86\,\text{kN}$

Along axis B: $V_{SW3} = 45.47 \times \dfrac{1.65}{7.32} = 10.25\,\text{kN}$

The designed shear bearing capacity of the shear wall is:[32]

$$V = f_d l = \left(f_{vd} k_1 k_2 k_3 \right) l$$

In this formula, $k_1 = k_2 = k_3 = 1.0, f_{vd} = 4.3 \times 1.25 = 5.38$ kN/m

Thus, $f_d = f_{vd} k_1 k_2 k_3 = 5.38 \times 1.0 \times 1.0 \times \dfrac{1.0}{0.85} = 6.32\,\text{kN}$. The wall along G has the largest shear input; thus, its capacity is,

$$V = 2 f_d l = 2 \times 6.32 \times 3.05 = 38.55 > 22.86\,\text{kN}$$

2 BOUNDARY ELEMENT CHECK

The boundary member of the shear wall is the stud at the end of the wall made of two 120 mm × 84 mm glubam elements. The design axial force of the boundary member is:

$$N_{f1} = \dfrac{22.86 \times 2.58}{3.05} = 19.34\,\text{kN}$$

The plybamboo is neglected. Therefore, the axial bearing capacity of the boundary member of the wall can be evaluated.
Tension ($f_t = 18.4$ MPa, based on Table 4.18):

$$N_t / 0.85 = 120 \times 84 \times 18.4 \times \dfrac{10^{-3}}{0.85} = 218.2\,\text{kN} > N_{f1} = 19.34\,\text{kN}$$

Compression ($f_c = 10.7$ MPa, based on Table 4.18): assume pinned ends in the out-of-plane direction bending and a length of 2.4 m. The compressive capacity is calculated according to the Chinese design code for glulam structures GB/T 50708,[34] using the material information: 120 mm × 84 mm × 2440 mm glubam elements, $f_{cd} = 10.7$ MPa, $E = 9400$ MPa.

For both pinned ends ($k_e = 1.0$), effective length: $L_{eff} = k_e L = 2440$ mm. Since the wall along axis G is connected to two walls at the ends in its orthogonal direction, thus the buckling is considered to occur with bending along its strong axis, or taking $b = 120$ mm (note, for weak axis bending, $b = 84$ mm).

$$f_{ce} = \frac{0.47E}{\left(L_{eff} / b\right)^2} = 10.68\,\text{MPa}$$

$$\text{Stability coefficient } \varphi = \frac{1+\left(f_{ce} / f_{cd}\right)}{1.8} - \sqrt{\left[\frac{1+\left(\dfrac{f_{ce}}{f_{cd}}\right)}{1.8}\right]^2 - \frac{\dfrac{f_{ce}}{f_{cd}}}{0.9}} = 0.756$$

Capacity: $F = \varphi A f_{cd} = 81.5\text{kN} > 19.34$ kN, → OK!

9.5.2.3 First-floor Shear Wall Design

I SHEAR CAPACITY

Similar to the second floor, the earthquake force is resisted by the walls along the axes J, G, and B. The total lateral shear force to be considered is,

$$V_{eq} = F_{2-eq} + F_{1-eq} = (34.98 + 15.34) \times 1.3 = 65.42 \text{ kN}$$

Along axis J: $V_{SW5} = 65.42 \times \dfrac{1.56}{7.32} = 13.94\,\text{kN}$

Along axis G: $V_{SW6} = 65.42 \times \dfrac{3.66}{7.32} = 32.71\,\text{kN}$

Along axis B: $V_{SW7} = 65.42 \times \dfrac{2.1}{7.32} = 18.77\,\text{kN}$

The largest shear input is at the wall along G. Based on Chinese code GB 50005:[32]

$$V = 2 f_d l = 2 \times 6.32 \times 3.05 = 38.55\text{kN} > 32.71 \text{ kN}$$

2 BOUNDARY ELEMENT CHECK

The elements of the shear walls are the double glubam studs with a section dimension of 120 mm × 84 mm. The design axial force is,

$$N_{f1} = \frac{22.86 \times 5.02 + 9.97 \times 2.58}{3.05} = 46.06\text{kN} < 81.5\text{KN}, \rightarrow \text{OK!}$$

9.5.2.4 Design of Second-floor Diaphragm (First-floor Roof)

The total length of the walls aligned along the transverse direction is shorter, thus governing the design, for a uniformly distributed design load of,

$$w_f = 1.3 \times \frac{15.34}{7.32} = 2.72 \,\text{kN/m}$$

This load needs to be transferred to the first-story walls along the axes B, G, and J, as shown in Figure 9.46 (b).

I SHEAR BEARING CAPACITY OF THE FLOOR DIAPHRAGM

The interface length of the wall along axis G is the shortest; i.e., 3.05 m. The interface of shear wall G and the diaphragm in the span of BG (length = 4.2m) is subjected to the largest shear force,

$$V_1 = 0.5 \times 2.72 \times 4.2 = 5.71 \,\text{KN}$$

The second-floor slab is constructed using 60 mm x 185 mm joints sheathed with 12.5 mm thick plybamboo panels. The designed shear bearing capacity of the roof is:[32]

$$V = f_d B = f_{vd} k_1 k_2 B$$

In this formula, $k_1 = k_2 = 1.0$; $B = 3.05$m; $f_{vd} = 5.4 \times 1.25 = 6.75$ kN/m (treated as type-4 diaphragm, and seismic adjustment coefficient 1.25, using Table P.0.3-2[32]).
   Thus, the designed shear bearing capacity along axis G is:

$$V_d = f_d \frac{B}{\gamma_{\text{Eh}}} = (6.75 \times 1.0 \times 1.0) \times \frac{3.05}{0.85} = 24.22 \,\text{kN} > V_1 = 5.71 \,\text{kN}$$

2 BEARING CAPACITY CHECK FOR BOUNDARY ELEMENTS

The boundary elements for the second-floor diaphragm are the same as those for the roof diaphragm. They are identified as safe, referring to the calculation for the roof diaphragm.

9.5.2.5 Design of Connection

Because the uplift force acting on the roof of the house (8.29 kN) is smaller than the weight of the roof (115 kN), there is no need to check the uplift force. The connecting structure between the wall and the floor is shown in Figure 9.48.

I CONNECTION OF THE SECOND WALL AND THE ROOF OF THE BUILDING

The designed shear bearing capacity of the 82 mm long nailed joint with a 3.66 mm diameter using normal steel nails is:

$$N_v = k_v d^2 \sqrt{f_c} = 10.2 \times 3.66^2 \times \sqrt{23} \times 10^{-3} = 0.66 \,\text{kN}$$

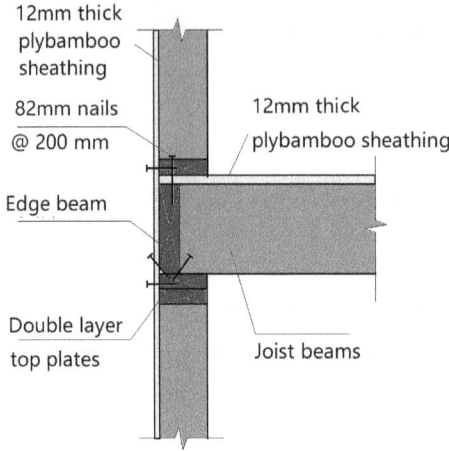

Figure 9.48 Details of connection between exterior wall and floor.

The design value of the horizontal seismic action load passed from the roof is:

$1.4 \times 34.98 = 48.97$ kN.

Thus, the number of nails needed is: $48.97/0.66 = 74.2$.
The total length of the second endwise wall is: 18.3 m.
Therefore, the nail pitch is:

$$\frac{18300}{74.2} = 247\,\text{mm},$$

This result is rounded down to 200 mm.
The calculating method for the broadwise nail pitch is the same as the above formula.

2 CONNECTION OF THE FIRST WALL AND THE ROOF OF THE BUILDING

The design value of the endwise horizontal seismic action load passed from the roof is:

$1.4 \times (34.98 + 15.34) = 70.45\,\text{kN}$

The number of nails needed is: $70.45/0.66 = 106.7$.
The total length of the first endwise wall is: 15.95 m.
Therefore, the nail pitch is:

$$\frac{15950}{106.7} = 150\,\text{mm}.$$

This result is rounded down to 100 mm.
The calculating method for the broadwise nail pitch is the same as the above formula.

3 CONNECTION OF THE WALL AND THE FOUNDATION

Choose anchor bolt M12 to connect the first wall and the foundation. The lateral designed bearing capacity of each single bolt is:

$$N_v = k_v d^2 \sqrt{f_c} = 5.5 \times 12^2 \times \sqrt{23} \times 10^{-3} = 2.96 \text{ kN}$$

The design value of the endwise horizontal seismic action passed from the roof is 70.45 kN.

The number of bolts needed is: 70.45/2.96 = 23.8.

The total length of the first endwise foundation is 15.95 m.

Therefore, the bolt pitch is:

$$\frac{15950}{23.8} = 670 \text{ mm},$$

This result is rounded down to 600 mm.

The calculating method of the endwise foundation bolt pitch is the same as the above formula.

9.5.2.6 Construction Measures

Several typical seismic details should be considered, similar to the common practice for timber structures in North America. At the corners of the windows and door openings, steel stripes should be nailed to prevent excessive separation of the elements joining at the corners. Since the structure is constructed using the platform procedure, steel stripes should also be tied between the stories to provide vertical tension continuity. The first story walls should be securely tied down to the foundation to prevent overturning or excessive uplift during earthquake loading, which is also effective in resisting severe wind.

9.6 Glubam Heavy Frame Structures

Single-story or multi-story frame buildings rely on joints with full or partial moment resistance to connect beams and columns to form frame systems. This type of timber structure is typically called a heavy frame structure, to distinguish it from the light-weight frame system. In many cases, the combined use of braces or shear walls in the frame system is more efficient for lateral force resistance. For relatively high-rise timber structures, the current trend is to use a hybrid system with a timber heavy frame, cross laminated timber walls/slabs, and concrete shear walls or steel frame/braces, such as the one shown in Figure 9.49. In this section, glubam heavy frame structures are discussed.

9.6.1 Glubam Single Story Frames

Feng et al.[35] experimentally studied glubam beam–column portal frame models with both push-over and cyclic loading, as shown in Figure 9.50. Four approximately 1:3 model frames were tested. The glubam material was the thin-strip type and the beam and the columns were connected at the corners of the portal frame using steel bolts, whereas the column feet were connected with the testing floor through pin connections.

Figure 9.49 Eighteen-story timber and concrete hybrid building
Source: Photo courtesy of Dr. X. Zhang.

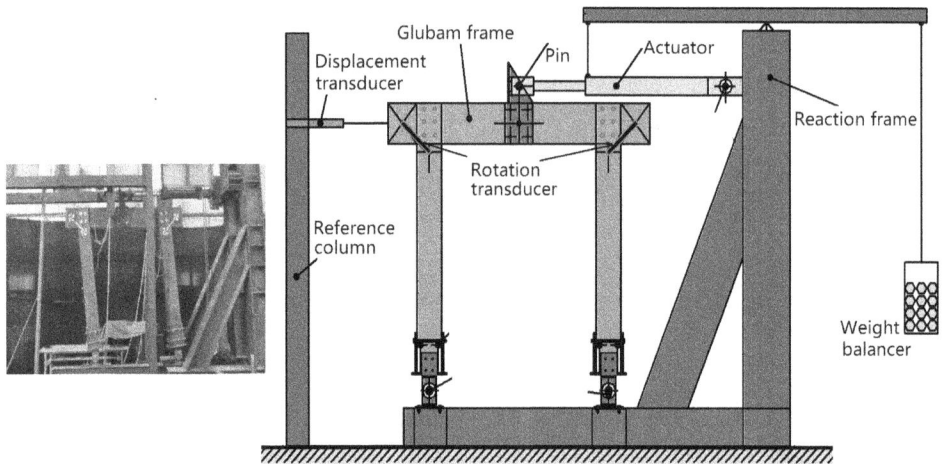

Figure 9.50 Feng et al.'s test on glubam portal frame.
Source: Feng et al. (2014). [35]

The test results show that the behavior of the portal frame was significantly affected by the behavior of the beam–column connections. The damage to the bolts was uneven depending on the location of each bolt. If the base material has adequate strength, the steel bolts can dissipate a reasonable amount of energy. The results also indicate that the connection design based on the current Chinese design code[32] yields a conservative estimation of the capacity of the portal frame.

The first glubam portal frame building was designed and constructed for an experimental laboratory at Hunan University in 2011. As shown in Figure 9.51 (a), the building has 13 glubam portal frames with a clear span of 6.3 m, spaced at 4.25 m. The beam length is 8.4 m, supported on two columns of different heights (4.5 m and 4.8 m tall, respectively) with a span of 6.3 m. The laboratory building is still in operation after 10 years. Another example shown in Figure 9.51 (b) is a waste treatment facility at the Zhejiang University International Campus. The portal frames had a span of 7.0 m, with external braces to the upper beam which is 9.2 m. The examples of portal frame buildings shown in Figure 9.51 (a) and (b) were made with thin-strip glubam and thick-strip glubam, respectively.

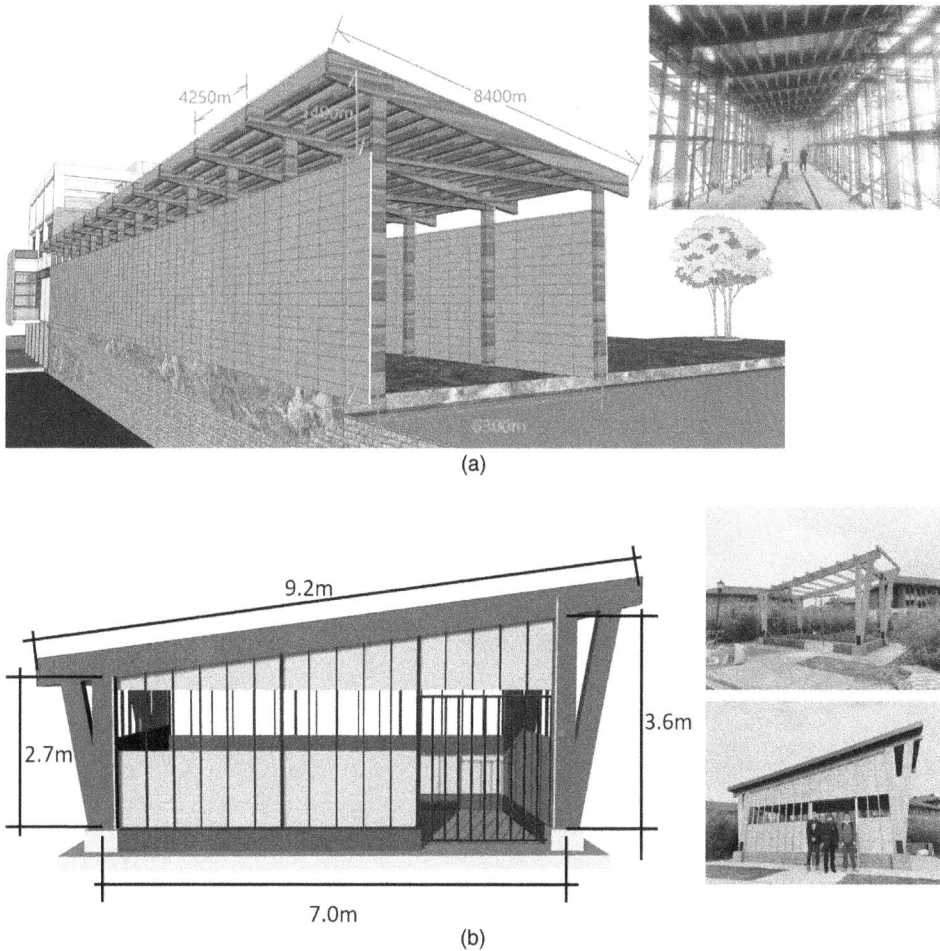

(a)

(b)

Figure 9.51 Examples of portal frame buildings: (a) Hunan University laboratory constructed in 2011; (b) a waste treatment building on Zhejiang University International campus built in 2019.

A more complex building constructed in 2011 is shown in Figure 9.52. The architectural design was provided by Prof. C.Y. Wei of Hunan University, while the author's group was responsible for the structural design and construction. The structure was probably the first engineered bamboo structure on such a scale in terms of component size and total usage of engineered bamboo. The largest girder was about 15 m long with a section size of 800 mm deep and 150 mm wide, cantilevered for 7.1 m, as shown in Figure 9.53 (a). In this structure, due to the complex roof system, essentially all the connections between columns and beams or girders are different with many skews in vertical and horizontal directions. Figure 9.53 (b) to (d) exhibit some of the details of the glubam frame structure during construction.

(a)

(b)

Figure 9.52 A complex glubam frame building at Meixi Lake, Changsha, Hunan Province: (a) bird's eye view effect; (b) current conditions in July 2021.

7.1m long cantilever

(a)

(b)   (c)   (d)

Figure 9.53 Glubam building with complex spatial frame system under construction: (a) large cantilever; (b) skewed connection; (c) column base connection; (d) glubam beam and concrete
column connection.

9.6.2 Design Example of a Multi-story Glubam Frame Building

An office building currently under design by Ma and Chen of the author's group is taken
as an example to illustrate the design process of multi-story glubam frame buildings
(Figure 9.54). The 3-story 2-span (8 m each) by 6-bay (5 m each) office building features
a common frame deployment, as shown in Figure 9.55. The story heights are 3 m, 4 m,
and 3.5 m from the first to the third story. The slope of the double-pitched roof is 1:8.

The design is based on Chinese codes with a 50-year service life, and a Class III
safety design level. The design intensity is considered level VII, corresponding to a basic
common earthquake acceleration of 0.08 g.[15] The site category is II, with a typical site
period of $T_g = 0.35$ s. Based on the Chinese loading code GB 50009,[14] the basic wind

Figure 9.54 A glubam building design.

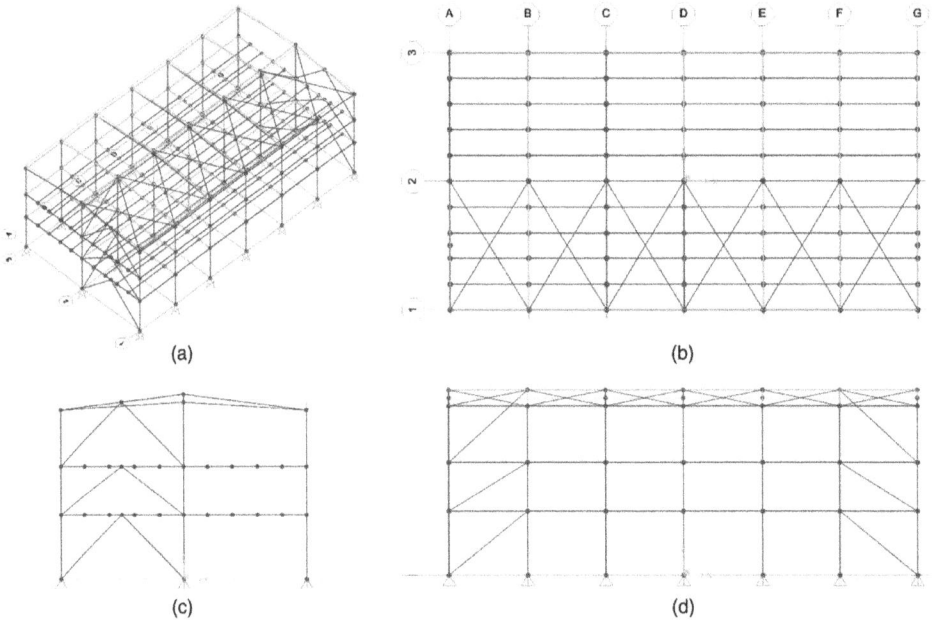

(a)

(b)

(c)

(d)

Figure 9.55 Multi-story glubam building: (a) three-dimensional model; (b) model plan; (c) frame A and B; (d) frame-2.

Table 9.3 Element list

	Material	CroSec	Width	Depth	Diameter	Length	Number
Slab	Lightweight concrete	Rect.	-	0.1	-	1.6	1500
Beam y	glubam	Rect.	0.3	0.6	-	8	28
Beam x	glubam	Rect.	0.24	0.45	-	5	36
2nd beam x	glubam	Rect.	0.12	0.36	-	5	80
Roof beam y	glubam	Rect.	0.12	0.48	-	8.062	18
Story-1 column	glubam	Rect.	0.3	0.5	-	4	21
Story-2 column	glubam	Rect.	0.3	0.5	-	3	21
Story-3 side column	glubam	Rect.	0.3	0.5	-	3.5	14
Story-3 middle column	glubam	Rect.	0.3	0.5	-	4.5	4
Story-1 support x	glubam	Rect.	0.36	0.36	-	6.403	2
Story-2 support x	glubam	Rect.	0.36	0.36	-	5.831	2
Story-3 support x	glubam	Rect.	0.36	0.36	-	6.403	2
Story-1 support y	glubam	Rect.	0.3	0.3	-	5.657	2
Story-2 support y	glubam	Rect.	0.3	0.3	-	5	2
Story-3 support y	glubam	Rect.	0.3	0.3	-	6.021	2
Cable x	650 steel	Circle	-	-	0.05	5	4
Cable y	650 steel	Circle	-	-	0.05	8.016	8
Cable z	650 steel	Circle	-	-	0.05	0.5	4
Cable xy	650 steel	Circle	-	-	0.05	9.487	8

Note: Density of materials: glubam = 750 kg/m³; lightweight concrete = 1200 kg/m³; steel = 7800 kg/m³; glass = 2500 kg/m³.

pressure for 50 years of return period is 0.35 kN/m²; basic snow pressure is 0.45 kN/m²; the floor live load is 2 kN/m² for the office usage; the roof live load is 0.5 kN/m².

The structural design is based on the current Chinese design code for timber structures, GB 50005.[32] The entire calculation process is encoded using python software for a fully automatic update upon change of specifications of elements so that an efficient design procedure is achieved. The final structural components chosen are listed in Table 9.3.

9.6.2.1 Load – Combination

The load is calculated by the maximum of the following, as per the Chinese loading code GB 50009:[14]

- $1.35D + 1.4 \times (0.7(L + Lr \text{ or } S) + 0.6W)$
- $1.2D + 1.4 \times (L + 0.7(Lr \text{ or } S) + 0.6W)$
- $1.2D + 1.4 \times (W + 0.7(L + Lr \text{ or } S))$
- $1.0D + 1.3E + 1.0L + 0.5S$

The upward wind load is also considered:

- $1.0D + 1.4W$

The standard load is calculated as:

- $1.0D + 1.0L + 0.7\ (Li + Lr\ \text{or}\ S) + 0.6W\ (i>1)$

where D – dead load; L – live load; Lr – roof live load; S – snow load; W – wind load; E – earthquake load.

For reference purposes, under the corresponding U.S. code (Section 1605.2 of the 2015 *IBC*):[36] the design load can be calculated by the maximum of the following:

- $1.4D$
- $1.2D + 1.6L + 0.5\ (Lr\ \text{or}\ S)$
- $1.2D + 1.6\ (Lr\ \text{or}\ S) + (f1L\ \text{or}\ 0.5W)$
- $1.2D + 1.0W + f1L + 0.5\ (Lr\ \text{or}\ S)$
- $1.2D + 1.0E + f1L + f2S$
- $0.9D + 1.0W$
- $0.9D + 1.0E$

The standard load is calculated by the following (Section 1605.3 of the 2015 *IBC*):[36]

- D
- $D + L$
- $D + (Lr\ \text{or}\ S)$
- $D + 0.75L + 0.75\ (Lr\ \text{or}\ S)$
- $D + (0.6W\ \text{or}\ 0.7E)$
- $D + 0.75\ (0.6W) + 0.75L + 0.75\ (Lr\ \text{or}\ S)$
- $D + 0.75\ (0.7E) + 0.75L + 0.75S$
- $0.6D + 0.6W$

9.6.2.2 Load – Earthquake Analysis

The earthquake response is analyzed under the following considerations:

- Both x and y directions;
- Both base shear method and response spectrum method (SRSS); note that the base shear method is applicable only to rigid frames, ignoring the difference between x and y directions.
- Each of the following hypotheses: rigid frame without braces, pinned joints with braces, and pinned joints with shear walls; however, only the braced frame design is briefly provided in this book.
- Common earthquake ($\alpha_1 = 0.08$ g) and rare earthquake ($\alpha_1 = 0.5$ g).

A total of 14 scenarios are calculated, the maximum of which is taken as the design value, with a few exceptions.

- Response spectrum method is employed in favor of base shear method.
- A rare earthquake scenario is only considered when designing joints and checking deformation between stories.

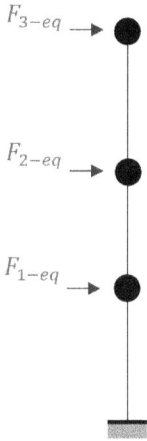

Figure 9.56 Simplified dynamics model.

Because the building is under 20 m in height, the lateral earthquake forces at each floor level can be calculated using the lumped mass stick model shown in Figure 4.56, according to the base shear method.[37]

The natural period of vibration of the buildings T is,

$$T = 0.05\left(h_n\right)^{0.75} = 0.302s$$

where, h_n is the height in meters. The horizontal force applied at each floor level is,

$$F_i = \frac{G_i H_i}{\sum\limits_{j=1}^{n} G_j H_j} F_{EK}\left(1-\delta_n\right) = \frac{G_i H_i}{\sum\limits_{j=1}^{n} G_j H_j} \alpha_1 G_{eq}\left(1-\delta_n\right)$$

where, α_1 represents the horizontal seismic influence coefficient, and $\alpha_1 = \left(\dfrac{T_g}{T}\right)^{\gamma} \eta_2 \alpha_{max}$, in which, $T_g = 0.35$, $\eta_2 = 1.0 \cdot 0.1 < T = 0.302 < T_g = 0.35$, so $\alpha_1 = \eta_2 \alpha_{max} = 0.08$. For low-rise buildings, the whiplash effect of the roof can be ignored, with the roof influence coefficient taken as $\delta_n = 0.0$; G_{eq} represents the equivalent structural gravity load. For the current building, the representative gravity load can be taken as 85% of the total vertical load combined, according to GB 50011[15]; i.e., $1.0D + 1.0L + 0.5S$.

$$G = \begin{bmatrix} 2404 \\ 2404 \\ 578 \end{bmatrix} kN$$

Such that the equivalent total gravity load is:

$$G_{eq} = 0.85 \times \| G_1 \| = 4578 \text{ kN}$$

Therefore, the story forces are:

$$F_{1-eq} = \frac{2.404 \times 4}{2.404 \times 4 + 2.404 \times 7 + 0.578 \times 11} \times 0.08 \times 4578 = 107.4 \text{ kN, applied at a height}$$

of 4 m above the top of the foundation.

$$F_{2-eq} = \frac{2.404 \times 7}{2.404 \times 4 + 2.404 \times 7 + 0.578 \times 11} \times 0.08 \times 4578 = 187.9 \text{ kN, applied at a height}$$

of 7 m above the top of the foundation.

$$F_{3-eq} = \frac{0.578 \times 11}{2.404 \times 4 + 2.404 \times 7 + 0.578 \times 11} \times 0.08 \times 4578 = 71.0 \text{ kN, applied at a height}$$

of 11 m above the top of the foundation.

9.6.2.3 Element – Secondary Beam under Gravity Load

The simply supported beam under bending is calculated, referring to section 5.2 of GB 50005.[32] The typical secondary beams (section of 120 mm wide and 360 mm deep) have a length of 5 m (for simplicity, the total length is conservatively used), and are aligned in the direction of x, spaced at 1600 mm. The load combination $1.2D + 1.4L$ yields a distributed design load and is transferred to a linear load of 8.342 kN/m on a typical secondary beam. In the following analysis, the elastic modulus is taken as 10 GPa for the thick-strip glubam material.

Stress check under bending, as per section 5.2.1 of GB 50005:[32]

$$\sigma = \frac{My}{I} = \frac{qL^2 y}{8I}$$

$$\sigma = 10.1 \text{MPa} \leq [\sigma] = 20.4 \text{MPa}, \rightarrow \text{Safe!}$$

The load combination $1.0D + 1.0L$ yields a distributed service load transferred to a linear load of 6.418 kN/m on a typical secondary beam. The deflection check, according to section 5.2.9 of GB 50005[32] is:

$$u = \frac{5qL^4}{384EI} < [u] = \frac{1}{250} L = 0.02 \text{ m}, \rightarrow \text{Satisfactory}$$

The corresponding U.S. code can be found in the National Design Specification.[33]

9.6.2.4 Element – Main Beam under Gravity Load

The main beams with a section of 600 mm depth and 300 mm width are aligned along the y-direction and have a total span of 8 m, which is used for the following calculations as the clear span of the simply supported beams, for simplicity and conservatism.

Vertical gravity design loads are mainly applied to a main beam from the secondary beams on two sides (for internal frames). For simplicity, the loads from the secondary beams are distributed as a linear design load $(1.2D + 1.4L)$ of 34.09 kN/m, and a service load $(1.0D + 1.0L)$ of 26.74 kN/m on a main beam.

Capacity check (section 5.2.1 of GB 50005[32]): no need to consider lateral torsional buckling because the beams are laterally restricted by the secondary beams.

$$\sigma = \frac{My}{I} = \frac{qL^2 y}{8I} = 15.15 \, \text{MPa} < [\sigma] = 20.4 \, \text{MPa}, \rightarrow \text{Safe!}$$

Deflection check (section 5.2.9 of GB 50005[32]):

$$u = \frac{5qL^4}{384EI} = 0.0264 \, \text{m} < [u] = \frac{1}{250} L = 0.032 \, \text{m}, \rightarrow \text{Safe!}$$

9.6.2.5 Element – Main Beams in x-Direction under Gravity Load and Earthquake Load

With the flexible-floor diaphragm assumption, a slab can be treated as equivalent to a simply supported beam, with the braced frames as the supports, as shown in Figure 9.57. Under such an assumption, beams in the x-direction (240 mm wide and 450 mm deep) are under combined bending and axial loading, with the axial loading derived from the entire frame structure, as the equivalent to a simply supported beam under a distributed lateral earthquake load, and the bending moment induced by the vertical gravity load.

The maximum distributed lateral earthquake load is on the second floor,

$$q = \frac{F_{2-eq}}{L} = \frac{187.9 \, \text{kN}}{30 \, \text{m}}$$

with the maximum bending moment at the center of the equivalent beam,

$$M = \frac{qL^2}{8} = 704.63 \, \text{kN} / \text{m}$$

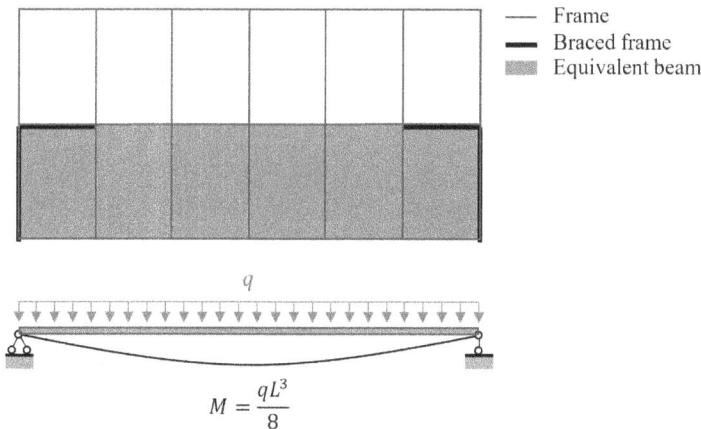

Figure 9.57 Slab diaphragm and the equivalent beam.

Then the axial tensile force of the beam on the edge is (assuming that only one of the two 8 m long spans is in action),

$$T = \frac{M}{8\,\text{m}} = 88.08\,\text{kN}$$

The load combination $1.0D + 1.0L + 1.3E$ yields the following axial load and bending moment,

$$N = 1.3T = 114.48\,\text{kN},$$

$$M = 1.0\left(D+L\right) = \left(3.2 + 5.72\right) \times \frac{5^2}{8} = 27.9\ \text{kN}/\text{m (Vertical direction)}$$

A stress check, as per section 5.3.1 of GB 50005,[32] for the member with combined bending and axial compression is performed as follows (for strength, axial stability, and lateral stability),

1 Stress check for the member under axial tension ($1.3E$) and bending ($1.0D + 1.0L$):

$$utilization = \frac{N}{A_n f_t} + \frac{M}{W_n f_m} = \frac{114.48e3}{0.108e6 \times \dfrac{32.3}{0.8}} + \frac{27.9e6}{8.1e6 \times \dfrac{38.4}{0.8}} = 0.0085 + 0.0718$$

$$= 0.08 < 1.0, \rightarrow \text{Safe!}$$

(Note, in the calculation, 0.8 is the seismic adjustment factor of bearing capacity of members γ_{RE}, as per section 4.2.10 of GB 50005.[32])

2 Stress check of beam under axial compression ($1.3E$) and bending ($1.0D + 1.0L$):

$$utilization = \frac{N}{A_n f_c} + \frac{M_0 + Ne_0}{W_n f_m} = \frac{114.48e3}{0.108e6 \times \dfrac{20.4}{0.8}} + \frac{27.9e6 + 114.48 \times \left(0.05 \times 0.45\right)e6}{8.1e6 \times \dfrac{38.4}{0.8}}$$

$$= 0.0135 + 0.0786 = 0.092 < 1, \rightarrow \text{Safe!}$$

3 Axial stability under combined bending and axial loading (as per section 5.3.2 of GB 50005,[32] however, neglecting adjustment factors for design strengths f_m, f_c).

$$k = \frac{M_0 + Ne_0}{W_n f_m\left(1 + \sqrt{\dfrac{N}{A f_c}}\right)} = \frac{27.9e6 + 114.48 \times \left(0.05 \times 0.45\right)e6}{8.1e6 \times \dfrac{38.4}{0.8}\left(1 + \sqrt{\dfrac{114.48e3}{0.108e6 \times \dfrac{20.4}{0.8}}}\right)} = 0.0653$$

$$k_0 = \frac{Ne_0}{W_n f_m\left(1 + \sqrt{\dfrac{N}{A f_c}}\right)} = \frac{114.48 \times \left(0.05 \times 0.45\right)e6}{8.1e6 \times \dfrac{38.4}{0.8}\left(1 + \sqrt{\dfrac{114.48e3}{0.108e6 \times \dfrac{20.4}{0.8}}}\right)} = 0.0056$$

Axial loading coefficient, as per section 5.1.4 of GB 50005,[32] (note, according to Chapter 4, the characteristic values, f_{ck} and E_k are taken as 47.7 MPa and 10 GPa, respectively. In addition, the earthquake load adjustment coefficients of the design strength are taken as 1.0 for f_c and f_m, for simplicity and conservatism).

Slenderness ratio (weak dir.): $\lambda c = 51.2$

Stability coefficient (weak dir.): $\varphi = 0.379$

The stability coefficient considering axial loading and initial bending moment,

$$\varphi_m = (1-k)^2 (1-k_0) = 0.808$$

$$\sigma = \frac{N}{\varphi\varphi_m A_0} = \frac{114.48e3}{0.379 \times 0.808 \times 0.108e6} = 3.438\,\text{MPa} \leq [\sigma] = 20.4\,\text{MPa},$$

$\rightarrow$ Safe!

4 Lateral stability under combined bending and axial compressive loading (as per section 5.3.3 of GB 50005[32]):

   Stability coefficient for the beam in vertical direction, as per section 5.1.4 of GB 50005:[32]

Slenderness ratio : $\lambda c = 51.2$

Stability coefficient: $\varphi_y = 0.844$

   Lateral stability coefficient for the beam with lateral supports at ends, as per section 5.2.3 of GB 50005:[32]

$\varphi_l = 1$

$$utilization = \frac{N}{\varphi_y A_0 f_c} + \left(\frac{M}{\varphi_l W f_m}\right)^2$$

$$= \frac{114.48e3}{0.844 \times 0.108e6 \times \dfrac{20.4}{0.8}} + \left(\frac{27.9e6}{1 \times 8.1e6 \times \dfrac{38.4}{0.8}}\right)^2$$

$$= 0.0493 + 0.00515 = 0.0544 < 1, \rightarrow \text{safe!}$$

The result is close to the corresponding check using the U.S. National Design Specification[33] (section 3.9).

9.6.2.6 Element – Column

In this building design, the lateral load is carried by the frames with lateral braces; thus the columns are designed as axially loaded elements with pinned ends. In reality, the columns are designed with steel plate connections with the foundation and beams; thus a significant moment resistance is provided for the column ends, primarily as a safety reserve.

Total vertical design load: $(1.0D+1.3E+1.0L+0.5S):combo = 6765.7\,\text{kN}$

Vertical design load of 1 column: $F = 563.81\,\text{kN}$

Considering the uneven load due to the third floor, the values above are each multiplied by 1.2.

Refer to GB 50005, Table 5.1.5, using the worst-case scenario (GB 50005, section 5.1.4).[32]

Slenderness ratio (300 mm x 500 mm section column with a length of 4000 mm, pinned at both ends): $\lambda c = 51.2$, and stability coefficient: $\varphi = 0.790$

Stress check under axial loading: (GB 50005,[32] section 5.1.2):

$F = 676.57\,\text{kN}$ (assuming only 10 columns to carry 6765.7 kN)

$$\sigma = \frac{F}{\varphi A} = \frac{676.57e3}{0.79 \times 0.15e6} = 5.7\,\text{Mpa} \le [\sigma] = 20.4\,\text{MPa}, \rightarrow \text{Safe!}$$

9.6.2.7 Element – Braces

Taking the y-direction as an example, the lateral earthquake force on the laterally braced frame is as shown in Figure 9.58, considering the accidental torsion. For simplicity, the brace elements are assumed to be pinned elements, as shown in Figure 9.59. The brace element is treated as an axially loaded column with pinned ends (refer to GB 50005,[32] section 5.1).

The y-direction brace under common earthquake conditions is, for instance, calculated as following (refer to GB 50005,[32] section 5.1.4) by considering a square section with 300 mm side length and a total length of 5657 mm (ignoring the sectional sizes, for conservatism and simplicity).

Slenderness ratio : $\lambda c = 78.27$

Stability coefficient of the brace member: $\varphi = 0.188$

Stress under axial loading (GB 50005,[32] section 5.1.2)

$$F_{b2} = 0.55 \times (107.9+187.4+71.0) = 201.5\,\text{kN}$$

$$N = \frac{F_{b2}}{2\cos\theta} = 142.44\,\text{kN}$$

$$\sigma = \frac{N}{\varphi A} = 8.41\,\text{MPa} \le [\sigma] = 25.5\,\text{MPa}, \rightarrow \text{safe!}$$

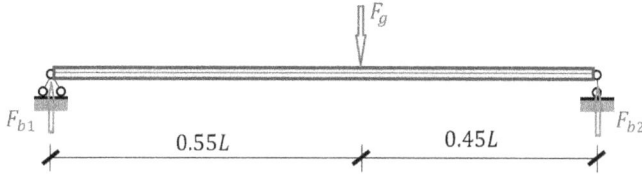

Figure 9.58 Lateral earthquake force on the laterally braced frame.

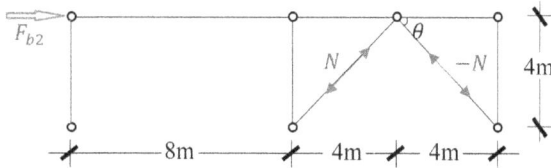

Figure 9.59 Brace elements in frame.

9.6.2.8 Design Check Using Structural Design Software

The structure with selected element specifications is then checked again using SAP2000 software. The results are highly consistent with the manual calculations.

The design specifications may have other requirements for the braced frame structure. For example, the current initial design is based on gravity columns for the frames with all the lateral seismic forces taken by the braces. In reality, the frames can also carry a certain amount of seismic load.

It should also be noted that the design and safety check for the connections are not provided for this example. The connections should be designed to ensure consistency between the analytical model and the actual design of the building.

9.6.2.9 Comments

The design of this heavy frame glubam building was completed in the summer of 2021, and the construction was planned for completion at the end of 2021. The author wishes to have the opportunity to update the report of the design and construction of the office building in future editions of this book.

References

[1] Zha, X.X. (2011). *Light-gauge steel mobile buildings* (in Chinese). Science Press, Beijing.
[2] Vigneshkannan, S., Bari, J.A., & Easwaran, P. (2017) A general study of light gauge steel structures – A review. *International Journal of Advanced Research Methodology in Engineering & Technology*, 1(4).
[3] Xiao, Y., Shan, B., & She, L.Y. (2010). A type of fast assemblable bamboo and timber mobile building. China Innovation Patent CN200810031732.1. Notice No. CN101338622B, released on June 9, 2010.

[4] Lu, G.J., Li, L.J., Peng, Z.J., Luo, Y.S., Lin, S.Y., & Zhang, S.J. (2010). Investigation and analysis on fire risks for temporary earthquake relief buildings. *Fire Protection Technology and Product Information*, 3, 3–5.

[5] Ariyanayagam, A.D., & Mahendran, M. (2018). Experimental study of non-load bearing light gauge steel framed walls in fire. *Journal of Constructional Steel Research*, 145, 529–551.

[6] Huang, X. (1959). *Bamboo roof structures*. Construction Engineering Press, Beijing.

[7] Ministry of Housing and Public Works. (2006). Bangladesh National Building Code (BNBC). Ministry of Housing and Public Works, Dacca.

[8] Lienhard, J.H. (1997). Balloon frame houses. No. 779. The Engines of our Ingenuity, https://uh.edu/engines/epi779.htm

[9] Xiao, Y., Huang, Z.Y., & Chen, G. (2007). Bamboo residential houses. China Innovation Patent CN200610136812. Notice N. CN1963056A, Released on May 16, 2007.

[10] Xiao, Y. & She, L. (2010). Two-by-four house construction using laminated bamboos. In *World Conference on Timber Engineering*. Trees and Timber Institute, National Research Council, Trentino, Italy.

[11] Xiao, Y., Chen, G., Shan, B., Yang, R.Z., & She, L.Y. (2010). Research and application of lightweight glue-laminated bamboo frame structures. *Journal of Building Structures*, 31(6).

[12] Chen, G. (2011). Experimental research and engineering application of light frame bamboo structures. Doctoral dissertation, supervised by Y. Xiao, Hunan University.

[13] Chen, G., Shan, B., & Xiao, Y. (2011). Aseismic performance tests for a lightweight glubam house (in Chinese). *Journal of Vibration and Shock*, 30(10), 136–142.

[14] Ministry of Housing and Urban–Rural Development of the People's Republic of China. (2019). GB 50009-2019: Load code for the design of building structures (in Chinese). China Architecture & Building Press, Beijing.

[15] Ministry of Housing and Urban–Rural Development, General Administration of Quality Supervision, Inspection and Quarantine of the People's Republic of China. (2019). GB 50011-2019: Code for seismic design of buildings (in Chinese). China Architecture & Building Press, Beijing.

[16] Wang, R., Xiao, Y., & Li, Z. (2017). Lateral loading performance of lightweight glubam shear walls. *ASCE Journal of Structural Engineering*, 143(6).

[17] Mi, H., Ni, C., Chui, Y.H., & Karacabeyli, E. (2006). Racking performance of tall unblocked shear walls. *ASCE Journal of Structural Engineering*, 132(1), 145–152.

[18] Wang, R. (2019). Research on lightweight Glubam frame structures. Doctoral dissertation, Hunan University.

[19] Xiao, Y., & Ma, J. (2012). Fire simulation test and analysis of laminated bamboo frame building. *Construction and Building Materials*, 34, 257–266.

[20] WFRA. (2003). *Literature review on the contribution of fire resistant timber construction to heat release rate*. Project Report 20633. Warrington Fire Research Consultancy and Testing, Surry Hills, NSW .

[21] AITC. (2003). *Superior fire resistance*. American Institute of Timber Construction, Revere Parkway, Centennial, CO.

[22] Forney, G.P. (2008). *Fire Smokeview (Version 5) – A tool for visualizing fire dynamics simulation data, Volume I: User's guide*. National Institute of Standards and Technology (NIST), Washington, DC.

[23] McGrattan, K., Hostikka, S., & Floyd, J. (2009). *Fire Dynamics Simulator (Version 5) technical reference guide*. National Institute of Standards and Technology (NIST) Press, Washington, DC.

[24] McGrattan, K., Hostikka, S., & Floyd, J. (2010). *Fire Dynamics Simulator (Version 5) user's guide*. National Institute of Standards and Technology (NIST) Press, Washington, DC.

[25] World Health Organization (WHO). (1983). *Indoor air pollutants: Exposure and health effects assessment*. Euro Reports and Studies Working Group Report No.78, Nördlingen. WHO, Copenhagen.

[26] Jones, A.P. (1999). Indoor air quality and health. *Atmospheric Environment*,33, 4535–4564.

[27] Wanner, H.U., & Kuhn, M. (1986). Indoor air pollution by building materials. *Environment International*, 12, 311–315.

[28] Xiao, S.B., Li, N.P., Li, J., Xiao, Y., & Shan, B. (2008). Measurement and analysis of the indoor air quality of the modern bamboo house in China (in Chinese). *Journal of Safety and Environment*, 8(6).

[29] Ministry of Housing and Urban–Rural Development, General Administration of Quality Supervision, Inspection and Quarantine of the People's Republic of China. (2020). GB 50325-2020: Standard for indoor environmental pollution control of civil building engineering (in Chinese). China Planning Press, Beijing.

[30] National Health Commission, and Standardization Administration of the People's Republic of China. (2014). GB/T 18204.2-2014: Examination methods for public places – Part 2: Chemical pollutants. Standards Press of China, Beijing.

[31] Ministry of Housing and Urban–Rural Development, General Administration of Quality Supervision, Inspection and Quarantine of the People's Republic of China. (2019). GB 50345-2012: Technical code for roof engineering. China Architecture & Building Press, Beijing.

[32] Ministry of Housing and Urban–Rural Development of the People's Republic of China. (2017). GB 50005-2017: Code for design of timber structures (in Chinese). China Architecture & Building Press, Beijing.

[33] American Wood Council. (2018). *National design specification for wood construction*. Leesburg, VA.

[34] Ministry of Housing and Urban–Rural Development, General Administration of Quality Supervision, Inspection and Quarantine of the People's Republic of China. (2012). GB/T 50708-2012: Technical code of glued laminated timber structures (in Chinese). China Architecture & Building Press, Beijing.

[35] Feng, L., Xiao, Y., Shan, B., Chen, J., & Shen, Y.L. (2014). Experimental study on bearing capacity of glubam beam-column bolted joints. *Journal of Building Structures*, 35(4).

[36] International Code Council. (2015). *International building code (IBC)*. International Code Council, Country Club Hills, IL.

[37] Ministry of Housing and Urban–Rural Development of the People's Republic of China. (2017). GBT 51226-2017: Technical standard for multi-story and high-rise timber buildings. China Architecture & Building Press, Beijing.

Index

For Product Safety Concerns and Information please contact our EU
representative GPSR@taylorandfrancis.com
Taylor & Francis Verlag GmbH, Kaufingerstraße 24, 80331 München, Germany